T0391222

Springer Series on

SIGNALS AND COMMUNICATION TECHNOLOGY

For further volumes:
http://www.springer.com/series/4748

Signals and Communication Technology

Multimodal User Interfaces
From Signals to Interaction
D. Tzovaras ISBN 978-3-540-78344-2

Human Factors and Voice Interactive Systems
D. Gardner-Bonneau, H.E. Blanchard (Eds.)
ISBN 978-0-387-25482-1

Wireless Communications
2007 CNIT Thyrrenian Symposium
S. Pupolin (Ed.) ISBN: 978-0-387-73824-6

Satellite Communications and Navigation Systems
E. Del Re, M. Ruggieri (Eds.)
ISBN 978-0-387-47522-6

Digital Signal Processing
An Experimental Approach
S. Engelberg ISBN 978-1-84800-118-3

Digital Video and Audio Broadcasting Technology
A Practical Engineering Guide
W. Fischer ISBN 978-3-540-76357-4

Three-Dimensional Television
Capture, Transmission, Display
H.M. Ozaktas, L. Onural (Eds.)
ISBN 978-3-540-72531-2

Foundations and Applications of Sensor Management
A.O. Hero, D. Castañòn, D. Cochran,
K. Kastella (Eds.) ISBN 978-0-387-27892-6

Digital Signal Processing with Field Programmable Gate Arrays
A.O. Hero, D. Castañòn, D. Cochran,
U. Meyer-Baese ISBN 978-3-540-72612-8

Adaptive Nonlinear System Identification
The Volterra and Wiener Model Approaches
T. Ogunfunmi ISBN 978-0-387-26328-1

Continuous-Time Systems
Y.S. Shmaliy ISBN 978-1-4020-6271-1

Blind Speech Separation
S. Makino, T.-W. Lee, H. Sawada (Eds.)
ISBN 978-0-4020-6478-4

Cognitive Radio, Software Defined Radio, and Adaptive Wireless Systems
H. Arslan (Ed.) ISBN 978-1-4020-5541-6

Wireless Network Security
Y. Xiao, D.-Z. Du, X. Shen
ISBN 978-0-387-28040-0

Terrestrial Trunked Radio – TETRA
A Global Security Tool
P. Stavroulakis ISBN 978-3-540-71190-2

Multirate Statistical Signal Processing
O.S. Jahromi ISBN 978-1-4020-5316-0

Wireless Ad Hoc and Sensor Networks
A Cross-Layer Design Perspective
R. Jurdak ISBN 978-0-387-39022-2

Positive Trigonometric Polynomials and Signal Processing Applications
B. Dumitrescu ISBN 978-1-4020-5124-1

Face Biometrics for Personal Identification
Multi-Sensory Multi-Modal Systems
R.I. Hammoud, B.R. Abidi, M.A. Abidi (Eds.)
ISBN 978-3-540-49344-0

Cryptographic Algorithms on Reconfigurable Hardware
F. Rodriguez-Henriquez
ISBN 978-0-387-33883-5

Ad-Hoc Networking
Towards Seamless Communications
L. Gavrilovska ISBN 978-1-4020-5065-7

Multimedia Database Retrieval
A Human-Centered Approach
P. Muneesawang, L. Guan
ISBN 978-0-387-25627-6

Broadband Fixed Wireless Access
A System Perspective
M. Engels, F. Petre
ISBN 978-0-387-33956-6

Acoustic MIMO Signal Processing
Y. Huang, J. Benesty, J. Chen
ISBN 978-3-540-37630-9

Algorithmic Information Theory
Mathematics of Digital Information Processing
P. Seibt ISBN 978-3-540-33218-3

Continuous-Time Signals
Y.S. Shmaliy ISBN 978-1-4020-4817-3

Interactive Video
Algorithms and Technologies
R.I. Hammoud (Ed.) ISBN 978-3-540-33214-5

Optical Transmission
The FP7 BONE Project Experience
A. Teixeira, G.M.T. Beleffi (Eds.)
ISBN 978-94-007-1766-4

António Teixeira · Giorgio Maria Tosi Beleffi
Editors

Optical Transmission

The FP7 BONE Project Experience

Editors
António Teixeira
DETI/Instituto de Telecomunicações
University of Aveiro
Aveiro
Portugal
teixeira@ua.pt

Giorgio Maria Tosi Beleffi
ISCTI, MSE Department of Communications
Viale America 201
00144 Rome
Italy
giorgio.tosibeleffi@sviluppoeconomico.gov.it

ISSN 1860-4862
ISBN 978-94-007-1766-4 e-ISBN 978-94-007-1767-1
DOI 10.1007/978-94-007-1767-1
Springer Dordrecht Heidelberg London New York

Library of Congress Control Number: 2011940128

Printed on acid-free paper

Springer is part of Springer Science+Business Media (www.springer.com)

Preface

Social inclusion, network management and neutrality, business and cost models are becoming key concepts in todays/tomorrows broadband optical communications design and development. Research has led to massive developments in the optical communications field, boosting the deployment of new technologies able to enhance bandwidth and services for the end users.

The Network of Excellence BONE, FP7 NoE led by Prof. Peter Van Daele IBBT, focused its activity in stimulating collaboration, research and know-how exchange through integration activities, bringing together several of the most relevant research labs in the field in Europe. Through the establishment of Virtual Centres of Excellence, the BONE-project observed and helped drawing the roadmaps for the "Network of the Future" recurring to training and sharing of knowledge, lab facilities and vies on technologies and architectures.

The BONE project VCEs are: VCE Network Technologies and Engineering, VCE Services and Applications, VCE Access networks, VCE Optical switching systems, VCE Transmission techniques, VCE In-building Networks.

The Virtual Centre of Excellence on transmission (VCE-T), led by Prof. Periklis Petropoulos, observed the optical high-speed transmission topic. Efficient transmission is a key prerequisite and enabler of future broadband networks and greatly affects networks in all its vectors, from arhcitectures to protocols. Part of the work and researchers in this center of excellence gathered together to generate this book on the topics related to transmission, and present their views on the topic.

This book starts with a description of the state of the art in transmission systems, chapter one, then dives into the topics related to signal processing, management and monitoring of these systems, Chapter two. In chapter three, simulation principles for long haul are discussed, and complemented with the recent achievements on the experimental side in chapter four. Finally, aiming at the economics behind a set of models and discussions are made on the models and strategies for the next generation optical networks.

This book was initiated by the self clustering of the authors, monitored and guided by the Technical Book Committee and specially by Chapter Leaders. Of course nothing of this would have been possible without the extremely interesting funding tool, the networks of excellence, instrument of the European commission, to promote, in a very inexpensive way, cooperation and unity within European research centres.

A special thanks goes to our parents and families that supported us in every moment (Michela, Rita, Emanuele, Pedro, Francesco, Anna Maria, Carla Maria, Abílio, Cília and Franklim). Last but not least we want to remember two key persons in the International optical scenario like Prof. Fabio Neri and Prof. Benedetto Daino that passed away in 2011.

Giorgio Maria Tosi Beleffi
António Luis Jesus Teixeira

Book Advisory Board

List of Authors

Achille Pattavina
António Teixeira
Armando Pinto
Carmen Vázquez
Christer
Christer Mattsson
Christophe Caucheteur
Christos Tsekrekos
Claus Larsen
David Bolt
David Larrabeiti
Davide Forin
Domenico Siracusa
Elisabeth Pereira
Francesca Parmegiani
Francesco Matera
Franco Curti
Gabriele Incerti
Gabriella Bosco
Gerald Franzl
Giorgia Parca
Giorgio Maria Tosi Beleffi
Guido Maier
Jawaad Ahmed
Jeroen Nijhof
Jiajia Chen
Jose A. Lazaro
José Lazaro
Josep Prat
Julio Montalvo

Karin Ennser
Kivilcim Yüksel
Lena Wosinska
Lucia Marazzi
Marcelo Zannin
Marco Forzati
Marco Tabacchiera
Marina Settembre
Michela Svaluto
Miroslaw Kantor
Morten Ibsen
Muneer Zuhdi
Paolo Monti
Periklis Petropoulos
Philippe Gravey
Pierluigi Poggiolini
Rebecca Chandy
Robert Keiley
Silvia Di Bartolo
Tatiana Loukina
Valeria Carrozzo
Veronique Moeyaert
Wladek Forysiak

List of Acronyms

3G	Third generation
AC	Asymmetrical clipping
ACO-OFDM	Asymmetrical clipping optical OFDM
ADC	Analogue-to-digital convertor
ADM	Add-drop multiplexer
ADSL	Asymmetric digital subscriber line
AM	Amplitude modulation
AOM	Acousto-optic modulator switch
AON	Active optical networks
AOWC	All optical wavelength conversion
APT	Automatic polarisation controller
AS	Autonomous systems
ASE	Amplified spontaneous emission
ASON	Automatically switched optical networks
AWG	Arrayed waveguide grating
BAL	Balanced photodetector
BER	Bit error rate
BERT	Bit error rate tester
BGP	Border gateway protocol
BOS	Blue ocean strategy
BPSK	Binary phase-shift keying
BTB	Back to back
CapEX	Capital expenditures
CBR	Constraint-based routing
CD	Chromatic dispersion
CFBG	Chirped fibre Bragg gratings
CMA	Constant modulus algorithm
CNR	Carrier to noise ratio
CO	Central office
CO-OFDM	Coherent ortogonal frequency division multiplexing

CP	Cyclic prefix
CRM	Customer relationship management
CRZ	Chirped return-to-zero
CW	Continuous wave
CWDM	Coarse wavelength division multiplexing
DAB	Digital audio broadcasting
DAC	Digital-to-analogue converter
DCF	Dispersion compensating fibres
DCU	Dispersion Compensation Unit
DD	Direct detection
DFB	Distributed-feedback lasers
DFG	Difference-frequency generation
DFT	Discrete Fourier transform
DGD	Differential group delay
DHT	Discrete Hartley transform
DM	Dispersion management
DMT	Discrete multi-tone modulation
DOCSIS	Data over cable service interface specifications
DPSK	Differential phase-shift keying
DQPSK	Differential quadrature phase shift keying
DS	Downstream
DSF	Dispersion shifted fibres
DSL	Digital subscriber line
DSP	Digital signal processing
DVB	Digital video broadcasting
DWDM	Dense wavelength division multiplexing
DWP	Distributed wavelength path provisioning
EB	Extender box
ECL	External cavity laser
ECOC	European conference on optical communication
EDF	Erbium doped fibre
EDFA	Erbium doped fibre Amplifier
EDWA	Erbium doped waveguide amplifier
EGP	Exterior gateway protocol
EPON	Ethernet PON
ERO	Explicit route object
FBG	Fibre Bragg gratings
FEC	Forward error correction
FF	Feeder fibre
FFT	Fast Fourier transform
FHT	Fast Hartley transform
FiOS	Fibre optic service
FIRF	Finite impulse response filters
FIR	Finite impulse response
FP	Fabry-Perot

FS	Frequency-shifter
FSC	Fibre switch capable
FTTB	Fibre to the building
FTTC	Fibre to the curb or cabinet
FTTH	Fibre to the home
FTTx	Fibre to the x
FUTON	Fibre optic networks for distributed, heterogeneous radio architectures and service provisioning
FWHM	Full width at half maximum
FWM	Four wave mixing
GDP	Gross domestic products
GMPLS	Generalised multi-protocol label switching
GPON	Gigabit passive optical network
GVD	Group velocity dispersion
HDP	Highly dispersive pulses
HDSL	High-speed digital subscriber line
HFC	Hybrid fibre coaxial
HNLF	Highly nonlinear fibre
HR	Home run
IA-RWA	Impairment-aware routing wavelength assignment
ICI	Intercarrier interference
ICT	Information and communications technologies
iDFT	Inverse discrete Fourier transform
IETF	Internet engineering task force
iFFT	Inverse fast Fourier transform
iFHT	Inverse fast Hartley transform
IGP	Internet gateway protocol
IL	Insertion loss
ILP	Integer linear programming
IM	Intensity modulation
IM-DD	Intensity modulation direct detection
IP	Internet protocol
IPTV	IP television
IQ	In-phase and quadrature
ISI	Intersymbol interference
ISIS	Intermediate system-intermediate system
IT	Information technology
ITU	International telecommunications union
L2SC	Layer two switching capable
LAN	Local area network
LCOS	Liquid crystal on silicon
LDP	Label distribution protocol
LFIB	Label forwarding information base
LLUB	Local loop unbundling
LMP	Link management protocol

LMS	Least mean square algorithm
LO	Local oscillator
LSA	Link state advertisement
LSC	Lambda switch capable
LSP	Label switched paths
LSR	Label switching router
mCBR	Multi-constraint-based routing
MDU	Multiple dwelling unit
MEM	Micro-electro-mechanical
MIMO	Multiple-input multiple-output
MMF	Multimode fibre
MPLS	Multi-protocol label switching
MPSK	Multi-phase shift keying
MRR	Micro ring resonator
MUSE	Multi service access everywhere
MZI	Mach–Zehnder interferometer
MZM	Mach–Zehnder modulator
NALM	Nonlinear amplifying loop mirror
NE	Network element
NF	Noise figure
NG-PON	Next-Generation PON
NLC	Nonlinearity compensation
NLSE	Nonlinear Schrödinger equation
NMS	Network management system
NMZ	Nested Mach–Zehnder Modulator
NO	Network Owner
NRZ	Non-return to zero
O/E/O	Optical/Electronic/Optical Conversion
OA	Optical Amplifier
OADM	Optical add drop multiplexer
OAM	Operations, administration, and maintenance
OBPF	Optical bandpass filter
OBS	Optical Burst Switching
OCC	Optical cross connect
OCE	Optical channel estimation
OCM	Optical channel monitoring
ODB	Optical duobinary
ODL	Optical delay line
ODN	Optical distribution network
OEO	Optical-electrical-optical
OFDM	Orthogonal frequency division multiplexing
OFDR	Optical frequency domain reflectometry
OGC	Optical gain clamping
OIF	Optical internetworking forum
OIM	Optical impairment monitoring

OLT	Optical line terminal
ONU	Optical network unit
O-OFDM	Optical orthogonal frequency division multiplexing
OOK	On-off keying
OPC	Optical phase conjugation techniques
OPEX	Operational Expenditures
OPM	Optical performance monitoring
OPS	Optical packet-switched
OSA	Optical spectrum analyser
OSI	Open systems interconnection
OSNR	Optical signal-to-noise ratio
OSP	Outside plant
OSPF	Open shortest path first
OTDR	Optical time domain reflectometry
OTN	Optical transport network
P2MP	Point-to-multipoint
P2P	Point-to-point
PAM	Pulse-amplitude modulation
PAPR	Peak-to-average power ratio
PBS	Polarisation beam splitter
PC	Post compensation
PCE	Path computation element
PCEP	Path computational element communication protocol
PDA	Photonic design automation
PDF	Probability density function
PDL	Polarisation-dependent loss
PDM	Polarisation-division-multiplexed
PDU	Protocol data units
PIEMAN	Photonic integrated extended metro and access network
PLC	Planar lightwave circuits
PLID	Physical layer impairment database
PM	Polarization multiplexed
PMD	Polarisation mode dispersion
PNNI	Private network-to-network interface
PON	Passive optical network
PPG	Pulse pattern generator
PPLN	Periodically poled lithium niobate
PR	Pre-compensation
PRBS	Pseudo-Random Binary Sequence
PSA	Phase-sensitive amplifiers
PSC	Packet switch capable
PSCF	Pure silica-core fibre
PSK	Phase-shift keying
PTF	Power transfer function
PXC	Photonic cross-connect

QAM	Quadrature amplitude modulation
QoS	Quality of service
QoT	Quality of transmission
QPSK	Quadrature phase-shift keying
R	Regenerators' regeneration stages
RBS	Rayleigh backscattering
RF	Radio frequency
RN	Remote node
ROF	Relaxation oscillation frequency
ROPA	Remotely pumped optical amplifier
ROS	Red Ocean Strategy
RPE	Reverse-proton-exchange technique
RR	Ring-resonator
RSVP-TE	Resource reservation-traffic engineering
RWA	Routing wavelength assignment
Rx	Receiver
RZ	Return to zero
SARDANA	Scalable Advanced Ring-based passive Dense Access Network Architecture
SB	Svenska Bostäder
SBS	Stimulated Brillouin scattering
SCP	Structure-conduct-performance
SDH	Synchronous digital hierarchy
SG	Sagnac
SHB	Spectral hole burning
SHG	Second-harmonic-generation
SLA	Service-level agreement
SMF	Single mode fibre
SNR	Signal-to-noise ratio
SOA	Semiconductor optical amplifiers
SONET	Synchronous optical network
SOP	State of polarisation
SP	Service provider
SPM	Self phase modulation
SRAM	Static random access memory
SRLG	Shared risk link group
SSB	Single side band
SSF	Split-step Fourier
SSFBG	Superstructured fibre Bragg gratings
SSMF	Standard single-mode fibre
TCAM	Ternary content addressable memory
TCO	Total cost of ownership
TDM	Time division multiplexing
TDMA	Time division multiple access
TE	Traffic engineering

TED	Traffic engineering database
TI	Transparency island
TLS	Tunable laser source
TRF	Transmission-reflection function
TWC	Tunable wavelength converters
Tx	Transmitter
ULAF	Ultra-large effective area fibre
US	Upstream
VA	Variable attenuator
VDSL	Very high-speed digital subscriber line
VOA	Variable optical attenuator
VPHG	Volume phase grating
VPLS	Virtual private lan services
WAN	Wide area network
WC	Wavelength conversion
WDM	Wavelength division multiplexing
WiFi	Wireless fidelity
WIMAX	Worldwide interoperability for microwave access
WLAN	Wireless local area network
WSS	Wavelength-selective-switch
XPM	Cross-phase modulation

Contents

Chapter 1
State of the Art on Transmission Techniques

Christos Tsekrekos, Wladek Forysiak, Robert Killey, Francesco Matera, Michela Svaluto Moreolo, and Jeroen Nijhof

1.1 Introduction

Enabling long-haul and high-speed transmission has been the premier feature of fibre optics that has rendered it an attractive and successfully deployed technology. Although optical fibres are being engaged in several applications, such as signal processing, lasers and sensors, transmission systems remain a core and continuously evolving application leading to demonstrations of several Tbit/s over hundreds of kilometres. To achieve such impressive performances, several techniques have been investigated and applied. WDM often combined with polarisation multiplexing,

C. Tsekrekos (✉)
Athens Information Technology Center, 19.5 klm, Markopoulo Ave. GR-19002, Peania Attikis, Greece
e-mail: tsekrekos@pn.comm.eng.osaka-u.ac.jp

W. Forysiak • J. Nijhof
Ericsson, UK

Ansty Park, Coventry, CV7 9RD, UK
e-mail: Wladek.Forysiak@oclaro.com; jeroen.nijhof@ericsson.com

R. Killey
Department of Electronic & Electrical Engineering, University College London, Torrington Place, London, WC1E 7JE, England
e-mail: r.killey@ee.ucl.ac.uk

F. Matera
Fondazione Ugo Bordoni, Viale America, 201, 00144 Roma, Italy
e-mail: mat@fub.it

M.S. Moreolo
Centre Tecnològic de Telecomunicacions de Catalunya (CTTC), Av. Carl Friedrich Gauss 7, 08860 Castelldefels, Barcelona, Spain
e-mail: michela.svaluto@cttc.es

A. Teixeira and G.M.T. Beleffi (eds.), *Optical Transmission: The FP7 BONE Project Experience*, Signals and Communication Technology,
DOI 10.1007/978-94-007-1767-1_1,

offers the means to exploit the large optical bandwidth. The aggregate transmission bit rate depends on the achievable bit rate per wavelength and the number of deployed wavelengths. The simplest and most straightforward approach based on IM-DD is giving its place to more advanced techniques based on phase modulation and coherent detection. Combination of optical fibre transmission with DSP has opened a new window in transmission systems, facilitating the use of coherent detection and the mitigation of nonlinear effects, as well as allowing the application of advanced transmission techniques such as OFDM. This chapter provides an overview of the state of the art of optical fibre transmission techniques.

First, in Sect. 1.2, an introduction to the optical fibre channel characteristics is presented. The objective is to briefly describe the main propagation effects and how they influence the transmission performance. Subsequently, Sect. 1.3 describes systems based on IM-DD, providing a comparison between NRZ and RZ encoding. The need for transmission systems using differential phase modulation is then explained, and modulation formats such as NRZ DPSK and RZ DQPSK are presented. Further enhancing the transmission performance requires the application of coherent techniques, combined with DSP and advanced modulation formats, such as multi-level QAM. The challenges and benefits of coherent transmission are analysed in Sect. 1.4. The application of optical OFDM (O-OFDM) has attracted much attention due to its scalability to very high bit rates and its resiliency to dispersion. In the last section of this chapter, Sect. 1.5, O-OFDM systems based on direct detection as well as coherent detection are described.

1.2 Fibre Channel Characteristics

The channel of an optical fibre transmission system can be seen as a cascade of several devices, the optical fibre being the key component. In the case of very simple systems, the channel coincides with a segment of optical fibre that can range from hundreds of metres (as in the case of access systems) up to one or two hundred kilometres. For longer distances, optical amplification is necessary to compensate fibre losses, and according to the network requirements, other devices can be included in the channel, such as optical equalisers, dispersion compensating fibres (DCFs), optical filters, optical add-drop multiplexers, and optical cross connects. In Fig. 1.1, an example of optical fibre transmission channel is shown that includes fibres, optical amplifiers and DCFs.

1.2.1 Linear Effects in Optical Fibres

An optical fibre is a very thin glass wire composed by a core and a cladding with a different refractive index (Agrawal 2007). In Fig. 1.2, an example of an optical fibre is shown.

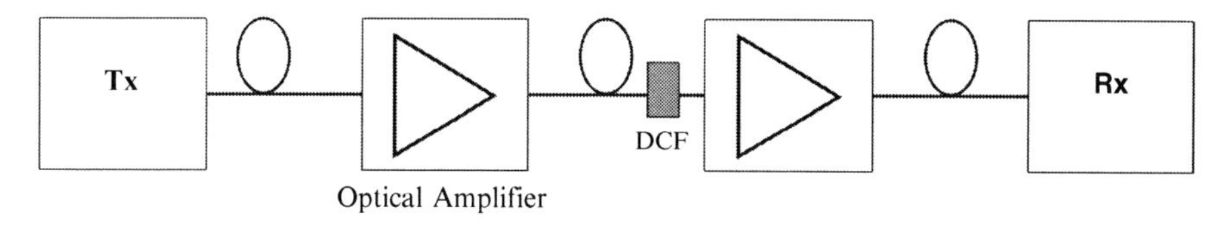

Fig. 1.1 Example of an optical fibre channel. DCF dispersion compensating fibre

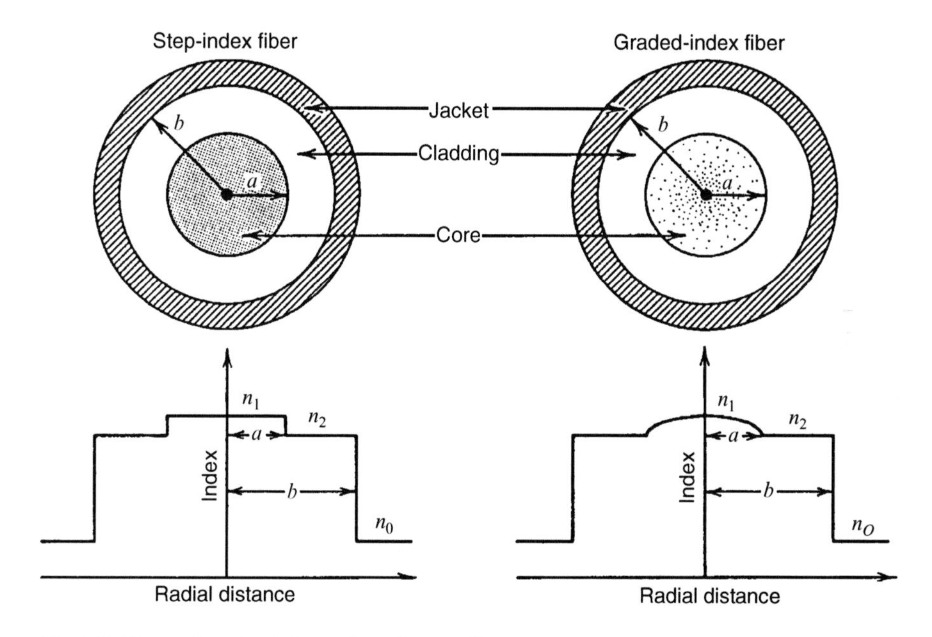

Fig. 1.2 Step and graded index optical fibre profiles

The propagation of the light in optical fibres is based on the principle of the total reflection as shown in Fig. 1.3.

Fibre attenuation is very low, and as can be seen in Fig. 1.4, three operational windows have been commonly used. The most used ones are at 1.300 nm and at 1.550 nm, where the attenuation value is only about 0.2 dB/km.

Optical fibres can be distinguished between SMFs and MMFs. For short-range applications, e.g. in in-building networks, MMFs are preferably used due to their low cost and ease of installation. On the other hand, for distances longer than a few kilometres, SMFs are employed.

Since fibre loss is very low and can be compensated by means of optical amplification, in SMF systems, the main degrading effect in signal transmission is chromatic dispersion, which is caused by the fact that the propagation constant β is a function of the optical frequency ω, and, therefore, the spectral components of a signal propagate with different velocities causing temporal pulse broadening. The chromatic dispersion is measured by the parameter $\beta_2 = \frac{d^2\beta}{d\omega^2}$.

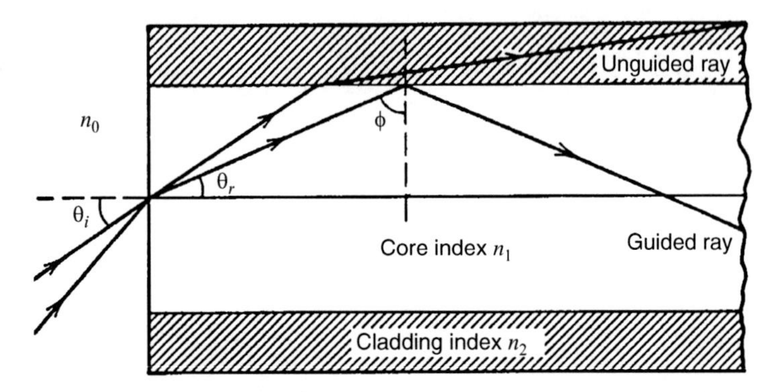

Fig. 1.3 Propagation in optical fibres

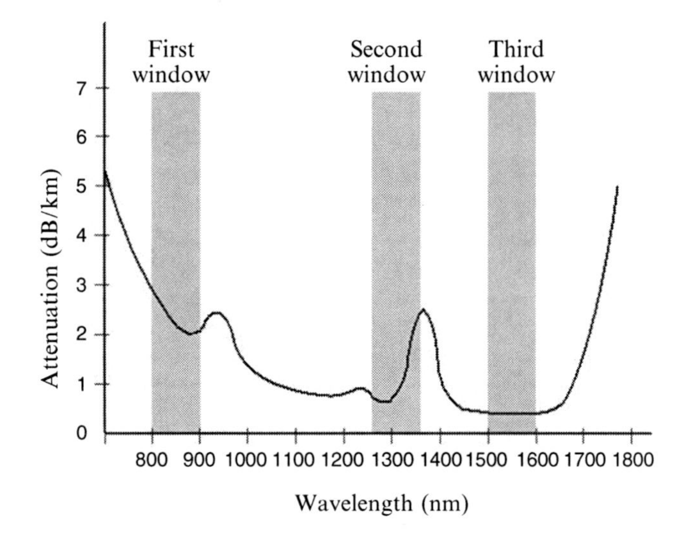

Fig. 1.4 Attenuation in optical fibres

If $A(0,t)$ is the envelope of a signal at the SMF input, the corresponding signal at distance z is given by

$$A(z,t) = \frac{1}{2\pi}\int_{-\infty}^{\infty} \tilde{A}(0,\omega)\exp\left(\frac{i}{2}\beta_2 z\omega^2 - i\omega t\right) d\omega \qquad (1.1)$$

where $\tilde{A}$ represents the Fourier transform of $A(t)$.

The signal evolution is particularly simple in case of a Gaussian pulse with time duration T_0 (Agrawal 2007). In fact, the pulse during propagation maintains the Gaussian shape and only changes its duration T according to the equation:

$$T = T_0\sqrt{1 + \left(\frac{z}{L_D}\right)^2} \tag{1.2}$$

where $L_D = \frac{T_0^2}{|\beta_2|}$ is defined as the dispersion length. The dispersion length is a measure of the effect of the chromatic dispersion on the transmission performance. This effect can be neglected when the dispersion length is much longer than the fibre segment under test.

Chromatic dispersion is often measured by means of the dispersion parameter D in units of ps/nm/km that is linked to β_2 through the relation $D = -\frac{2\pi c}{\lambda 2}\beta 2$, based on the derivative with respect to the wavelength. The D parameter allows to simply estimate the signal degradation due to chromatic dispersion in terms of the signal bandwidth $\Delta\lambda$.

It has to be pointed out that the term 'single mode' is not strictly correct since an SMF always supports two polarisation modes, which in ideal conditions (perfect circular symmetry, no twisting or stress) are degenerate modes, which means they have the same propagation constant β. However, in practice, due to several imperfections in the fibre and in the corresponding cable, the two modes lose the degeneration, and they manifest different propagation constants. Such an effect is responsible for the PMD that induces signal pulse broadening due to the fact that the two polarisation components of the signal travel with different group velocities. Obviously, if the PMD can be neglected, the fibre can be assumed as a purely single-mode one. PMD can be a detrimental effect for systems operating at a bit rate higher than 10 Gbit/s over long distances (>500 km).

Among SMFs, we can distinguish different ITU-T standards that characterise the fibre mainly in terms of chromatic dispersion. Here we report the main G.65x standards:

- G.652: step-index fibres with zero dispersion at 1.300 nm and a parameter D at 1.500 nm around 17 ps/nm/km.
- G.653: dispersion-shifted fibres with zero dispersion at 1.550 nm. The zero dispersion at the referred wavelength (1.550 nm) will enhance the phase matching conditions and therefore nonlinear effects impact.
- G.655: non-zero dispersion-shifted fibres created to avoid the nonlinear effects in G.652. For G.655 fibres, the chromatic dispersion at 1.550 nm is a few ps/nm/km (normally less than 5 ps/nm/km).
- G.657: fibres for local and in-house applications.

1.2.2 Nonlinear Effects in Optical Fibres

Even though optical fibre could be considered as an example of linear medium, due to the fact that a signal can propagate for very long distances, some nonlinear effects can be manifested causing signal degradation. Furthermore, nonlinear impairments are much more evident when the fibre channel hosts many wavelength channels. Therefore, we can say that an index of the nonlinear degradation is given by the product: (fibre link) × (number channel) × (channel power).

It is not easy to describe the nonlinear impairments in optical fibre transmission, since in a nonlinear regime, the intuitive rule of 'summing the effects' is not valid. One of the consequences is the fact that impairments induced by nonlinearities also depend on the type of the transmitted signal. For conventional transmission systems, the Kerr nonlinearity is the main nonlinear effect, and it can be manifested in three different ways depending on the transmission scenario.

In the case of single channel systems, the intensity $I(t)$ of the same signal induces a phase variation $\Delta\phi$ in the signal given by

$$\Delta\phi = \frac{2\pi n_2 I(t) z}{\lambda_0} \tag{1.3}$$

where n_2 represents the nonlinear Kerr index and λ_0 the wavelength of the carrier signal. Therefore, in this case, the Kerr effect causes a phase variation, depending on the power variation over time, which is manifested as SPM. The main degradation induced by SPM, in the case of zero chromatic dispersion, is spectral broadening of the signal. The combination of the Kerr effect and chromatic dispersion can yield different signal distortions. In the case of a pulse signal in the normal chromatic dispersion regime ($\beta_2 > 0$), the signal manifests a wider pulse broadening, with respect to the absence of the Kerr effect, that increases with the power. Conversely, in the anomalous regime ($\beta_2 < 0$), the behaviour is more complex, and the pulse can also experience spectral narrowing. It has to be pointed out that when particular conditions are satisfied in terms of initial power, pulse time duration, pulse shape and chromatic dispersion, the pulse can maintain its initial shape and duration. When this condition is satisfied, the pulse is called *soliton*, and it will be described more thoroughly in Chap. 4.

The Kerr effect manifests itself in a very different way in the presence of two or more wavelengths that propagate and carry high power. In this case, spurious frequencies are generated, and for this reason, the effect of Kerr nonlinearity is referred to as FWM. The last manifestation of the Kerr nonlinearity is when two or more pulses propagate with different carrier frequencies. In this case, when the pulses overlap, the phase of a pulse is modified by the power variation of the other pulse. In such a case, the effect of the Kerr nonlinearity is called XPM. As it will be shown in Chap. 4, to exactly evaluate the signal evolution in optical fibre, including the Kerr effect, it is necessary to numerically solve the nonlinear Schrödinger equation.

To conclude this section on the fibre channel, two other nonlinear effects have to be mentioned: Raman and Brillouin scattering. Both phenomena are a consequence of power transfer from the signal to other signals. In particular, in Brillouin scattering, a part of the signal power is converted to a signal that counter-propagates at a frequency that is shifted by 11 GHz with respect to the input one. Brillouin scattering arises when the input power at a certain frequency is higher than several milliwatts. Therefore, given the current presence of in-line optical amplifiers that can reduce the input power requirements as well as transmission of signals with wide bandwidth, the Brillouin effect can be negligible in most of the operating systems.

Conversely, the Raman effect manifests itself in the presence of many high power channels distributed in a wide bandwidth (high dense WDM). Hence, the link power budget has to be carefully evaluated in order to avoid Raman degradation. It has to be pointed out that the Raman effect can be also used as a means of amplification, since the power of pump signals can be transferred to amplify weak signals. Therefore, in some links, high power pump signals can be injected at the end of the fibre span to achieve an amplification of signals in the fibre region where their power is lower.

1.3 Intensity and Differential-Phase Modulation Formats

The state of the art in modulation for optical communications has advanced quickly during the past 10 years, and increasingly complex formats promise to be adopted in commercial dense WDM (DWDM) systems as the demand to increase capacity in the relatively narrow spectrum supported by EDFA continues. This rapid evolution is illustrated by comparing the leading systems' demonstrations at the annual conference on Optical Fibre Communications (OFC) over the past decade. In 2000, the highest capacity demonstrations of 3 Tbit/s, spanning both C-band and L-band spectral regions (1,530–1,600 nm), were achieved using optical intensity modulation, also known as OOK (Nielsen et al. 2000; Ito et al. 2000), whereas in 2010, more than 64-Tbit/s capacity was demonstrated in a similar bandwidth by groups working with advanced polarisation-division-multiplexed (PDM) formats, modulating both field amplitude and phase, PDM-16-QAM and PDM-36-QAM (Sano et al. 2010; Zhou et al. 2010). During this 10-year period, therefore, the state of the art in spectral efficiency increased from 0.4 bits/s/Hz to 6–8 bits/s/Hz, approaching the so-called Shannon limit to information capacity (Essiambre et al. 2008; Ellis et al. 2010), and, consequently, the research community has acknowledged that practical considerations on the deliverable OSNR in terrestrial networks will force it to consider other techniques to increase capacity, including new transmission media (Ellis et al. 2010).

The adoption of new technology in commercial DWDM systems tends to lag the state of the art in research by a few years, so while the underlying 'per-channel' data rates in the year-2000 demonstrations were 20–40 Gbit/s (those in 2010 were 107–170 Gbit/s), commercial deployments at that time were limited to 10 Gbit/s at a spectral efficiency of 0.2 bit/s/Hz. Today, in 2010, while deployments in the DWDM backbone are mostly of 10 Gbit/s, with the 40-Gbit/s share growing rapidly as pricing becomes competitive, the first generation of 100-Gbit/s interfaces has just become available from a couple of leading vendors (Ruhl et al. 2010; Faure et al. 2010). Meanwhile, the almost universal adoption of the ITU-T standard 50-GHz grid, driven partly by requirements for increased optical transparency, flexibility and routing via WSS technology (Ma and Kuo 2003), has meant that current deployments are limited to spectral efficiencies of 0.2–2 bit/s/Hz and likely to remain that way for a few years while the necessary cost reduction for widespread deployment of 100 Gbit/s is achieved.

During the past decade, a wide range of modulation formats was proposed and investigated at 10 and 40 Gbit/s, as a suitable compromise among simplicity (cost), performance and spectral efficiency was sought by equipment suppliers and operators. Indeed, it has been argued that the wide range of technological choices thrown up by the research community helped in part to create an uncertain environment and fragmented supply chain, which led to a delay in the take up of 40 Gbit/s (Schmidt and Hong 2009). Excellent reviews covering the wide range of advanced modulation formats investigated in research laboratories exist already (Winzer and Essiambre 2008), so here we focus on just two OOK formats for 10-Gbit/s transmission and two PSK formats for 40 Gbit/s which we consider the most common options deployed today. For 10 Gbit/s, these are NRZ and RZ, while for 40 Gbit/s, they are DPSK and DQPSK. All these techniques rely on external modulation via MZMs, direct detection via a single, balanced, or dual balanced detectors, and each provides excellent transmission performance for long-haul DWDM applications, with reaches exceeding 3,000 km at 10 Gbit/s and approaching 1,500 km at 40 Gbit/s.

1.3.1 Intensity Modulation

The simplest and still the most widely used modulation format in commercial DWDM systems today at 10 Gbit/s remains OOK, and in particular NRZ. In OOK, the digital signal to be transmitted is encoded into a series of symbols where binary ones (1's) and zeroes (0's) are represented by the presence and absence of light. The name 'NRZ' stems from the fact that the optical power does not return to zero between successive 1's, as shown in Fig. 1.5. Although 10-Gbit/s NRZ signals can be generated via direct modulation of the laser current, and indeed a large class of small form factor pluggable transceivers (XFPs) is based on this approach (XFP MSA), for the best transmission performance, external modulation via an MZM is used to minimise chirp and produce the most compact spectra.

The basic MZM structure is shown schematically in the composite modulators of Fig. 1.6. The incoming light is split into two paths at the input coupler, and one (or both) paths are equipped with phase modulators which can be driven with electrical signals, $V_{1,2}(t)$, so that the two optical fields acquire a phase difference relative to one another. On combination at the output coupler, the two fields interfere and combine, either constructively or destructively as a function $V_{1,2}(t)$, so leading to intensity modulation. If the phase modulation depends linearly on the drive voltage, the amplitude of the incoming light, $A(t)$, becomes

$$A'(t) = A(t)\cos\left[\pi(V_{\text{bias}} + V(t)/V_{\pi})\right] \tag{1.4}$$

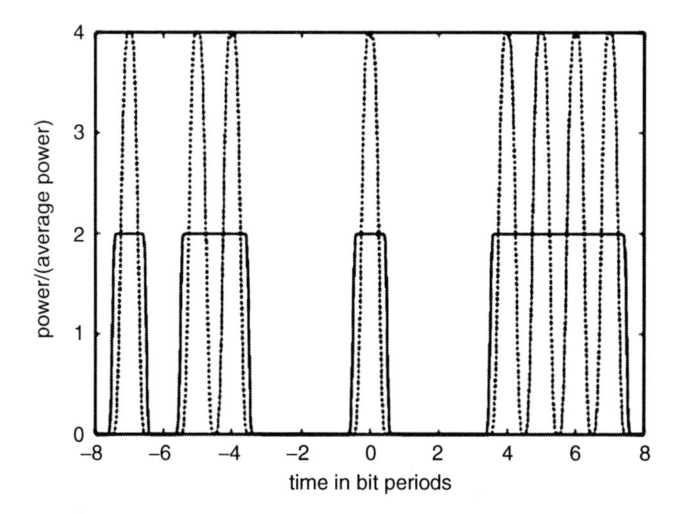

Fig. 1.5 Optical intensity for NRZ- (*solid*) and RZ-(*dotted*) modulated signals encoding the data pattern 1011000100001111

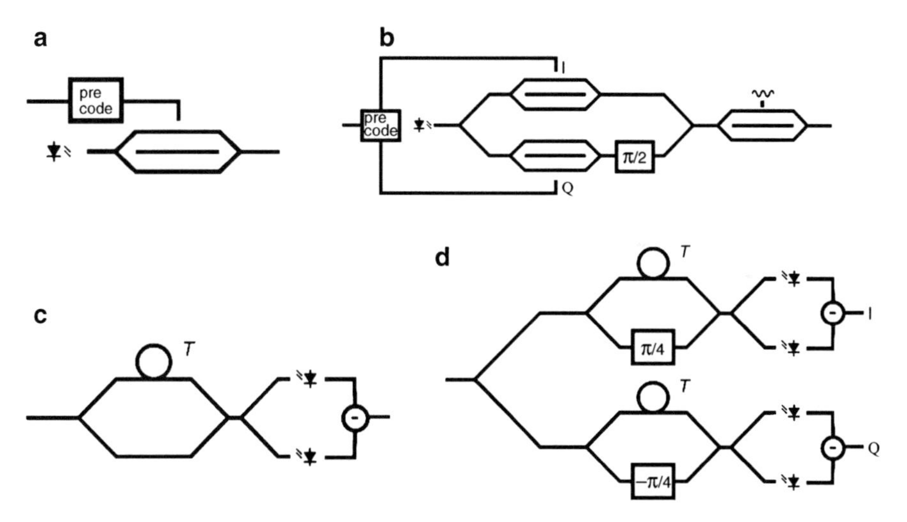

Fig. 1.6 Schematics of MZM-based transmitters for (**a**) NRZ-DPSK and (**b**) RZ-DQPSK, and delay-interferometer- and balanced-detector-based receivers for (**c**) NRZ-DPSK and (**d**) RZ-DQPSK

where $V(t) = V_1(t) - V_2(t)$, V_{bias} is a DC bias voltage, and V_π is the modulation voltage required to change the phase difference between the two arms by π, and so switch the output between full transmission and complete extinction.

To generate an NRZ signal, the MZM is biased at 50% transmission and driven from minimum to maximum transmission with a voltage swing of V_π. Figure 1.7

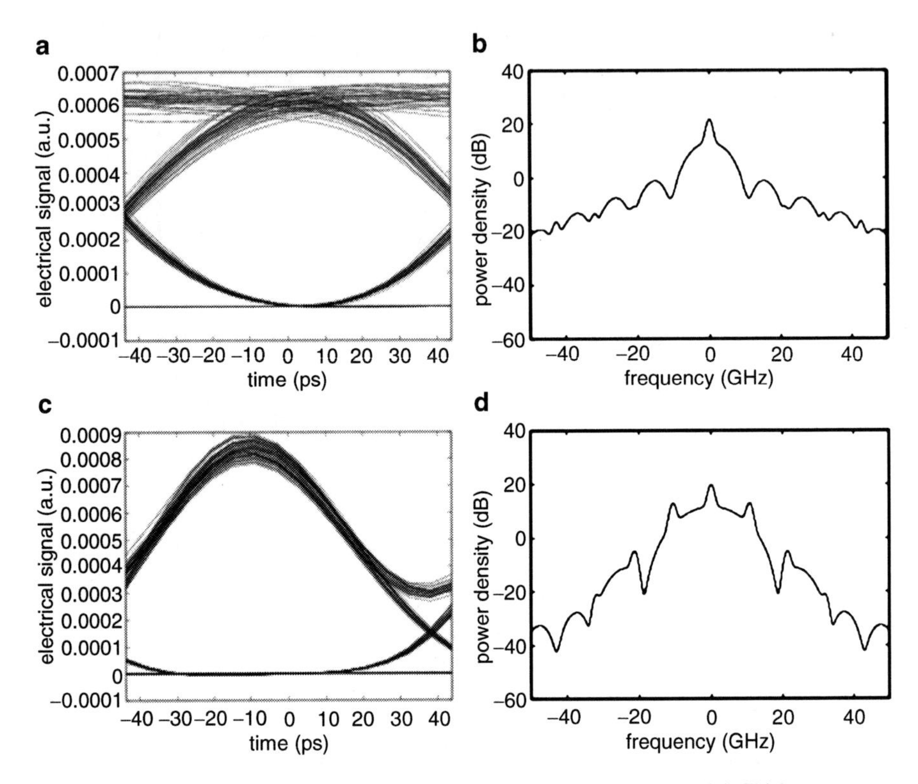

Fig. 1.7 Eye diagrams and optical spectra for NRZ (**a**, **b**) and RZ (**c**, **d**) at 10.7 Gbit/s

shows a typical simulated, and therefore idealised, 10.7-Gbit/s NRZ eye diagram, together with its optical spectrum. The eye diagram shows the basic shape of the NRZ 'pulse' together with transitions between 1s and 0s and the 'rails' at the top and bottom corresponding to sequences of 1s and sequences of 0s, respectively. The NRZ optical spectrum (drawn here with a resolution of 0.01 nm) is made up of a strong carrier at the channel centre, with a continuous portion dependent of the shape of the individual NRZ pulses (rise and fall times), and some weak, residual tones at multiples of the bit rate.

RZ modulation is the simplest variant of OOK which can be generated to offer an improvement in system performance. As the name suggests, in RZ modulation, the underlying waveform is given a pulse-like shape, and for a sequence of data 1s, the optical power falls to zero, or close to zero, at the start and end of each symbol, as shown in Fig. 1.5. Various forms of RZ can be generated by a series of two modulators, one imparting the data in NRZ form and the second 'carving' the RZ pulses. Three basic options resulting in RZ pulses with 33%, 50% and 67% duty cycle are possible by biasing and driving the MZM appropriately. For 50% RZ, the MZM is biased at 50% transmission and driven sinusoidally at the data rate between the maximum transmission and extinction points, resulting in pulses of half of the bit duration, while for 33% or 67% RZ, the bias is changed to full transmission or extinction, respectively, and driven through $2V_{\pi}$ at half the data rate. In each case, some variation of the pulse width can be achieved by adjusting the modulator bias and decreasing the modulator swing, at the cost of reduced extinction ratio or increased modulator insertion loss. Note that, today, because the electrical drivers and MZM for 10 Gbit/s have sufficient bandwidth, a single modulator can be used by generating the appropriate RZ waveforms electronically (Kao et al. 2003) and thus providing savings in component cost.

Figure 1.7 shows a simulated 10.7-Gbit/s 50% RZ eye diagram, alongside its optical spectrum. The pulse shape is evident, together with the rail representing sequences of data 0s, although a full extinction between symbols is absent because of pulse broadening due to optical and electrical filtering. The optical spectrum shows a strong central carrier component, a broader continuous portion due to the narrow RZ pulses and slightly more prominent residual tones at multiples of the bit rate. As for 10-Gbit/s NRZ, the optical spectrum is sufficiently compact to fit into a 50-GHz spectral slot and indeed to pass comfortably through more than a dozen nodes equipped with WSSs with negligible filtering penalty (Tibuleac and Filer 2010).

At the cost of slightly increased complexity and a broader optical spectrum, RZ modulation offers a number of performance advantages over NRZ. Firstly, typical end-of-life OSNR requirements for back-to-back error rates of 10^{-5} – before FEC – are typically 2 dB better for 50% RZ. Secondly, the tolerance to PMD is greater, allowing a penalty allocation of order 1 dB for a peak DGD of 40 ps, as opposed to 30 ps. Thirdly, and perhaps most importantly, the RZ waveform, via its broader spectrum, offers increased resilience to nonlinear impairments, typically allowing the launch of at least 2 dB more power per channel over long terrestrial systems, depending on the fibre type, channel spacing and the dispersion management of the optical line. When combined together, these factors can result in significantly greater reach for RZ transceivers versus NRZ, easily exceeding 3,000 km in typical terrestrial deployments, whereas NRZ connections are usually limited to around 2,000 km. However, although a significant minority of 10-Gbit/s transceivers sold today are RZ, since most connections of terrestrial operators are 1,500 km or below, the majority of sales are based on NRZ modulation.

It is worth noting that although simple NRZ and 50% RZ modulation make up most of the 10-Gbit/s market today, various alternatives were developed and even productised over the past decade. In particular, carrier-suppressed RZ, which gives 67% RZ with π-phase shifts between adjacent bits, was thoroughly investigated due to its perceived improved nonlinear tolerance and lower demands on electrical drives (Sano and Miyamoto 2010). Similarly, chirped RZ attracted the attention of the submarine community because a tuneable chirp could be generated to offset significant various amounts of uncompensated residual dispersion (Bergano 2005). On the other hand, ODB modulation, sometimes called phase-shaped binary transmission, appeared to offer the prospect of a narrower spectrum and increased dispersion tolerance but suffered from implementation difficulties, higher OSNR requirements by 1–2 dB and reduced nonlinear tolerances (Pennickx et al. 1997). Finally, to improve OSNR sensitivity, nonlinear tolerance and therefore reach, optical phase-shift-keying (Gnauck and Winzer 2005) was also explored at 10 Gbit/s, and although commercially not realised in terrestrial markets, this work formed the basis for today's 40-Gbit/s technology.

1.3.2 Differential Phase Modulation

NRZ and RZ modulation formats were also investigated at 40 Gbit/s, and ODB was productised since its narrow spectrum makes it readily compatible with deployment on a 50-GHz spaced grid, but none of these options were found to have sufficient performance to warrant widespread deployment in the terrestrial backbone. Instead, it was gradually realised that the better OSNR tolerance of the simplest PSK formats could be used to offset some of the increased requirements for operating at 4× the data rate, without losing too much in overall system reach.

In contrast to IM, or amplitude-shift keying, where only the electric field amplitude is modulated and its square detected, PSK keeps the field amplitude constant and modulates only the phase of the electric field. Thus, BPSK uses two values for the optical phase $[0, \pi]$ corresponding to normalised field amplitudes of $[1, -1]$, and QPSK uses four values $[0, \pi, \pi/2, -\pi/2]$ at a symbol rate of half the total bit rate. More generally, where the symbols representing the data bits can take on complex values in the two-dimensional signal space of amplitude and phase, the modulation format is known as QAM. Note that QPSK and 4-QAM are equivalent.

The simplest optical detection method using a single photodiode can only detect the field intensity, so this cannot be used for phase-modulated formats. Phase information can be recovered with a more complex receiver incorporating a delay interferometer in which one branch provides a delay of one symbol, so that the split and recombined signal recovers the phase difference between consecutive symbols. It would be possible to reconstruct the absolute phase by adding the recovered phase differences, but this is sensitive to cycle slips and error propagation. Therefore, it is more convenient to use differential pre-coding and choose the transmitted phases so that the data is encoded in the phase differences of consecutive symbols, e.g. a 1 is encoded a phase change of π and a 0 as no phase change, rather than the absolution phase. The resulting modulation format is known as differential PSK or DPSK.

Figure 1.6 shows schematically the most conventional DPSK transmitter and receiver. Optical phase modulation can be performed using a straight-line phase modulator, but since this modulates the phase along the unit circle in the complex plane, the speed of phase transitions is limited by the bandwidths of the electrical drive and phase modulator, and any overshoots or ringing produces phase distortions. If instead, an MZM is driven symmetrically around its transmission zero, the modulation takes place along the real axis through the origin of the complex plane, which always produces exact π phase changes, at the cost of residual intensity dips at the locations of these transitions. Figure 1.8 shows the 42.7-Gbit/s NRZ-DPSK eye diagram, together with its optical spectrum.

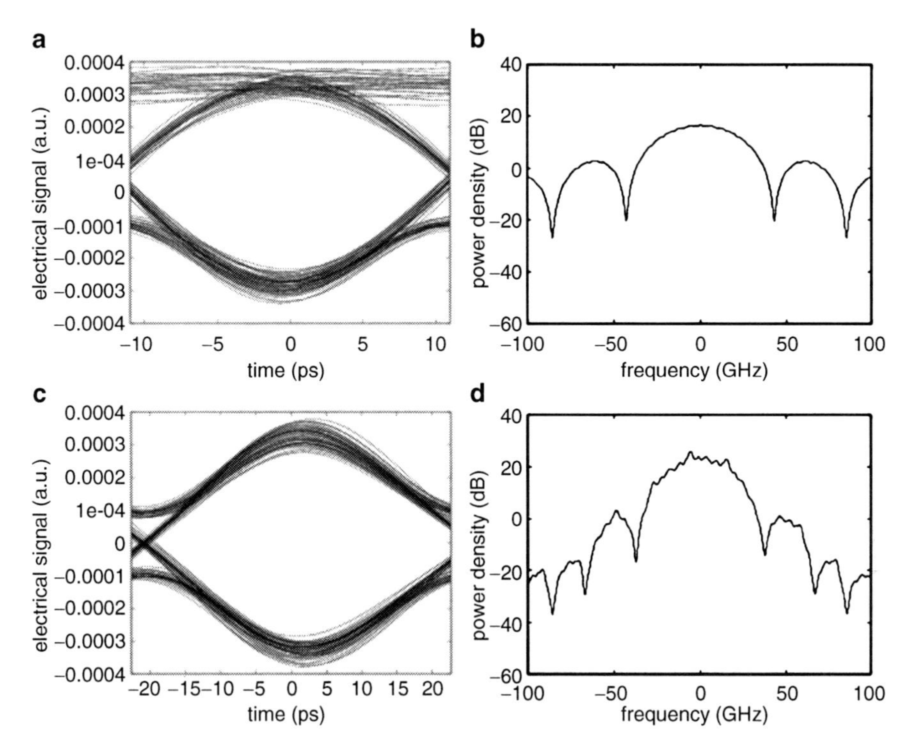

Fig. 1.8 Eye diagrams and optical spectra for NRZ-DPSK (**a**, **b**) and RZ-DQPSK (**c**, **d**) at 42.7 Gbit/s

To convert the phase-modulated signal into an intensity-modulated signal, the transmitted signal is passed through a balanced DPSK receiver, as shown in Fig. 1.6b, which includes a 1-bit delay interferometer. Thus, the signal interferes with a delayed version of itself to give constructive interference if the phase difference between adjacent bits is zero and destructive interference if it is π. Clearly, the two ports of the delay interferometer produce logically inverted outputs, and although only one port is sufficient for single-ended detection, a balanced receiver based on the difference between the two gives rise to a 3-dB sensitivity advantage. Note that the intensity dips due to the Mach-Zehnder phase modulation are visible on the bottom rail of the eye diagram.

The NRZ-DPSK optical spectrum of Fig. 1.8b contains no prominent central carrier, as expected of a PSK modulation format, and is broad compared to the 10-Gbit/s spectra of Fig. 1.7b, d, with still significant power appearing ± 25 GHz from the central frequency. At first sight, therefore, it would seem that NRZ-DPSK at 40 Gbit/s is incompatible with operation on a 50-GHz grid. However, with some pre-filtering of the individual optical spectra due to optical multiplexers and interleavers, together with a modification in the demodulation technique, known as partial demodulation (Mikkelson et al. 2006; Malouin et al. 2007; Govan and Doran 2007), NRZ-DPSK can be transmitted successfully with 50-GHz neighbours, albeit with a higher requirement in the OSNR of approximately 2 dB for the same pre-FEC BER at 10^{-5}, compared to operation at 100-GHz spacing. As with 10-Gbit/s OOK, an additional pulse carver could be used to convert NRZ-DPSK into RZ-DPSK, but at 40 Gbit/s, this is rarely done because the resulting spectrum would be too wide.

DQPSK is the second significant modulation format for the 40-Gbit/s market today (Griffin and Carter 2002). In fact, although more complex to implement, as can be seen in the transmitter and receiver schematics shown in Fig. 1.6, it is currently poised to overtake NRZ-DPSK as the dominant modulation format, because of its attractive performance characteristics, if the expected price reductions are met. In particular, RZ-DQPSK is being favoured over NRZ-DQPSK because of its increased nonlinear tolerance and greater resilience to PMD.

As shown in Fig. 1.6, a DQPSK transmitter is usually implemented by two nested MZM operated as DPSK modulators, with a $\pi/2$-phase shift between the upper and lower arms to generate the I and Q components. At the receiver, the signal is split into two parts, and two balanced receivers with differently balanced delay interferometers are used to demodulate the two binary data streams. Figure 1.8 shows a simulated 42.7-Gbit/s RZ-DQPSK eye diagram, alongside its optical spectrum. The shape of optical spectrum is similar to that of the NRZ-DPSK, though it is somewhat narrower due to the halved symbol rate, but not by a factor of 2 due to the 50% RZ modulation. This narrowed spectrum is important to achieve full compatibility with 50-GHz channel spacing and leads to very modest filtering penalties for multiple passes through typical WSSs (Tibuleac and Filer 2010). The RZ-DQPSK eye diagram shows both I and Q components.

At the cost of increased complexity, which can be mitigated with photonic integration and is partly offset by the ability to use lower speed components, RZ-DQPSK offers some performance advantages over NRZ-DPSK, although at first sight, the increased OSNR requirement of RZ-DQPSK over NRZ-DPSK of about 1.5 dB would seem to point to the opposite conclusion. Firstly, DQPSK is properly compatible with a 50-GHz infrastructure, and the partial demodulation required to operate DPSK in restricted bandwidths actually increases its OSNR requirements, eliminating the intrinsic gain (which can be utilised in 100-GHz systems). As a result, RZ-DQPSK suffers from less filtering penalties due to intermediate WSS-based nodes than NRZ-DPSK. Secondly, the nonlinear tolerance of RZ-DQPSK, though sensitive to fibre type, channel spacing and dispersion map, is sufficient to allow channel powers of 2–3 dBm over 10–15 spans and therefore sufficient OSNR over distances exceeding 1,000 km. Although it is expected that RZ-DQPSK is more susceptible to nonlinear phase noise (Gordon and Mollenauer 1990), a number of field trials have shown RZ-DQPSK to be robust to the presence of 10-Gbit/s OOK neighbours and transmission over lumped dispersion maps (Fuerst et al. 2006, 2008; Nijhof and Forysiak 2006; van den Borne et al. 2006). Thirdly, and most importantly, the PMD tolerance due to the halved symbol rate increases the DGD tolerance for a 1-dB penalty allocation from 6 to 8 ps to around 17 ps. Although this is smaller than the 10-Gbit/s OOK formats, it is a useable budget for terrestrial fibre of reasonable quality, together with any PMD impairment due to DWDM transmission equipment, whereas the much smaller budget of NRZ-DPSK is quickly exhausted.

In comparing 10-Gbit/s NRZ with these 40-Gbit/s formats, we note that the increased end-of-life OSNR requirements for commercial modules are typically 3–4 dB worse. Together with the increased sensitivity to PMD, this results in a reach reduction to 1,000 km for NRZ-DPSK and 1,500 km for RZ-DQPSK on typical terrestrial networks.

1.3.3 Summary

It is expected that OOK modulation formats will continue to dominate the 10-Gbit/s market, and RZ-DQPSK will become the dominant 40-Gbit/s modulation format soon. In the near future, coherent detection and DSP will be required for 100-Gbit/s transmission with PDM-QPSK as the most likely long-term prospect, although it is possible this format will also take part of the 40-Gbit/s market because it offers increased resilience to OSNR, greater reach and almost unlimited tolerance to PMD. Looking further forward still, in terms of likely commercial deployment, the picture is still emerging, although it is clear increasingly advanced modulation formats, already used in other communication systems, such as 16-QAM, are the likeliest candidates, together with other advances in transmission media to enable a more linear communication channel.

1.4 Field Modulation and Coherent Detection

A rapid growth in capacity and spectral efficiency of WDM transmission systems has taken place over recent years, resulting from the use of high spectral efficiency signal formats, polarisation-division multiplexing, coherent detection and high-speed DSP (Li 2009). WDM channel rates are increasing from 10–40 Gbit/s to 100 Gbit/s through the use of PDM-QPSK, with high tolerance to transmission impairments such as chromatic dispersion and PMD, and continued use of the ITU-T standard 50-GHz grid spacing of the WDM channels. In the future, the use of higher-order modulation formats such as QAM, either single carrier or subcarrier multiplexed, will allow encoding of more than 4 bits per symbol and spectral efficiencies beyond 2 bit/s/Hz. Figure 1.9 shows the spectral efficiency achieved in notable WDM transmission experiments over the last decade, with spectral efficiencies of up to 7 bit/s/Hz being reported in 2,000-km transmission experiments in 2010 (Liu et al. 2010).

Coherent optical communications was the subject of a major research effort during the 1980s due to the improved sensitivity it offers over conventional direct detection (Okoshi and Kikuchi 1988). However, interest in this technique waned in

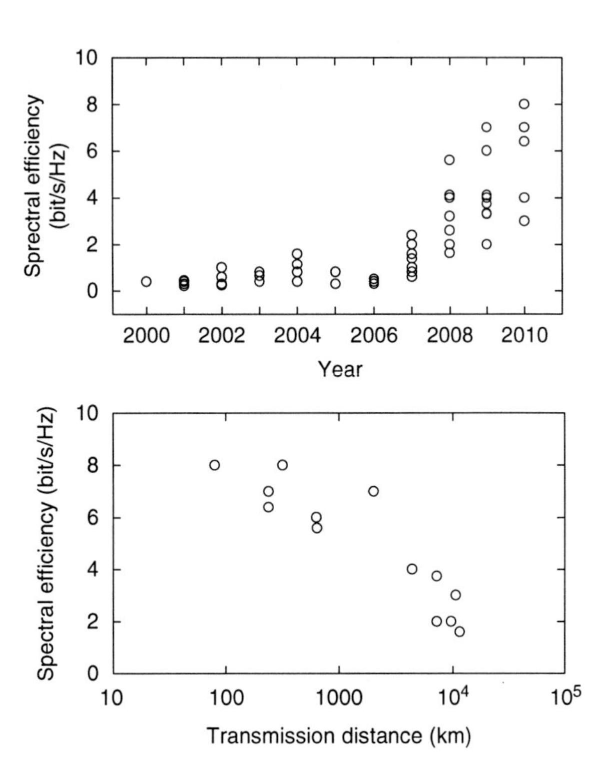

Fig. 1.9 Spectral efficiencies achieved in WDM transmission experiments over the last decade

the 1990s as a result of the introduction of EDFAs and the difficulty of phase and polarisation management with coherent detection. In a coherent transmission system, the information is encoded onto the phase and/or amplitude of the electric field of the light wave, and to measure the complex field, the incoming signal interferes with the output of a *LO* laser in a 90° hybrid as shown in Fig. 1.10 (Noe 2005). The two outputs of the hybrid are detected, giving electrical signals representing the real and imaginary parts of the optical signal. The phase of the transmitter laser drifts, and the polarisation fluctuates in the fibre due to environmental effects. In early implementations of coherent receivers, schemes to track the phase and polarisation of the signal were developed, requiring phase locking of the *LO* laser and fast polarisation controllers, which were difficult to implement in practise (Kazovsky 1986; Barry and Kahn 1992). However, the continuing advances in the power of DSP, as CMOS technology continues to scale following Moore's law, have now made it possible to implement effective phase estimation and polarisation tracking in the electronic domain, following detection of the optical signal with a free-running, fixed polarisation *LO* laser (Sun et al. 2008). Besides this, DSP offers the possibility of adaptive compensation of large values of dispersion and PMD (Savory 2008). For these reasons, interest in coherent transmission has re-emerged in the research community, and commercial exploitation of the technology is now underway.

Fig. 1.10 Schematic of a coherent receiver (Li 2009)

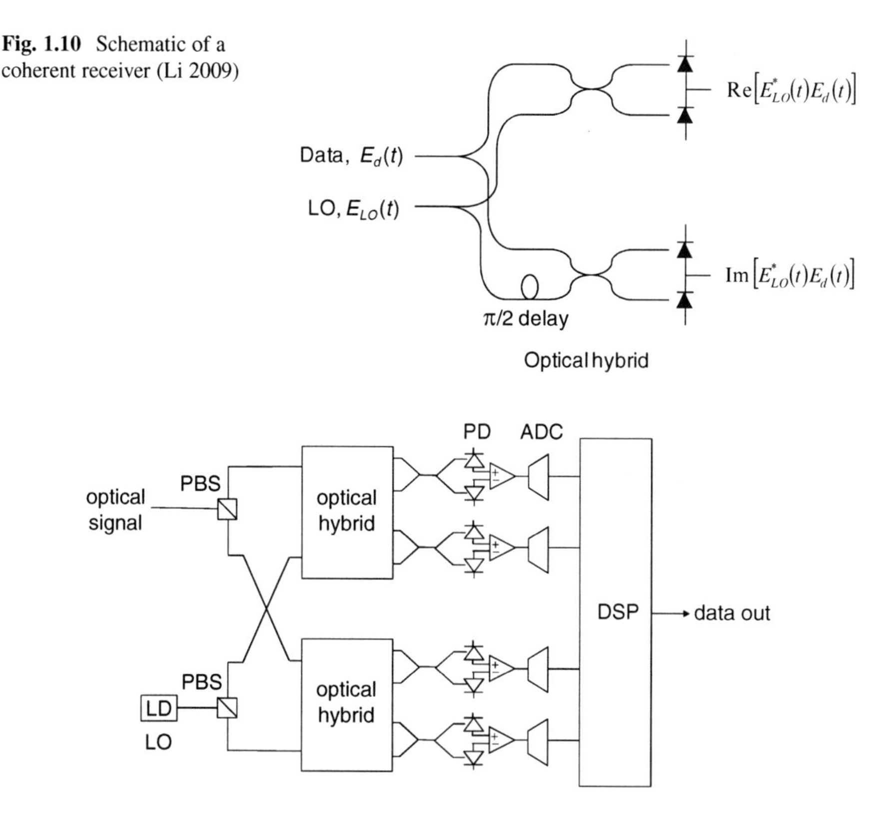

Fig. 1.11 Polarisation-diverse coherent optical receiver. *PBS*: polarisation beam splitter

In this section, an overview of coherent receiver architectures, and the constituent DSP blocks will be presented. Digital carrier phase estimation, polarisation tracking, chromatic dispersion and PMD compensation will be discussed, recent key experiments will be described and progress on compensation of fibre nonlinear effects will be reviewed. Finally, future higher-order modulation formats, moving from QPSK to QAM, will be discussed. This review will include transmitter architectures and implications for receiver algorithms.

1.4.1 Coherent Receiver Architecture

Figure 1.11 shows a schematic of a polarisation-diverse coherent receiver (Li 2009). The function of the receiver is to map the optical field into four electrical signals corresponding to the real and imaginary parts of the optical field for each of the two polarisations and to recover the information encoded onto the field. The incoming signal is split into two polarisations, each of which is combined with the aligned

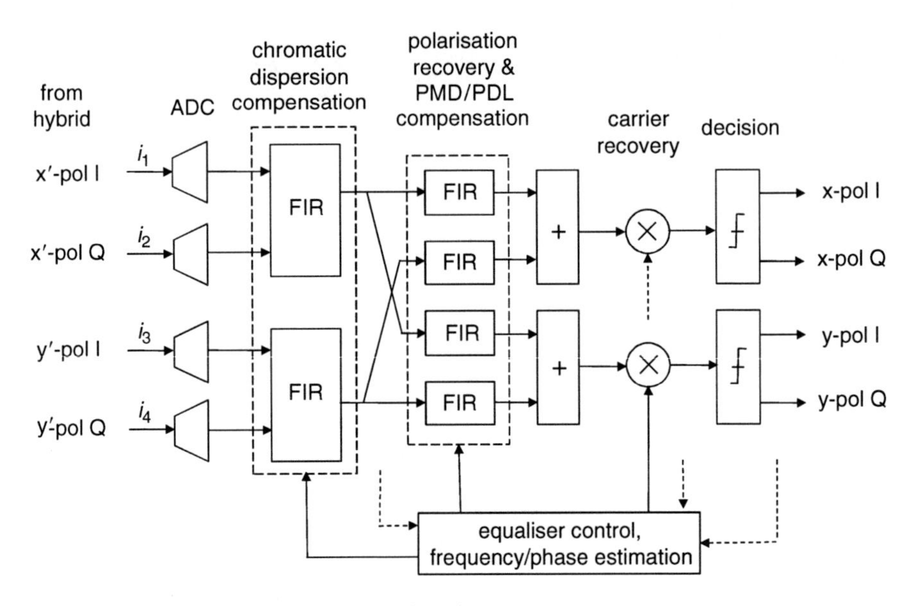

Fig. 1.12 DSP blocks in the coherent PDM receiver

polarisation component of the optical *LO* output using a 90° hybrid. The outputs of the hybrids are detected, and ADCs are used to convert the electrical signals into digital representations.

A schematic showing the main DSP blocks typically used is shown in Fig. 1.12 (Laperle et al. 2008; Savory 2008). Following the ADCs, clock extraction and retiming are carried out to obtain an integer number of samples per symbol. Normalisation and orthogonalisation are used to compensate for imperfections of the hybrids and differing responsivities of the photodetectors. Digital filtering is used to compensate for polarisation rotation, chromatic dispersion and PMD. Finally, carrier frequency and phase estimation is carried out, followed by symbol estimation and forward error correction. In the following sub-sections, the function of some of these DSP blocks will be described in more detail.

1.4.1.1 Carrier Phase Estimation

The requirement for phase locking the *LO* is avoided by using phase estimation in the electronic domain. To illustrate the process of phase estimation with a constant-amplitude phase-shift-keyed signal, a QPSK signal is considered. The phase estimation algorithm is depicted in Fig. 1.13.

The electric field of the signal can be represented as

$$E(t) = Ae^{j(\theta_S(t)+\theta_C(t))} \tag{1.5}$$

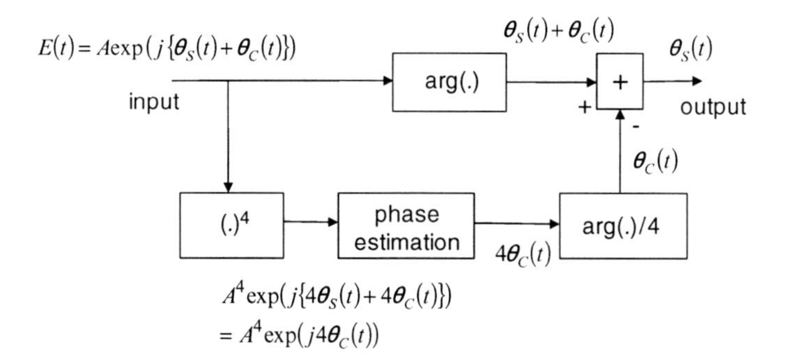

Fig. 1.13 Viterbi and Viterbi phase estimation algorithm

where A is the amplitude (a constant real value), θ_c is the phase of the transmitter laser relative to the *LO* phase, and the data phase θ_s is used to encode two bits per symbol, using four phase values: 0, $\pi/2$, π and $-\pi/2$ rad. By raising the received signal to the fourth power, the data phase is stripped off (Viterbi and Viterbi 1983; Noe 2005):

$$(E(t))^4 = \left(Ae^{j(\theta_S(t)+\theta_C(t))}\right)^4 = A^4 e^{j4\theta_C(t)} \tag{1.6}$$

since $e^{j4\theta_C(t)}$. The relative phase difference between the transmitter and *LO* is then known and can be subtracted from $E(t)$.

In the case of a noisy signal, the phase estimation is no longer accurate. Errors in the estimated value of θ_c occur with a magnitude which is inversely proportional to the OSNR. In this case, the impact of noise on the phase estimation error can be reduced by averaging the estimated phase over a sequence of symbols by filtering the estimates through filter, for example an equal tap weight finite transversal filter (Noe 2005; Ly-Gagnon et al. 2006):

$$\theta_{\text{c,estimated}} = \frac{1}{4}\arg\left(\sum_{k=1}^{N} E_k^4\right) \tag{1.7}$$

where E_k is the kth sample of the optical field, and N is the number of samples over which the phase estimates are averaged. The choice of the value of N is a trade-off determined by the noise level and the rate at which θ_c varies (determined by the transmitter and *LO* linewidths). The optimum phase estimation algorithm makes use of a Wiener filter to average the phase estimates (Taylor 2009).

While the algorithms described here apply to the case where transmitter laser and *LO* are nominally at the same frequency, since the frequency difference between transmitter and *LO* can also be estimated in the digital domain, DSP-based intradyne and heterodyne receivers, in which the transmitter and *LO* are at different frequencies, can also be realised, with the signal being recovered using frequency estimation (Leven et al. 2007).

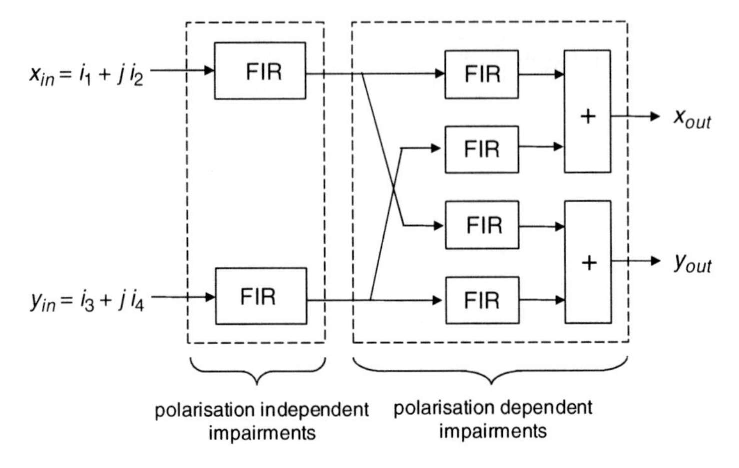

Fig. 1.14 Digital equalisation (Savory 2008)

1.4.1.2 Dispersion Compensation

Figure 1.14 shows the architecture of the digital filtering stage, compensating for polarisation rotation, chromatic dispersion and PMD (Savory et al. 2007). The first block deals with polarisation-independent impairments, while the second compensates polarisation-dependent effects. For linear filtering, the order in which the two blocks operate can be reversed.

The second order group chromatic dispersion of the fibre link results in a transfer function given by (Agrawal 2007)

$$H_{\mathrm{d}}(\omega) = e^{-\frac{j}{2}\beta_{2,\mathrm{c}}\omega^2} \tag{1.8}$$

where $\beta_{2,\mathrm{c}}$ is the cumulated group velocity dispersion of the link of length L

$$\beta_{2,\mathrm{c}} = \int_0^L \beta_2(z)dz \tag{1.9}$$

The dispersion can be compensated by a filter with a transfer function given by $H_{\mathrm{comp}}(\omega) = H_{\mathrm{d}}^*(\omega)$. Such a filter can be realised in the time domain, for example, using a FIR transversal filter with tap weights obtained by sampling the impulse response given by the inverse Fourier transform of $H_{\mathrm{d}}{}^*(\omega)$ (Savory 2008). Alternatively, the compensation can be carried out in the frequency domain, by taking discrete Fourier transforms (DFTs) of the incoming signal and applying the phase correction across the spectrum, in techniques such as overlap-and-add or overlap-and-save (Poggiolini et al. 2009).

1.4.1.3 Polarisation Tracking and PMD Compensation

Polarisation-dependent effects can be described by the Jones matrix. The output electric field, with x- and y-polarisation components, is related to the input field by

$$\begin{pmatrix} E'_x \\ E'_y \end{pmatrix} = \begin{pmatrix} J_{xx} & J_{xy} \\ J_{yx} & J_{yy} \end{pmatrix} \begin{pmatrix} E_x \\ E_y \end{pmatrix} = J \begin{pmatrix} E_x \\ E_y \end{pmatrix} \tag{1.10}$$

In polarisation-multiplexed signal transmission, two independent channels are transmitted on orthogonal polarisations. However, the polarisation in the fibre varies with time due to environmental effects. The receiver is required to track the polarisation rotation, and this is carried out in the second block in the digital compensation architecture in Fig. 1.14, using the lattice structure (Savory 2008). Blind estimation of the Jones matrix, i.e. without requiring training sequences, is possible by using statistical properties of the signal to control the settings of the filter. In the case of constant-intensity modulation formats such as QPSK, the estimation makes use of the fact that, provided the filter characteristics are correctly set, the modulus of the output signal should be constant. An estimation of the Jones matrix is obtained by minimising the mean squared errors $\left\langle \varepsilon_{x,y}^2 \right\rangle$ of the quantities $\varepsilon_x = 1 - |E'_x|^2$ and $\varepsilon_y = 1 - |E'_y|^2$. The matrix elements can be updated in an iterative way by using a stochastic gradient algorithm as described in Savory 2008. In addition to this approach, a decision-directed algorithm, based on monitoring the pre-FEC error rate, can also be used.

The PMD of the fibre can be modelled by the Jones matrix, by using time-dependent elements, where J_{ij} is the response at the ith output polarisation resulting from an impulse applied to the jth input polarisation. The PMD compensation can be carried out using linear FIR filters with impulse responses h_{ij} in the lattice structure, as shown in the digital filtering architecture in Fig. 1.14.

1.4.2 Recent Experimental Studies

Since the first publication of the use of coherent detection of QPSK signals with DSP by Taylor in 2004, the technology has developed rapidly, and the first experimental demonstrations and commercial implementations of 111-Gbit/s transceivers, following the 100-Gbit Ethernet standardisation (Optical Internetworking Forum 2010), have taken place over the last few years.

In one of the earliest reported experiments (Taylor 2004), a 10.66-Gbit/s BPSK signal was successfully detected using offline signal processing. The received signal was digitised and stored using a digital storage oscilloscope, and the subsequent signal processing was carried out offline, an approach that has been subsequently adopted in the majority of experiments on this topic. Dispersion compensation of 1,470 ps/nm was also achieved in this early experiment.

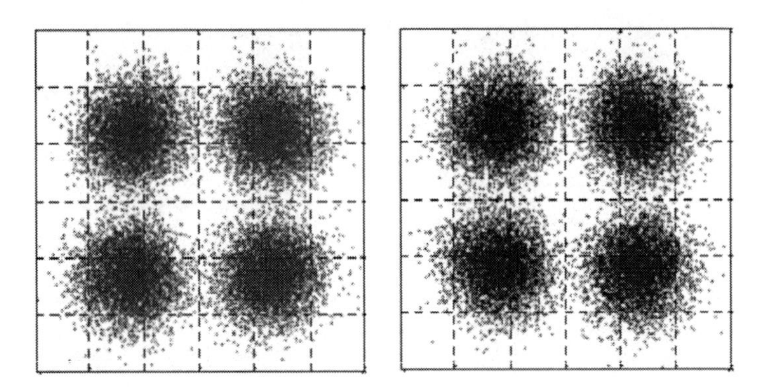

Fig. 1.15 Received signal constellations for *x*- and *y*-polarisations; 42.8-Gbit/s PDM-QPSK after transmission over 6,400 km of uncompensated standard SMF (Savory et al. 2007)

In 2007, single channel 42.8-Gbit/s PDM-QPSK transmission experiments over 6,400 km of uncompensated standard SMF were performed using a recirculating fibre loop, in which the 107,424 ps/nm of dispersion was compensated digitally (Savory et al. 2007). Figure 1.15 shows the recovered signal constellations for the *x*- and *y*-polarisations. Using a fixed differential delay and a loop synchronous polarisation scrambler, effective compensation of PMD with a mean DGD of up to 186 ps was demonstrated (Savory 2008).

A WDM PDM-RZ-DQPSK transmission experiment with a channel rate of 85.6 Gbit/s and spectral efficiency of 1.6 bit/s/Hz was reported in 2007 (van den Borne et al. 2007). The link used consisted of 1,700 km of standard SMF with in-line dispersion compensating fibre. An RZ signal format was employed, and the study included a comparison between interleaved and bit-aligned pulses in the two polarisations, the former being found to improve the tolerance to fibre nonlinearity, though offering lower DGD tolerance.

A transmission distance of 3,200 km over uncompensated standard SMF was achieved with a 4-dB margin with 40-Gbit/s WDM PDM-QPSK, described in (Laperle et al. 2008). Measurements of the impact of polarisation-dependent loss (PDL) of the link, which changes the state of polarisation and degrades the degree of orthogonality of the polarisation-multiplexed signal, were made, showing an OSNR penalty of less than 1.5 dB for mean PDL of up to 2 dB (Fig. 1.16).

Transmission using a channel data rate of 111 Gbit/s has been reported in a number of recent papers, e.g. in Fludger et al. (2008), Charlet et al. (2009), and Cai et al. (2010). In Fludger et al. (2008), a spectral efficiency of 2 bit/s/Hz was achieved in 1-Tbit/s WDM transmission over 2,375 km of standard SMF with in-line dispersion compensation. The optical filtering effect of five add-drop nodes along the line was emulated in these experiments. The possibility of using 1 sample-per-symbol signal processing combined with additional electrical filtering, offering the advantage of lower-cost ADCs, was also demonstrated.

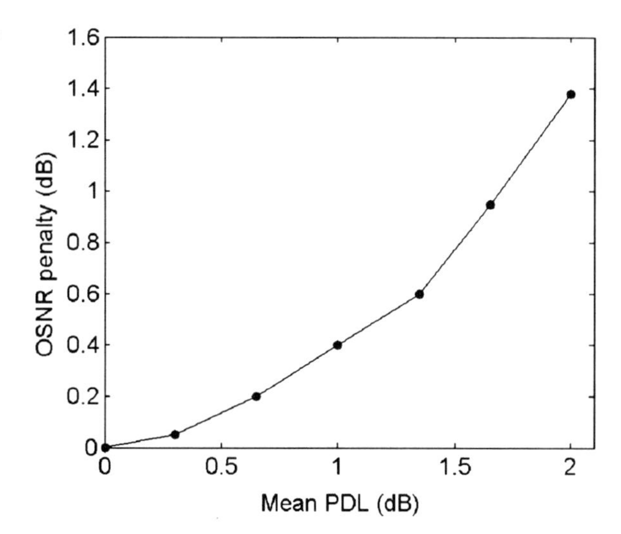

Fig. 1.16 $OSNR_{0.1nm}$ penalty versus mean link PDL (Laperle et al. 2008)

Coherent detection and DSP enabled a record capacity-distance product of 41.8 Petabit/s·km in (Charlet et al. 2009); 164 WDM channels transmitted over 2,550 km. To achieve this result, spans of + D/–D/+D high effective area fibre, and distributed and lumped Raman amplification were employed. Transoceanic distances (up to 10,608 km) with a spectral efficiency of 3 bit/s/Hz were spanned using PDM-RZ-QPSK at 112 Gbit/s per channel in experiments reported in (Cai et al. 2010). To achieve the high spectral efficiency, strong optical pre-filtering of the signals was performed, to minimise inter-channel cross talk, and the resulting intra-channel ISI was mitigated using a maximum a posteriori probability detection algorithm. Signal distortion due to fibre nonlinearity was minimised through the use of high effective area fibre ($A_{eff} = 150\ \mu m^2$).

1.4.3 Future Developments

Among the topics of ongoing research in the field of coherent transmission and high-speed DSP, two major ones are: (1) the compensation of fibre nonlinearity, to allow higher launch powers and hence increased OSNR, and (2) increasing spectral efficiency, using higher-order modulation formats such as 8- and 16-QAM, and reduced WDM channel spacing.

1.4.3.1 Nonlinearity Compensation

As described earlier in this chapter, nonlinear distortion occurs as a result of the Kerr effect, the intensity dependence of the optical fibre's refractive index. This results

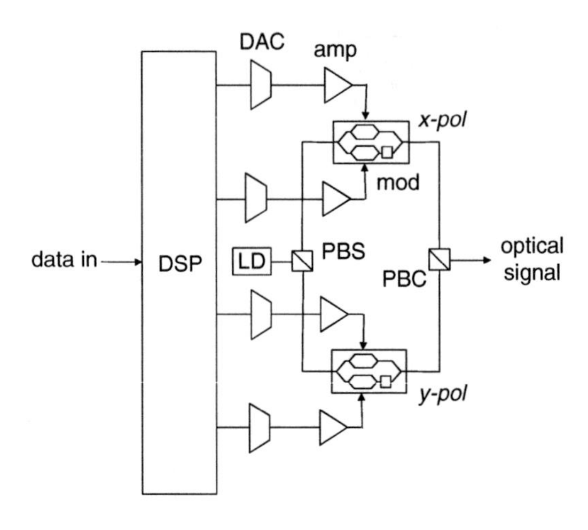

Fig. 1.17 DSP-based PDM transmitter for electronic predistortion and high-order modulation signal generation

in nonlinear phase noise which affects carrier phase estimation and increases the error vector magnitude in the recovered signal constellation. DSP can be used to mitigate the effects of nonlinearity, carried out either at the transmitter or receiver (Roberts et al. 2006; Killey et al. 2005, 2006; Watts et al. 2007; Waegemans et al. 2009; Yamazaki et al. 2007; Li et al. 2008; Mateo et al. 2010). Transmitter-based compensation takes advantage of error-free knowledge of the bit sequence to be transmitted (Roberts et al. 2010). A transmitter architecture to implement this is shown in Fig. 1.17. Signal processing is performed on the bits to be transmitted, in order to calculate the modulator drive signal waveforms, which are converted to analogue form using DACs operating at one or more samples per symbol. Besides the additional DSP and DACs, the design of the transmitter is identical to that of the conventional PDM-QPSK transmitter.

Nonlinearity compensation (NLC) can also be implemented at the receiver (Li et al. 2008), which allows tracking of the time-varying impact of PMD on the nonlinear distortion (Roberts et al. 2010). In receiver-based NLC, the algorithms must be robust to the additional noise on the signal.

Due to the memory of the nonlinearity (the chromatic dispersion causing symbol spreading and hence nonlinear interaction between multiple symbols within the channel), the task of SPM compensation is challenging. For inter-channel effects, XPM and FWM, the challenge is even greater, since the algorithm has to deal with a larger optical bandwidth, and the channel walk-off due to dispersion increases the memory depth.

An approach that has been investigated by a number of groups is digital back-propagation, split-step calculations, based on the nonlinear Schrödinger equation, of the propagation of the signal backwards through the link from receiver to transmitter, thereby reversing the combined effects of nonlinearity and dispersion (Ip and Kahn 2010). SPM compensation by back-propagation has been investigated in single channel 42.7-Gbit/s PDM-QPSK transmission experiments over uncompensated standard SMF in (Millar et al. 2009). The maximum transmission distance of 7,780 km was increased to 10,670 km. However, in a subsequent experimental study on NLC in WDM transmission at 112 Gbit/s (Savory et al. 2010), it was shown that the benefit of back-propagation, in which only intra-channel nonlinearities are taken into account, was significantly reduced due to XPM and FWM from the neighbouring channels. The reach extension achievable with NLC was reduced from 46% for single channel transmission to just 3% in the case of WDM transmission with a 50-GHz channel spacing. For NLC to be effective in long-haul systems with spectral efficiencies of 2 bit/s/Hz or more, it is clear that inter-channel nonlinearities must be taken into account.

Digital compensation of inter-channel XPM and FWM is the subject of recent research described in (Mateo et al. 2010), in which an advanced split-step method is proposed for the digital back-propagation method, using coupled nonlinear Schrödinger equations to compensate for XPM. A 2.5-times increase in transmission distance was achieved in simulations of a 100-Gbit/s per-channel 16-QAM WDM system, while the NLC complexity was kept to a minimum through factorisation of the dispersive walk-off effect between channels, which allows a significant increase in the allowable back-propagation step size.

1.4.3.2 Higher-Order Quadrature Amplitude Modulation

The use of higher-order modulation formats such as *m*-PSK or QAM, in which more than 2 bit/symbol per polarisation are transmitted, will allow further increases in spectral efficiency. Figure 1.18 compares the constellations of QPSK, 8-PSK and 16-QAM, while Fig. 1.19 quantifies the back-to-back required OSNR versus baud rate for a variety of coherent and direct-detection signal formats at a fixed bit rate of 107 Gbit/s, at a BER of 3×10^{-3} (which can be tolerated by FEC with 7% overhead) (Sun et al. 2009). It shows that the baud rate can be reduced, and hence WDM spectral efficiency increased, at the expense of OSNR requirement. For example, the baud rate of PDM-16-QAM is 13.4 GHz, half that of PDM-QPSK at 26.8 GHz, although the required OSNR is increased from 11.8 to 15.4 dB. Consequently, there is significant interest in developing higher-order PSK and QAM transceivers to maximise the capacity of links in which the OSNR is relatively high, e.g. short links, or links employing distributed Raman amplification.

A general-purpose transmitter architecture, able to generate all higher-order polarisation-multiplexed signal formats in addition to implementing electronic predistortion, is that shown in Fig. 1.17. It makes use of DSP and DACs to generate the multi-level modulator drive waveforms. Alternative transmitter architectures, specifically designed to implement particular formats, make use of optical techniques, e.g. parallel modulators or interferometers (Sakamoto et al. 2008; Makovejs et al. 2010).

WDM transmission experiments at a spectral efficiency of 6.2 bit/s/Hz over a distance of 630 km using PDM-16-QAM were reported in (Winzer et al. 2010). Since QAM signals do not have constant amplitude, more advanced DSP algorithms to achieve equalisation and phase estimation may be required. For example, in

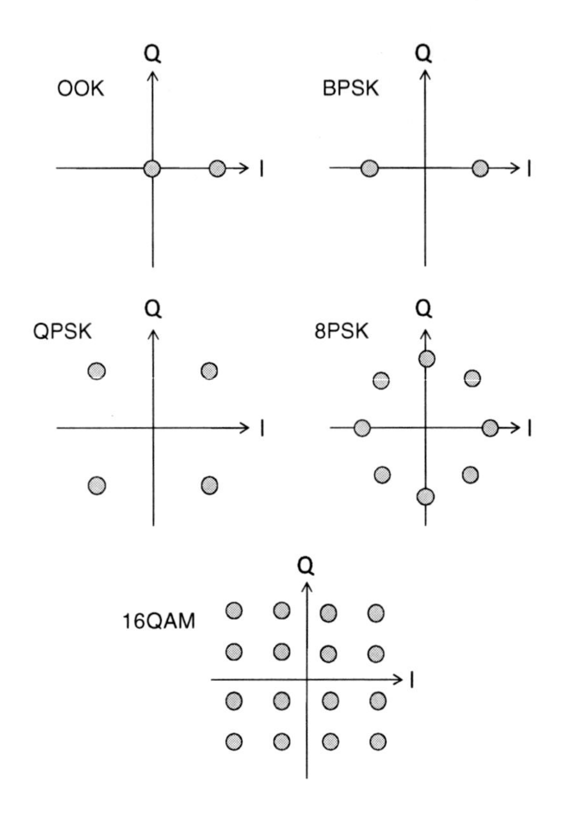

Fig. 1.18 Signal constellations of a variety of signal formats (Sun et al. 2009)

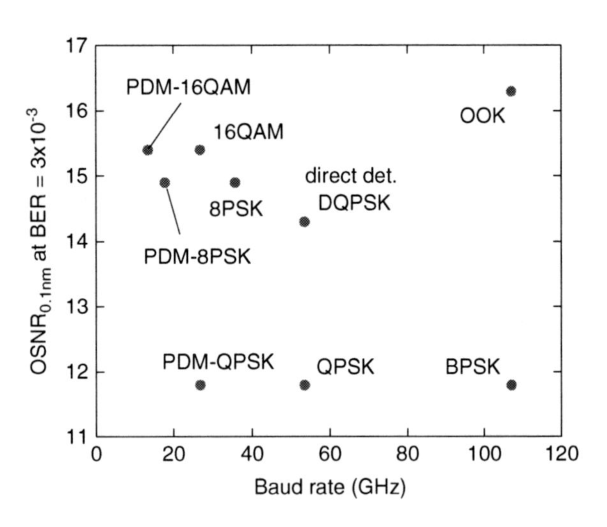

Fig. 1.19 Theoretical values of required OSNR versus baud rate for a variety of signal formats at 107 Gbit/s (Sun et al. 2009)

the case of equalisation, the constant modulus algorithm can be used initially for pre-convergence, followed by a radius-directed (or multi-modulus) decision-aided algorithm to take into account the fact that the symbol points in the constellation lie on multiple rings (Sethares et al. 1989; Fatadin et al. 2009; Winzer et al. 2010; Makovejs et al. 2010).

1.5 Optical Orthogonal Frequency Division Multiplexing

1.5.1 Direct-Detection Optical OFDM

OFDM is a multiplexing scheme used as multi-carrier modulation technique in broadband wireless and radio communications. It has found applications in several telecommunications standards, such as asymmetric, high-speed and very high-speed digital subscriber line (ADSL, HDSL and VDSL); digital audio and video broadcasting (DAB and DVB); IEEE 802.11 wireless local area network (WLAN) and IEEE 802.16 worldwide interoperability for microwave access (WiMAX).

OFDM was first introduced by Chang in 1966 (Chang 1966a) and patented at Bell Labs (Chang 1966b) as a transmission technique using multiple orthogonal carriers to cope with dispersive fading channels. The initial idea of using a bank of subchannel modulators was replaced by the more efficient solution of implementing it with the DFT, thanks to the work of Weinstein and Ebert (Weinstein and Ebert 1971).

The early stage of OFDM system applications was exclusively military. In the 1980s, after Cimini's contribution on OFDM for mobile communications (Cimini 1985), practical OFDM systems started to be considered and implemented for wireless applications.

The first paper on O-OFDM has appeared in the literature in 1996 (Pan and Green 1996), but its resilience to optical channel dispersion impairments was recognised for the first time only in 2001 (Dixon et al. 2001). Starting from 2005, it has found applications in optical fibre communications, and it has rapidly become a hot research topic due to the ever increasing demand for high-data rate transmission (Armstrong 2009).

OFDM represents an attractive technology for next generation optical networks for its scalability to high-speed transmission and its robustness against channel dispersions. In fact, as a multi-carrier technique, OFDM allows transmitting the signal over several orthogonal lower-rate subchannels. Therefore, its use in optical networks meets the twofold requirement of extending the attainable distance without in-line compensation and providing high-data rate transmission (Lowery et al. 2008; Jansen et al. 2008b, 2009b). Since its resilience to dispersion impairments can reduce the conventional per-span compensation, this technology offers an alternative electronic dispersion compensation method to traditional optical techniques (Lowery 2008).

Currently, it is debated whether single carrier transmission technique or O-OFDM will preferably enable the design of 100-GbE optical transport systems (Jansen et al. 2009a, b).

OFDM has rapidly gained attention in the optical community, also due to the progress in electronic DSP for optical communications. In fact, the signal processing in the OFDM transmitter/receiver takes advantage of the efficient algorithm of FFT. On one side, it enables the use of the mature technology and capabilities of DSP; on the other side, it gives a complex and bipolar signal that must be transmitted on an optical link. To do that, different solutions have been proposed: DD and coherent schemes can be used (Schmidt et al. 2008; Shieh et al. 2008a, b). DD systems have lower cost and complexity, whereas coherent OFDM allows implementing high spectral efficiency systems suitable for long-haul transmission with ultimate performance in receiver sensitivity and robustness against dispersions (Jansen et al. 2009b; Yang et al. 2009). DD O-OFDM has been also demonstrated for long-haul systems and, due to its simpler design and cost-effective implementation, the range of applications is wider, including MMF systems and transmission over plastic optical fibre links (Shieh and Djordjevic 2010).

This section is organised as follows: First, the principles of OFDM are given; then, DD O-OFDM systems are described for different types of implementation. A variant of the OFDM scheme, based on an alternative transform, is also discussed. Finally, we illustrate intensity-modulated DD O-OFDM systems for cost-sensitive applications. We describe two possible implementation techniques, comparing the performance in terms of power efficiency.

1.5.1.1 Principles of OFDM

OFDM is a multi-carrier transmission technique where the signal data stream is transmitted over N lower-rate subchannels, whose subcarriers are orthogonal to each other, and their spectra are allowed to overlap.

As recognised by (Weinstein and Ebert in 1971), DFT can be used to perform the OFDM modulation because the transform can be seen as a bank of modulators, whose narrowband channels have mutually orthogonal subcarriers. The subcarrier set is given by the transform kernel. Each one carries a data symbol encoded into a constellation point. QAM is usually used. PSK could also be used, but especially for large-size constellations, it is not suitable, due to the reduced distance between symbols.

The symbol duration T_s is related to the number N of subcarriers: Each QAM symbol has an N times longer duration, compared to single carrier modulation. Thus, OFDM is more robust against channel-induced dispersions.

As indicated in Fig. 1.20, the bit stream is parallelised and mapped into QAM constellation; the resulting vector of N complex elements $\mathbf{x} = [x_0, x_1, \ldots x_{N-1}]^T$ is modulated by using the inverse DFT (iDFT) of order N

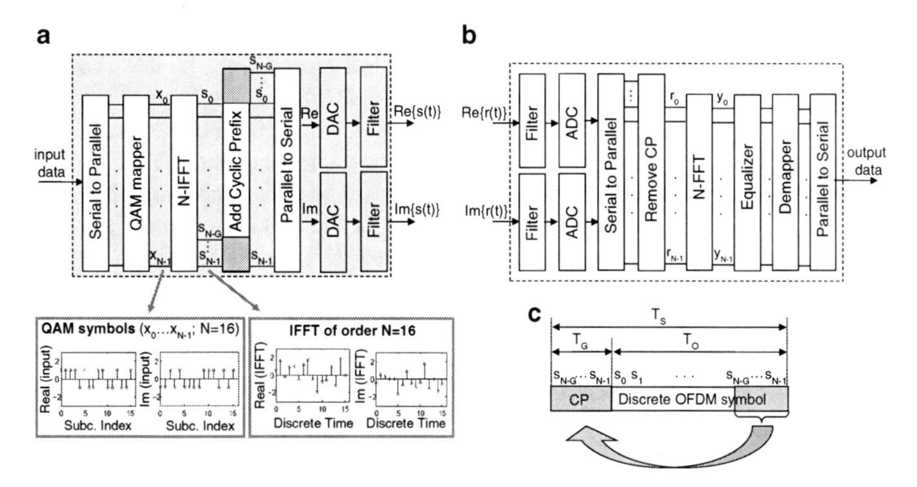

Fig. 1.20 Electrical OFDM (**a**) transmitter and (**b**) receiver. (**c**) CP extension for a discrete OFDM symbol of duration T_O

$$s_k = \frac{1}{\sqrt{N}} \sum_{n=0}^{N-1} x_n \exp\left(\frac{j2\pi kn}{N}\right) \quad k = 0, 1, \ldots, N-1. \tag{1.11}$$

At the receiver side, the discrete received signal $\mathbf{r} = [r_0, r_1, \ldots r_{N-1}]^T$ is demodulated by using the forward DFT

$$y_n = \frac{1}{\sqrt{N}} \sum_{k=0}^{N-1} r_k \exp\left(\frac{-j2\pi kn}{N}\right) \quad n = 0, 1, \ldots, N-1. \tag{1.12}$$

Note that the forward and inverse DFTs have the same normalising factor $1/\sqrt{N}$, so that the input vector and its transform have the same total energy.

The DFT and iDFT are implemented in the DSP by fast algorithms [FFT and iFFT] in order to lower the computational complexity. To do that, N is required to be a power of 2. The well-known radix-2 Cooley and Tukey algorithm reduces the required complex multiplications from N^2 to $N/2\log_2 N$ and the complex additions from $N(N-1)$ to $N\log_2 N$ (Cooley and Tukey 1965).

If neither noise nor distortion, due to the channel or introduced by the transmitter and receiver, is taken into account, the received signal equals the signal at the output of the transmitter ($r = s$). Therefore, the vector of QAM symbols is perfectly recovered, resulting $y = x$, for the orthogonality of the Fourier transform.

Let us now consider the delay introduced by a dispersive channel. Different subcarriers experience different delays, and for a fixed DFT window, two consecutive OFDM symbols can interfere generating ISI. Moreover, due to the delay spread, the signal components carried by the slower subcarriers appear incomplete in the DFT window, destroying the carrier orthogonality and resulting in an ICI. In order to mitigate both ISI and ICI, a cyclic redundancy is introduced in the OFDM symbol, called CP. The elements in the tail of the discrete OFDM signal vector s are copied and pasted as a prefix (see Fig. 1.20c). This represents an overhead in the transmitted signal, which reduces the supported data rate. The total symbol duration T_s is given by the duration of the OFDM symbol T_O and the additional component of the CP length T_G. Generally, the CP is a small fraction of the OFDM symbol, but to be effective, it should be longer than the delay spread. As the FFT order increases, the impact of CP on the data rate becomes less significant. Generally, a 10% CP is considered in practical systems. It has been demonstrated that large dispersion tolerance is achieved using long OFDM symbol lengths (Jansen et al. 2009a, b).

The use of CP combined with equalisation allows a correct recovery of signals distorted by a linear dispersive channel. Channel estimation is required for the equalisation, and it is enabled by periodically inserting training symbols. At the expense of this additional overhead [typically between 0.2% and 4% in coherent systems (Jansen et al. 2009a, b)], amplitude and phase errors in the y vector can be corrected with a simple *single tap* equaliser, performing one complex multiplication for each element.

The overhead amount, including oversampling, impacts on the spectral efficiency of the OFDM system. Oversampling is often applied to relax the filtering requirements, at the expense of the number of subcarriers carrying data.

The spectral efficiency of OFDM can be enhanced by using multi-level modulations, such as m-QAM; m typically varies between 4 and 64.

The digital complex OFDM signal is converted to analogue by using two DACs, one for the real and one for the imaginary part, respectively. Similarly, at the receiver side, two ADCs are also required. DAC and ADC are critical elements in the design of O-OFDM systems due to the required high resolution, especially when large constellations are applied.

1.5.1.2 Transmitter and Receiver Configurations

Transmitting the complex OFDM signal on an optical link by using conventional DD system is a challenging issue, which partially motivates the recent introduction of this technology in optical communications. In fact, standard OFDM systems require coherent detection: An *LO* is needed to recover the OFDM signal, whereas in DD system no laser is present at the receiver. Indeed, as illustrated in Fig. 1.21, a simple photodetector is used, which can be modelled with a square-law characteristic. Furthermore, only real signal can be modulated by using a single MZM. However, with a suitable choice of the transmission scheme, OFDM can be applied to this simple and cost-effective system that can be implemented by using commercial components. Different solutions are possible for the transmitter (depending on the application) while maintaining the same detection scheme (Schmidt et al. 2008).

A first solution is based on the electrical transmitter of Fig. 1.20a. The real and imaginary part of the baseband signal $s(t)$, at the output of the transmitter, must be

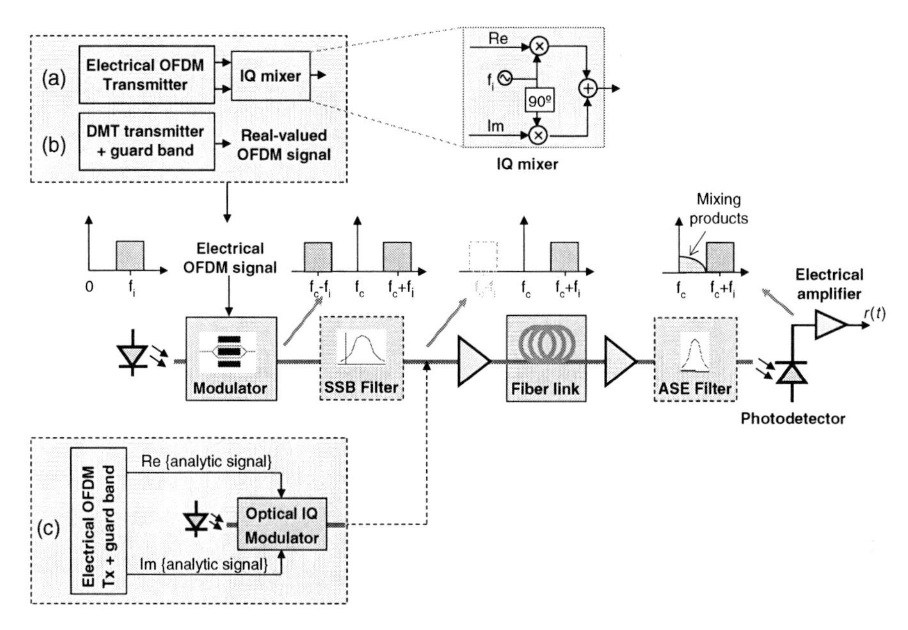

Fig. 1.21 Three different transmission architectures for O-OFDM using DD scheme. SSB modulation with guard interval is illustrated. (**a**) Electrical OFDM with IQ electrical mixer, (**b**) Disctrete multi-tone modulation, (**c**) Electrical OFDM with IQ optical modulator

first electrically up-converted, using a RF mixer, as indicated in Fig. 1.21a. Then the RF up-converted signal can be optically modulated by a single MZM. The MZM creates a double side band spectrum with respect to the optical carrier (f_c), as shown in Fig. 1.21.

Due to the lack of an *LO* at the receiver, in DD systems the optical carrier should not be completely suppressed. The carrier to signal power value is chosen to optimise the system performance. It has been observed that an optimal choice consists in allocating the same amount of power for both the carrier and the OFDM signal (Lowery and Armstrong 2006). The MZM bias point should also provide linear field modulation. In fact, the effectiveness of DD depends on the system linearity, and only a correct mapping between the RF OFDM signal and the optical signal enables a correct detection. In order to ensure that the OFDM subcarriers are represented only once by the optical frequencies and avoid chromatic dispersion fading, SSB modulation can be adopted. In this case, an optical filter is required to transmit only one side of the optical spectrum.

The photocurrent is given by the OFDM signal term and unwanted mixing products, since it is proportional to $|r(t)|^2$. The intermodulation distortions can be minimised, at the expense of the spectral efficiency, by inserting a guard band between the OFDM signal and the optical carrier. This can be easily obtained by properly selecting the intermediate RF frequency f_i of the IQ mixer. Typically it is 1.5 times the signal bandwidth; for example, if the OFDM signal has a 5-GHz bandwidth, f_i should be 7.5 GHz.

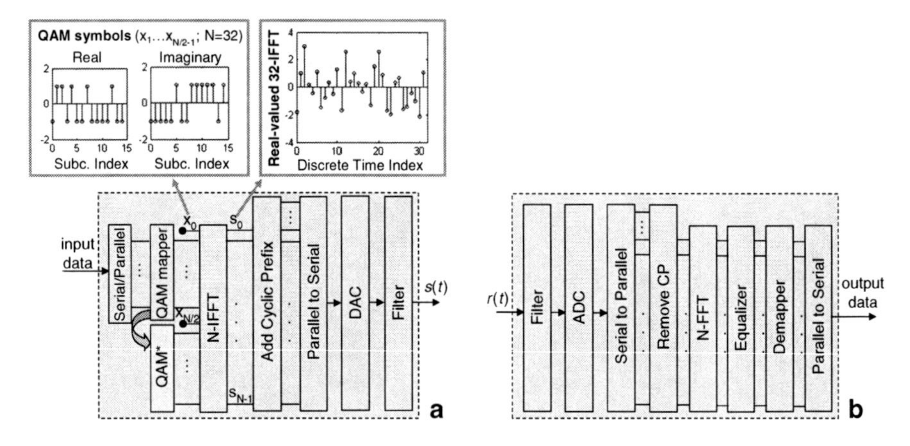

Fig. 1.22 DMT (**a**) transmitter and (**b**) receiver. The QAM data symbols supported by 32 subcarriers and the corresponding real-valued iFFT are indicated

Similarly to the transmitter scheme, an electrical IQ mixer is required at the receiver, after DD, to recover the baseband signal and perform the digital processing. To reduce the amplified spontaneous emission noise of the optical amplifiers, a narrow band filter can be added before the photodetector.

This DD scheme has been experimentally demonstrated for long-haul transmission (400 km of SMF) at 10 Gbit/s using 4-QAM modulation (Schmidt et al. 2008).

A simpler and cost-effective implementation of the OFDM technique is possible if the OFDM signal is real-valued. This scheme furnishes a second transmitter architecture solution. To generate real OFDM symbols, the input signal mapped into a complex constellation is forced to have Hermitian symmetry. In fact, the DFT of a real signal satisfies this property. When the OFDM signals are real-valued, the multi-carrier transmission technique is considered a special case of OFDM: In the literature, it is referred as discrete multi-tone modulation (DMT) (Lee et al. 2007, 2008). The complexity of the electronic design is reduced: Only one DAC and no IQ modulation onto an RF carrier are required at the transmitter, and after DD, the received signal $r(t)$ is converted to digital by one single ADC, without IQ mixing.

An electrical transmitter able to generate real-valued OFDM signals is described in Fig. 1.22a. The serial-to-parallel converted bit sequence is mapped into a 4-QAM constellation. Only half of the subcarriers support data symbols, since to obtain a real-valued vector at the output of the iFFT, half the carriers are required to support the complex conjugate. An example for a real-valued *32*-order iFFT is given in Fig. 1.22. The vector at the input consists of $N/2 - 1 = 15$ complex information symbols and their complex conjugate values (indicated in the figure with QAM*). The DC and Nyquist frequencies (x_0 and x_{16}) are set to zero to ensure Hermitian symmetry; alternatively, they could also be real-valued.

Since in DMT systems there is no need for up-conversion at an intermediate frequency, the guard band can be created at the expense of the number of subcarriers supporting data. For a guard band equal to the signal bandwidth, $N/2$ subcarriers of an N-order iFFT must be set to zero, reducing the information symbols to $N/4$ ($N/4$ are required for the Hermitian symmetry constraint).

SSB can be also applied to this second scheme (Fig. 1.21b), enabling the transmission at 20 Gbit/s with 16-QAM over 320 km of SMF, as experimentally demonstrated in 2007 (Schmidt et al. 2007). Ali et al. have numerically shown that up to 40 Gbit/s over 640 km of uncompensated standard SMF can be transmitted (Ali et al. 2009). They have also theoretically shown that by properly selecting the MZM bias point, the guard band can be reduced to enhance the system spectral efficiency at the expense of the receiver sensitivity.

However, in case of cost-sensitive applications, where the robustness against dispersion is a more relaxed requirement (e.g. in short-reach applications), double side band OFDM can be adopted.

A third transmission scheme is based on the concept of the Hilbert transform, described in Fig. 1.21c. An optical signal with non-negative frequency components is generated without the need of an optical filter. This can be done by modulating the real and imaginary parts of an analytic signal by using a complex optical modulator (IQ optical modulator). The iFFT output gives an analytic signal if half of the elements of the input vector x are set to zero. As for the second transmission scheme, a guard band can be generated by setting to zero the required input vector elements. Schmidt et al. have demonstrated that 12-Gbit/s transmission with 32-QAM is possible over an SMF length of 400 km (Schmidt et al. 2007). Peng et al. have also proposed an alternative spectral efficient implementation in which the symbols and zeros are interleaved (Peng et al. 2009). They have demonstrated that the optical carrier can be suppressed if the detection is assisted by a pilot tone, achieving better sensitivity and chromatic dispersion tolerance than baseband SSB O-OFDM; 10 Gbit/s is transmitted with 8-QAM over 260 km of standard SMF.

1.5.1.3 Alternative Transform for OFDM Modulation/Demodulation

Fourier transform always implies complex processing, and the phase carries fundamental information, while a real trigonometric transform is particularly attractive for the processing of real signals, and it can find applications in DMT systems. Discrete Hartley Transform is a real transform, similar to the DFT, differing in the fact that the even and negative odd parts of the DHT coincide with the real and imaginary parts of the DFT, respectively (Bracewell 1983). Similarly to DFT, DHT can be seen as a bank of modulators, whose mirror-symmetric sub-bands ensure subcarrier orthogonality, and enable to carry the data symbols for the parallel processing (Wang et al. 2005). Therefore, fast Hartley transform (FHT) can replace FFT algorithm to furnish an alternative OFDM scheme, as indicated in Fig. 1.23.

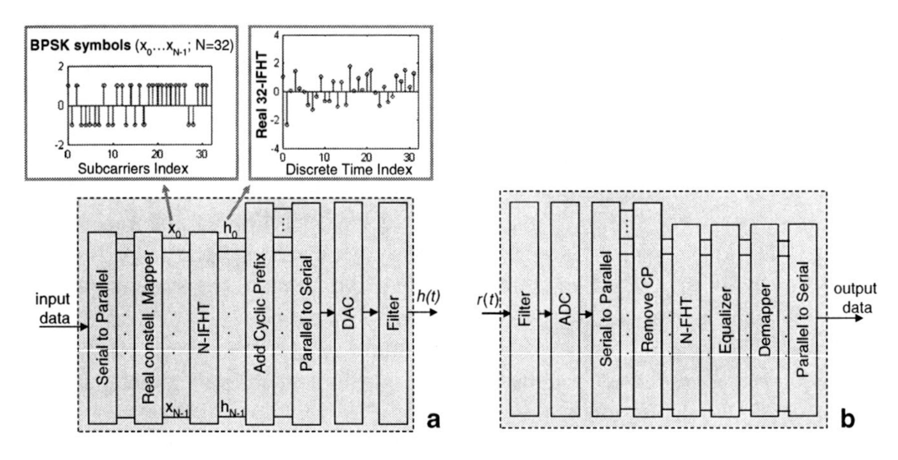

Fig. 1.23 FHT-based OFDM (**a**) transmitter and (**b**) receiver. In the insets, 32 BPSK input symbols and corresponding real-valued iFHT

According to the definition of DHT (Bracewell 1983), the OFDM symbol is given by

$$h_k = \frac{1}{\sqrt{N}} \sum_{n=0}^{N-1} x_n \left[\cos\left(\frac{2\pi k n}{N}\right) + \sin\left(\frac{2\pi k n}{N}\right) \right] \quad k = 0, 1, \ldots, N-1, \tag{1.13}$$

where x_n indicates the nth constellation symbol, and N is the transform order. The DHT kernel is real, and if a real constellation [e.g. BPSK, M-PAM (pulse-amplitude modulation)] is used for the subcarrier modulation, the OFDM symbol h_k is real.

In DMT systems, only half of the iFFT points are used to process the information data symbols (independent complex values); the second half is required to process the complex conjugate vector (see Fig. 1.22). DHT directly deals with real signals without the need for Hermitian symmetry constraint: If the input vector is real, the inverse FHT (iFHT) is real-valued, and the number of subcarriers carrying information symbols (independent real-valued values) coincides with the DHT points, as shown in Fig. 1.23a. Therefore, in order to transmit the same data signal, a lower constellation size is required, compared to the system of Fig. 1.22, and no additional resources are needed for the calculation of the complex conjugate vector, as in standard DMT (Svaluto et al. 2010).

DHT, being a real transform, is self-inverse; thus, the same FHT routine can be applied to calculate the inverse and forward transforms. The computational complexity is lower than FFT algorithms for complex data (e.g. the Cooley and Tukey algorithm), since only real multiplications have to be calculated, and no complex algebra has to be applied (Hou 1987). However, compared to optimised algorithms for the FFT of real-valued sequences, FHT requires about the same numbers of multiplications, but more additions. For decimation-in-time or decimation-in-frequency radix-2 algorithm, $N-2$, more additions are required to evaluate an N-order DHT, compared to the real-valued DFT of the same order (Sorensen et al. 1985). Similarly, for radix-4, split radix, prime factor and Winograd transform algorithms, the number of additions required by the FHT slightly exceeds the ones required by the FFT of a real-valued sequence (Sorensen et al. 1985). To improve the DSP speed, FHT algorithm with minimum arithmetic complexity can be applied. In this case, FHT only requires two more additions than the faster real-valued FFT (Duhamel and Vetterli 1987): The number of multiplications is $(N\log_2 N - 3N + 4)/2$, and the additions are $(3N\log_2 N - 5N)/2 + 6$.

As in the standard implementation of OFDM, a CP can be used to mitigate ISI and ICI. It has been demonstrated that sinusoidal transforms (including Hartley transform) with a symmetric extension may outperform DFT, when the delay spread due to the channel is longer than the CP duration (Merched 2006).

Similarly to DMT systems, as indicated in Fig. 1.23, only one single DAC at the transmitter and one single ADC at the receiver are required.

1.5.1.4 DD O-OFDM Using Intensity Modulation: AC and DC-Biased Solutions

Intensity-modulated DD O-OFDM represents an attractive solution for short-reach applications using SMF (Tang and Shore 2006), for applications using MMF, such as optical local area networks (Lee et al. 2008), or even for plastic optical fibre link (Lee et al. 2009). In fact, when IM is used, the system is less robust against dispersion impairments due to the nonlinear mapping between the OFDM baseband signal and the optical field, but it is characterised by low cost.

Direct laser modulation can replace external modulator to further lower the overall system cost, and high-capacity systems can be provided by using higher-size constellation. Adaptively modulated O-OFDM technique, where different subcarriers support different modulation formats, can be adopted to optimise the system performance (Tang and Shore 2007). This way, flexible and robust systems can be designed, for example, to upgrade installed 1-Gbit/s Ethernet backbones to 10 Gbit/s and beyond (Tang and Shore 2007).

Since in IM systems the OFDM signal $s(t)$ is represented by the optical intensity and not by the optical field, $s(t)$ must be unipolar, i.e. real and not negative. A positive signal can be obtained by adding a DC bias to the real OFDM signal generated by a DMT transmitter. Usually, the bias value B_{DC} is at least twice the signal standard deviation, resulting in an inefficient solution in terms of optical power. A power efficient technique to transmit unipolar signals is the asymmetrical clipping (AC), proposed in 2006 by Armstrong and Lowery (2006).

When AC is adopted, only the odd subcarriers are modulated, and the OFDM signal can be clipped at zero level without losing information. In fact, if only the symbols x_n carried by the odd subcarriers are non-zero, the OFDM signal is redundant: $s_k = -s_{k+N/2}$, N being the transform order. The non-redundant signal $s_{\mathrm{c}}(t)$ can be obtained by clipping at zero level $s(t)$

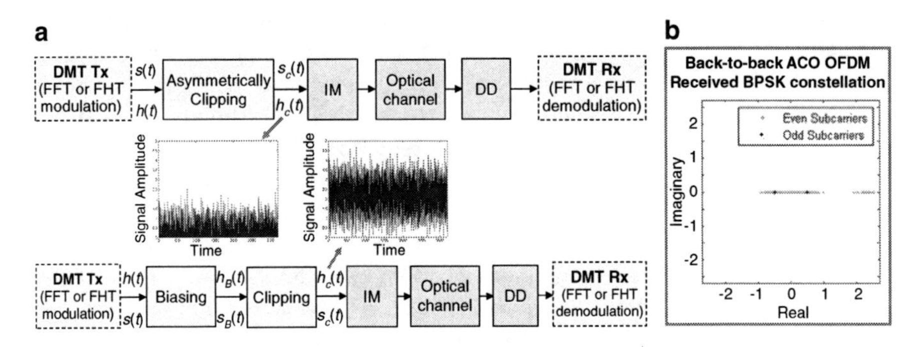

Fig. 1.24 (**a**) ACO (*top*) and DCO-OFDM (*bottom*) for FFT- and FHT-modulated signals (indicated with s(t) and h(t), respectively). In the insets, simulation of 50 symbols using BPSK and 64-FHT, for ACO and DCO. (**b**) Received constellation for FHT-based ACO-OFDM back-to-back system, using BPSK. For 4-QAM constellation FFT-based approach, see Fig. 3 of (Armstrong and Lowery (2006)

$$s_c(t) = \begin{cases} s(t) & s(t) > 0 \\ 0 & s(t) \leq 0 \end{cases} \tag{1.14}$$

The resulting unipolar OFDM signal is intensity modulated and transmitted in the optical channel. The symbol sequence is recovered, by using DD, from the odd subcarriers; the constellation points have half of the original values. All the clipping noise falls into the even subcarriers that can be easily discarded (Armstrong and Lowery 2006).

Clipping is a memoryless nonlinearity. For $N \geq 64$, the OFDM signal can be assumed to have a Gaussian distribution, so Bussgang's theorem can be applied, and clipping results in the attenuation of the input sequence and in the presence of additive noise.

This is still valid for OFDM signals modulated by DHT, as demonstrated in Svaluto et al. (2010). Figure 1.24a shows the AC and DC-biased systems (ACO and DCO-OFDM) for both FFT- and FHT-based DMT transmitters. The first inset presents the clipped FHT-modulated OFDM signal for BPSK-mapped symbols, carried by the odd subcarriers. The constellation received in a back-to-back system is reported in Fig. 1.24b. According to the Bussgang's theorem statement, the even subcarriers give the noisy component of the signal, and the odd subcarriers furnish the half-valued constellation points.

AC and DC-biased techniques trade power and bandwidth efficiency. Indeed, when AC is applied, the optical power, which is proportional to the electrical OFDM signal, is substantially reduced, compared to the DC-biased signal (see the insets of Fig. 1.24a). However, with DCO-OFDM all the subcarriers can be modulated. The choice of a fixed bias value trades power efficiency and additional noise. A minimum bias value should be twice the signal standard deviation, as considered in the system of Fig. 1.24a. Due to the high PAPR of OFDM signals, residual negative peaks can be present in the biased OFDM signal $s_B(t) = s(t) + B_{DC}$ (or $h_B(t) = h(t) + B_{DC}$, if FHT is used for the modulation) that have to be clipped at zero level for IM. Therefore, an additional noise component affects the signal, and depending on the clipping level, it can severely degrade transmission.

Here we report the performance simulations of AC and DC-biased solutions in additive white Gaussian noise channel to furnish an analysis and a comparison of the two approaches. No CP is considered, and the impulse response of the optical channel is assumed to be unitary. A Gaussian noise source is added to the systems of Fig. 1.24a, in the electrical domain at the receiver side, after DD.

For the analysis of DCO-OFDM, the bias is defined as $B_{DC} = k\sqrt{E\{s^2(t)\}}$, where $E\{s^2(t)\}$ is the signal variance, and the bias value in dB is $10\log_{10}(k^2+1)$. A 7-dB bias (corresponding to twice the signal standard deviation) and a larger bias value of 13 dB are considered for the simulations, according to the clipping levels considered in Armstrong and Schmidt (2008) and Tang and Shore (2007). Figure 1.25a shows the BER as a function of the bit electrical energy normalised to the noise power spectral density. The curves have been obtained for FHT-based O-OFDM systems using BPSK and 4-PAM modulation formats (Svaluto 2010). The same performance is achieved with the FFT-based DC-biased systems reported in (Armstrong and Schmidt 2008) using 4- and 16-QAM, respectively. In fact, in order to compare DHT-based and FFT-based O-OFDM systems transmitting the same signal at the same bit rate, the real M-ary constellation must be $M = 2^{\log_4 L}$, where L is the QAM constellation size (Svaluto et al. 2010).

The additional noise due to the clipping of residual negative peaks decreases with the increase of the bias value, at the expense of the electrical and optical power. The bias choice depends on this trade-off. However, when a large-size constellation is considered, a larger bias is needed (Tang and Shore 2007; Lee et al. 2008); otherwise, the system performance can be severely degraded and BER floor occurs above the target BER of 10^{-3} (Armstrong and Schmidt 2008). This BER value ensures error-free transmission if forward error correction is applied.

The power efficiency of AC technique is superior to DC-biased solution, as shown in Fig. 1.25b for FHT-based O-OFDM using BPSK, 4- and 8-PAM. The same BER curves can be obtained with FFT-based systems using 4-, 16- and 64-QAM (Armstrong and Schmidt 2008). In fact, in AC technique, only $N/4$ of the N FFT subcarriers support data symbols, while the available subcarriers of an N-order DHT are $N/2$ (all the odd-indexed carriers), since it does not require Hermitian symmetry.

Here we have analysed the performance in terms of electrical energy. According to the Bussgang's theorem and the considerations reported in Armstrong and Lowery (2006) and Armstrong et al. (2006), the relation between electrical and optical power can be easily derived for both ACO and DCO-OFDM systems. The electrical-to-optical conversion in dB can be obtained by using the formulas (Armstrong and Schmidt 2008)

$$E_{b(opt)}/N_0 = 10\log_{10}(1/\pi) + E_{b(elec)}/N_0 \tag{1.15}$$

$$E_{b(opt)}/N_0 = 10\log_{10}\left(k^2/(k^2+1)\right) + E_{b(elec)}/N_0 \tag{1.16}$$

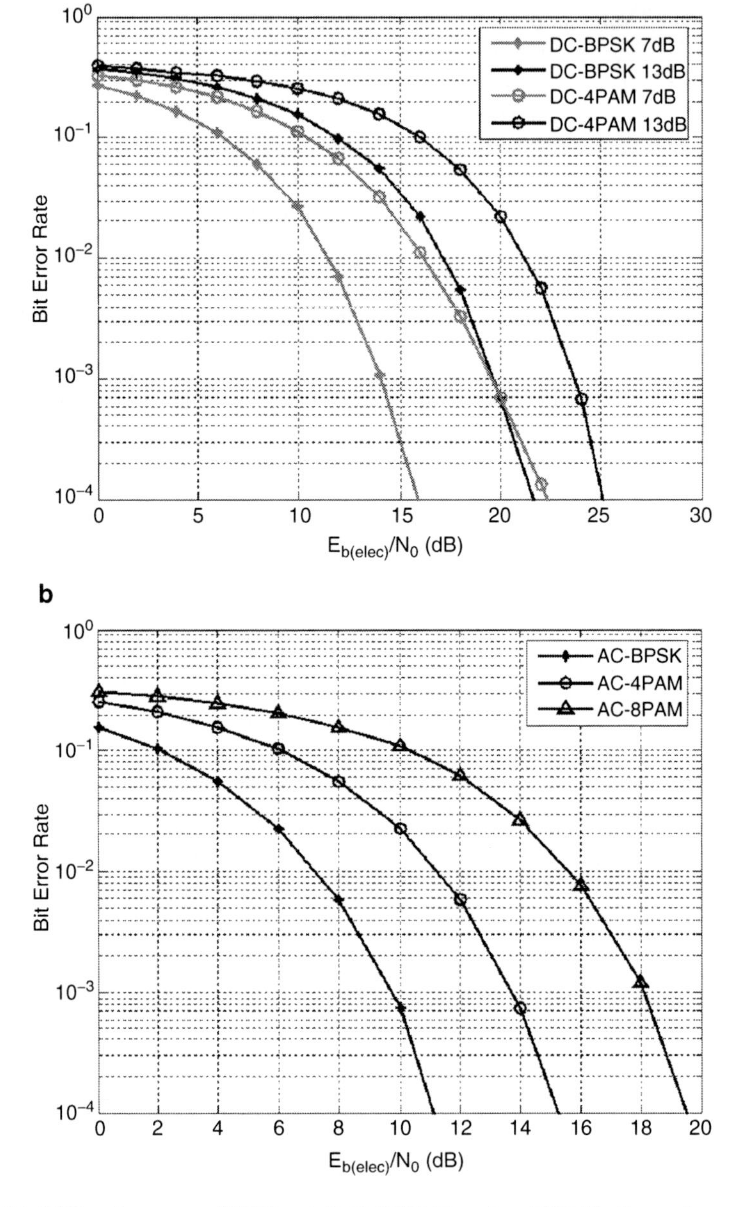

Fig. 1.25 BER performance of DCO and ACO-OFDM based on FHT. Similar performance can be achieved with FFT-based OFDM using L-QAM constellations (Armstrong and Schmidt 2008)

for AC and DC-biased systems, respectively. Equations 1.12 and 1.13 hold for OFDM signals obtained by either FFT or FHT modulation.

As an example, we calculate the normalised optical power per transmitted bit for a BER of 10^{-3}. If AC is considered, $E_{b(opt)}/N_0 = 4.8$ dB using either FFT with 4-QAM modulation or FHT with BPSK modulation (Armstrong and Schmidt 2008; Svaluto et al. 2010). When the system is based on DCO-OFDM, $E_{b(opt)}/N_0 = 13$ dB, considering a DC bias of 7 dB.

1.5.2 Coherent-Detection Optical OFDM

OFDM is a multi-carrier multiplexing scheme, where individual modulated subcarriers are transmitted in a multiplexed setup in parallel. Contrary to the classical frequency division multiplexing approach, the individual spectra of adjacent subcarriers may overlap. To avoid cross talk between the subcarriers, the orthogonality condition has to be fulfilled. At the frequency of the nth subcarrier, all other subcarriers have zeros (Fig. 1.26). This orthogonality is achieved if the subcarrier frequencies are equidistant in frequency domain leading to the frequency of the kth subcarrier (Bahai 1999)

$$f_k = f_0 + k \cdot f_s, \tag{1.17}$$

where f_0 is the lowest frequency of the OFDM signal, f_S is the frequency spacing, and k is the subcarriers index with $k = 0 \ldots N_{SC} - 1$.

1.5.2.1 Coherent Optical OFDM Transmitter and Receiver

For the generation of an O-OFDM signal, first, the equivalent baseband representation of the signal is calculated by DSP. Second, by driving an optical IQ modulator with real (I) and imaginary (Q) constituents of this time domain signal, the equivalent baseband signal is up-converted to the optical carrier frequency. The DSP calculates the time domain signal for one OFDM symbol period by applying the iDFT at the vector of complex subcarrier amplitudes. The amplitude vector is obtained by mapping the data bits to the subcarriers' constellations, i.e. the modulation of the subcarriers. The iDFT output signal corresponds to equally spaced subcarriers according to formula (1.1). For the modulation of subcarriers, QAM is typically applied in optical transmission.

For demodulation and de-multiplexing of the orthogonal subcarriers, a matched filter bank is required, realised by the DFT in the digital domain. In practise the DFT and iDFT are realised by FFT and iFFT algorithms exhibiting an effective realisation for computation. Figure 1.27 shows the schematics of typical OFDM transmitter and receiver setups. In the transmitter, the incoming serial data are re-organised into blocks and subsequently coded for the designated modulation scheme. The iFFT

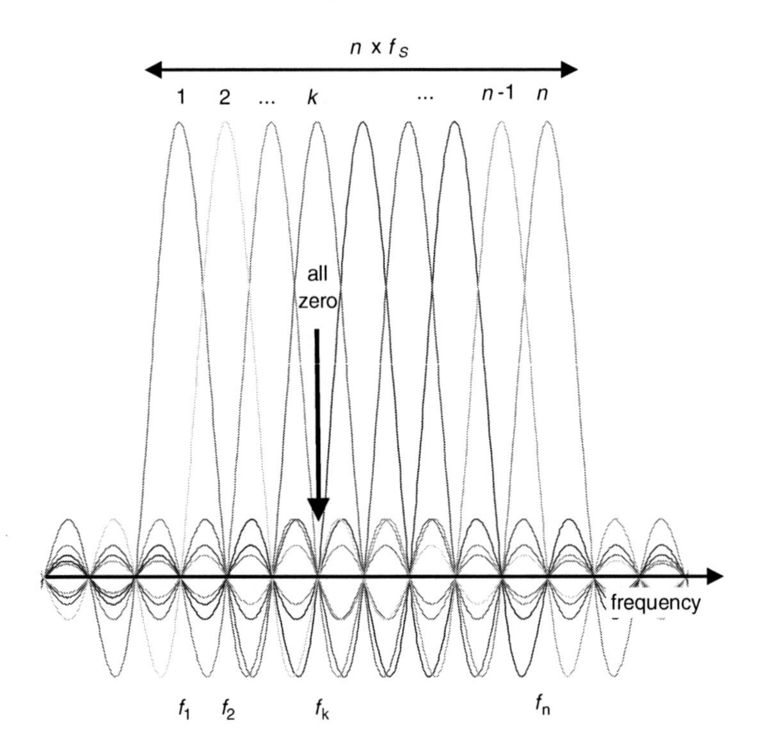

Fig. 1.26 Spectrum of an OFDM signal with overlapping spectra of the individual subcarriers

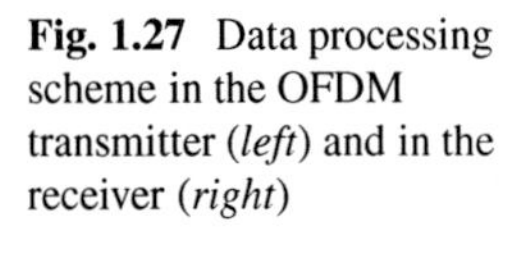

Fig. 1.27 Data processing scheme in the OFDM transmitter (*left*) and in the receiver (*right*)

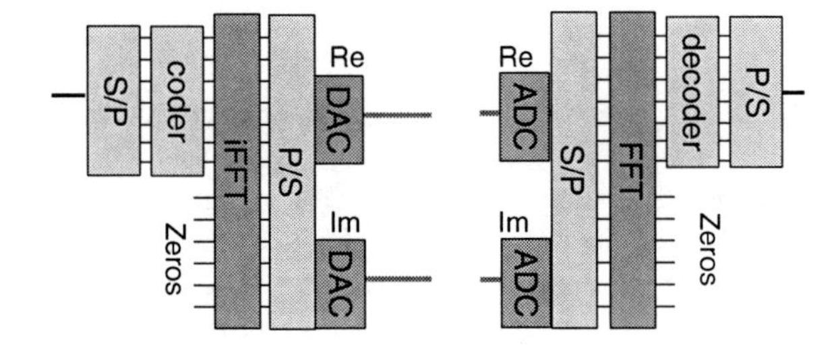

performs the frequency multiplexing of the modulated subcarriers. The resulting time domain signal vector at the iFFT output is re-converted into a serial stream, which drives the two DACs for the OFDM signal's real and imaginary constituents. In the receiver, the inverse procedure is applied to regenerate the individual subcarriers and the transmitted data.

OFDM signals are analogue and complex valued data signals. To modulate an optical carrier with a complex valued signal, an optical IQ modulator can be used (Fig. 1.28). In the transmitter's electro-optic converter, the driving voltage modulates directly the optical field. For the commonly used electro-optic modulators based on nested MZMs for I and Q, respectively, the real and imaginary

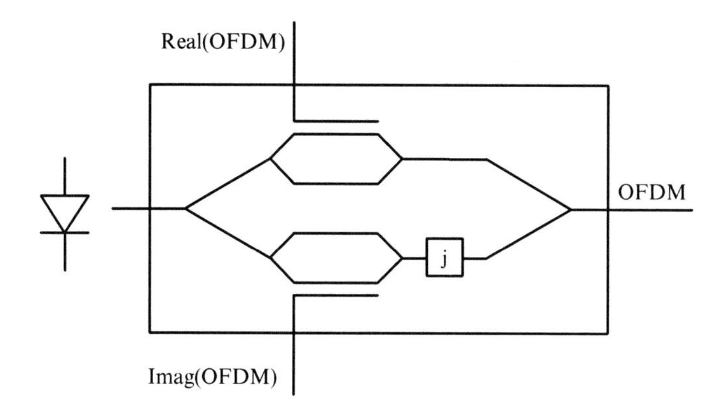

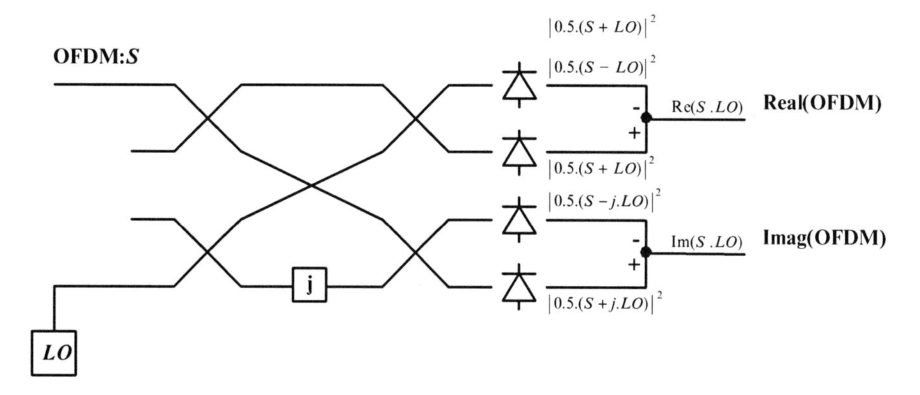

Fig. 1.28 Setups for CO-OFDM transmitter (*upper*) and receiver (*lower*)

constituents of the optical field are proportional to the sine of the driving voltages. In a limited operation range, proportionality between the electrical driving voltage and the optical field can be assumed. Thus the electrical OFDM signal is linearly converted into an O-OFDM signal. The O-OFDM signal is transmitted over the optical fibre to the optical receiver. For reception of optical signals, photodiodes are typically applied. Since in the case of direct detection of the OFDM signal by a photodiode the optical phase information is not preserved due to the square-law-detection process of the photodiode, two different reception schemes are used to convert the O-OFDM signals back to the electrical domain:

1. In direct-detection O-OFDM systems, an optical carrier is transmitted combined with the OFDM signal somewhat apart from the OFDM band. In the photodiode, the optical carrier mixes with the O-OFDM signal leading to a dominating part of the photodiode current corresponding to a linear conversion of the optical field into an electrical signal. Hence, the OFDM signal is down-converted to a low intermediate carrier frequency in the electrical domain containing complete phase information of the optical carrier.

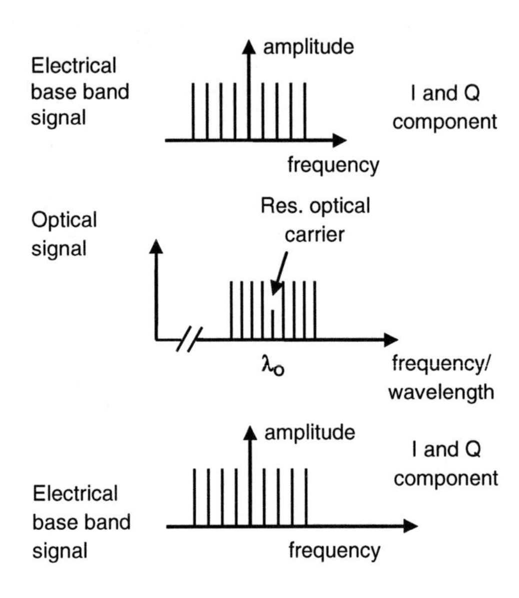

Fig. 1.29 Spectra of the OFDM signals in electrical domain (*Tx*), in optical domain (*optical fibre*) and in the receivers' electrical domain

2. Alternatively, the coherent O-OFDM (CO-OFDM) scheme is applying an optical *LO* in the receiver and thus eliminates the need for the transmission of an optical carrier. The *LO* mixes the OFDM signal from optical light frequencies down to an intermediate frequency close to the baseband. A down-conversion directly to the baseband, i.e. to an intermediate frequency of zero, requires a demanding frequency locking of the *LO* to the received OFDM signal, which is practically impossible due to *LO* tuning accuracy and frequency stability. Therefore, intradyne detection is commonly applied, which means a down-conversion to an intermediate frequency within the spectral width of the signal, often in the range of 1 GHz or less. In the receiver, the optical *LO* is superimposed to the received O-OFDM signal in an optical 90° hybrid. It superimposes the OFDM signal with the quadratural states associated with the *LO* signal in the complex field space. The optical hybrid delivers the four quadratural OFDM signal *S* and the *LO* to two pairs of balanced photodiodes. One pair of detectors receives the superpositions $S \pm LO$, and the other pair the superpositions $S \pm j \cdot LO$ with *LO* in quadrature. After square-law detection and balancing the output currents of the photodiodes, only the products of OFDM signal and *LO* signal remain, namely the real part $\mathrm{Re}(S \cdot LO)$ and the imaginary part $\mathrm{Im}(S \cdot LO)$ of the product of *S* and *LO*. Assuming an *LO* frequency equal to the frequency of the optical carrier, after balancing, the photodiode output signals are proportional to the real and imaginary parts of the O-OFDM signal and can be further processed in the receiver's signal processor.

The respective spectra of the OFDM signal are shown in Fig. 1.29. In the transmitter, the OFDM signal is generated in the baseband. After modulation by the

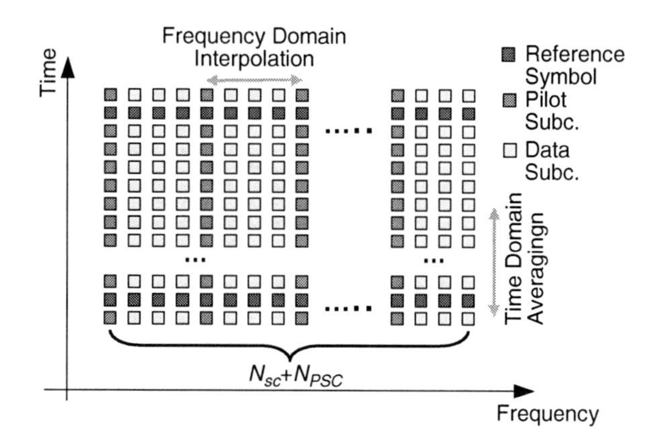

Fig. 1.30 Simple pattern for allocation of reference symbols and pilot subcarriers for channel estimation

IQ modulator, the OFDM signal is up-converted to the optical carrier frequency. In the receiver, after down-conversion, the OFDM signal is available in the baseband in the electrical domain. Therewith a full linear electro/optic/electro transition is demonstrated.

1.5.2.2 Signal Processing

In OFDM systems, the DSP is distributed between the transmitter and the receiver (Fig. 1.27). In the transmitter, further signal processing blocks are applied in addition to the iFFT. At first, few subcarriers are inserted for pilot assistance in the receiver. They are modulated with a known payload and distributed within the OFDM spectra. After iFFT, a cyclic extension is applied to prevent intersymbol interference between consecutive symbols. Therefore, the OFDM symbol is extended at the beginning and at the end of the symbol by copying samples from the end of the symbol to the beginning and from the beginning to the end. One up to a few training symbols are included periodically to train the receiver to the transmission channel. Finally, the OFDM signal is organised on a frame basis; each frame consists of a predefined number of training symbols and payload symbols (Fig. 1.30).

For reliable detection of OFDM signals (1) the start of an OFDM frame (2) and the start of an OFDM symbol (training and payload symbols) need to be detected. For this purpose, a first training symbol is utilised, which differs from all other training and payload symbols. The first training symbol has the same duration as a data symbol, but it consists of two identical sub-symbols with half length each. The received serial stream is processed in a sliding time window of one symbol duration (Shieh et al. 2008a, b; Moose 1994). By correlating samples y_d and y_{d+L} with distance L and the search of a maximum for $|P(d)|^2$ with

$$P(d) = \sum_{m=0}^{L-1} (y^*_{d+m} \cdot y_{d+m+L}) \tag{1.18}$$

the exact position of the reference symbol can be detected. Furthermore, the position of subsequent training and payload symbols is due to the definition of the overall OFDM frame consisting of training and payload symbols as defined in the transmitter.

To estimate a frequency shift of the *LO* frequency to the optical carrier, the two halves of the training symbol are transformed separately into the frequency domain by two FFTs (Jansen et al. 2007a, b; Buchali et al. 2008). By calculation of the mean phase shift of the second set of subcarriers with respect to the first set, the frequency offset of the OFDM signal can be derived. The mandatory frequency down-conversion is performed in the time domain by generating an appropriate carrier and multiplication of this carrier with the received OFDM signal on a sample by sample basis.

After down-conversion, the data samples belonging to cyclic pre- and post-fix are removed, so the original OFDM symbols are recovered. The remaining samples of each OFDM symbol are passed through an FFT, the outputs carrying directly the data information under consideration of the channel response. Each received subcarrier can be represented as the transmitted subcarrier multiplied by a channel coefficient for this subcarrier, which is in general a complex valued number. Therefore, dividing the received subcarrier by the channel coefficient is sufficient to compensate for the linear distortions of amplitude and phase of the subcarriers. To obtain the channel coefficients for all the subcarriers, a channel estimation based on training symbols can be used. By comparing the received training symbols with the transmitted training symbols whose subcarriers are known, the channel coefficient for each subcarrier can be obtained. The training symbols can be periodically inserted into the OFDM data symbol sequence so that the channel estimation can be performed periodically in order to track the dynamic behaviour of the channel. In addition, by using pilot subcarriers that are inserted in every OFDM symbol, fast time variations of the channel can be captured on a symbol by symbol basis (Fig. 1.30). The pilot subcarriers are equally distributed over the OFDM spectrum, and their state of modulation is known at the receiver. Since the pilot carriers are covering only a few points in the OFDM spectrum, the channel estimation between these points is done by interpolation in frequency domain. To improve the accuracy of the channel estimation in a noisy channel, one can average the channel coefficients obtained from multiple training symbols for each subcarrier. The improvement can also be obtained by averaging the channel coefficients in the frequency domain over multiple subcarriers (Buchali et al. 2009).

1.5.2.3 Transmission Characteristics

The performance of a coherent O-OFDM transmission system ought to be demonstrated for an experimental single polarisation transmission system (Dischler and Buchali 2009a) at 12.6–113 Gbit/s data rate, as presented in Fig. 1.31. Due to unavailability of transmitter and receiver terminals, especially of digital signal

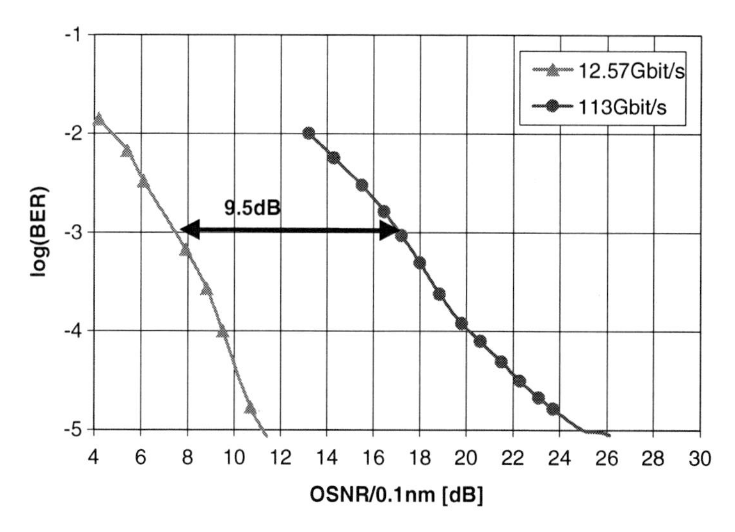

Fig. 1.31 BER versus OSNR for a 12.6 and a 113 Gbit/s in a back-to-back scheme

processors, the experiments have been performed using offline data processing in both terminals. In the transmitter, the pre-processed OFDM signal has been stored in the memory of an arbitrary waveform generator and transmitted periodically. The built-in DAC outputs drive the state of the art optical IQ modulator generating the O-OFDM signal. After transmission over optical fibre, the reception of the OFDM signal has been performed using a high-speed real-time oscilloscope. It performs the analogue-to-digital conversion. The subsequent DSP has been implemented offline in a PC including the complete algorithm. The required OSNR to achieve a BER of 10^{-3} is 7.5 dB for the 12.6-Gbit/s system and 9.5-dB increase for the 113-Gbit/s system. The difference in required OSNR of 9.5 dB corresponds totally to the ninefold increase in data rate (Fig. 1.31).

It has to be mentioned that coherent systems require the polarisation alignment of the received signal and the *LO* to enable superposition of both. To avoid the polarisation alignment, the application of a dual polarisation receiver consisting of two parallel coherent receivers is advantageous. The *LO* is applied in two orthogonal polarisations down-converting the polarisation aligned fraction of the received OFDM signal, respectively. The resulting two electrical signals may be processed jointly in the receiver to recover the complete OFDM signal. If doubling the receiver, the transmission of a polarisation-multiplexed signal is a next reasonable step. In this case, two OFDM transmitters are operated in parallel and their signals are superimposed in two orthogonal polarisations. In the receiver, the alignment between the polarisation of the received OFDM channels and the *LO* is random. Therefore, both transmitted channels have to be separated in the polarisation de-multiplex stage of the digital signal processor (Jansen et al. 2007a, b; Ma et al. 2008). Applying appropriate training symbols, a low penalty for polarisation de-multiplexing can be achieved leading to a system sensitivity similar to the single polarisation case (Dischler and Buchali 2009b). The spectral efficiency of CO-OFDM signals using QPSK subcarrier modulation is below the theoretical value of 2 bit/s/Hz due to the OFDM overhead and reaches about 1.7 bit/s/Hz in practise using a single polarisation transmission. For polarisation-multiplexed systems, a doubled spectral efficiency of 3.3 bit/s/Hz has been demonstrated (Dischler and Buchali 2009b) at 113-Gbit/s bit rate. The spectral width of this 113-Gbit/s signal is 34 GHz, fitting well into the 50-GHz WDM channel grid; therefore, O-OFDM is a very suitable technique for 100-G WDM transmission systems (Jansen et al. 2008a, b, c; Bao and Shieh 2007; Buchali et al. 2009).

OFDM signals consist of a high number of superimposed subcarriers. If the maximum values of many subcarriers occur at the same time, high peak power values are generated. Typically, PAPR values in the range of 10–15 dB are well known for OFDM systems. The probability for occurrence of high peak power values decreases with the power, the highest values occurring very seldom. The optical SMF degrades the transmitted signal if the optical power exceeds a threshold value (nonlinear threshold). In case of OFDM with high PAPR values, first the peak values are degraded nonlinearly starting already at a lower average power. The nonlinear threshold for 100-G OFDM systems has been investigated in detail for dispersion compensated and non-dispersion compensated fibre links (Buchali et al. 2009). In case of dispersion compensation, low nonlinear thresholds have been found, whereas in case of non-dispersion compensation, same nonlinear thresholds are present as has been found for single carrier 100-G systems.

References

Agrawal, G.P.: Nonlinear Fibre Optics, 4th edn. Academic, Boston (2007)

Ali, A., Leibrich, J., Rosenkranz, W.: Spectral efficiency and receiver sensitivity in direct detection optical-OFDM. In: Proceedings of the Optical Fibre Communication Conference 2009, OMT7, San Diego, California, USA, 22–26 March (2009)

Armstrong, J.: OFDM for optical communications. IEEE/OSA J. Lightwave. Technol. **27**, 189–204 (2009)

Armstrong, J., Lowery, A.J.: Power efficient optical OFDM. Electron. Lett. **46**, 370–372 (2006)

Armstrong, J., Schmidt, B.J.C.: Comparison of asymmetrically clipped optical OFDM and DC-biased optical OFDM in AWGN. IEEE Commun. Lett. **12**, 343–345 (2008)

Armstrong, J., Schmidt, B.J.C., Kalra, C., Suraweera, H.A., Lowery, A.J.: Performance of asymmetrically clipped optical OFDM in AWGN for an intensity modulated direct detection system. In: Proceedings of the IEEE Global Telecommunication Conference 2006, San Francisco, California, USA, 27 November – 1 December (2006)

Bahai, A.R.S.: Multi-Carrier Digital Communications. Kluwer, New York (1999)

Bao, H.C., Shieh, W.: Transmission of wavelength-division-multiplexed channels with coherent optical OFDM. IEEE Photon Technol. Lett. **19**, 922–924 (2007)

Barry, J.R., Kahn, J.M.: Carrier synchronization for homodyne and heterodyne-detection of optical quadriphase-shift keying. IEEE/OSA J. Lightwave Technol. **10**, 1939–1951 (1992)

Bergano, N.S.: Wavelength division multiplexing in long-haul transoceanic transmission systems. IEEE/OSA J. Lightwave Technol. **23**, 4125–4139 (2005)

Bracewell, R.N.: Discrete Hartley transform. J. Opt. Soc. Am. **73**, 1832–1835 (1983)

Buchali, F., Dischler, R., Xiao, X., Tang, Y.: Improved frequency offset correction in coherent optical OFDM systems. In: Proceedings of the European Conference Optical Communication 2008, Brussels, Mo.4.D.4, Brussels, Belgium, 21–25 September (2008)

Buchali, F., Dischler, R., Liu, X.: Optical OFDM: a promising high speed optical transport technology. Bell Labs Tech. J. **14**, 125–146 (2009)

Cai, J.-X., Cai, Y., Davidson, C., Foursa, D., Lucero, A., Sinkin, O., Patterson, W., Pilipetskii, A., Mohs, G., Bergano, N.: Transmission of 96 × 100 G Pre-filtered PDM-RZ-QPSK Channels with 3000% Spectral efficiency over 10,608 km and 400% Spectral efficiency over 4,368 km. In: Proceedings of the Optical Fibre Communication Conference 2010, PDPB10, Turin, Italy, 19–23 September (2010)

Chang, R.W.: Synthesis of band-limited orthogonal signals for multi-channel data transmission. Bell Syst. Tech. J. **46**, 1775–1796 (1966)

Chang, R.W.: Orthogonal frequency multiplex data transmission systems. US Patent 3488445 (1966)

Charlet, G., Renaudier, J., Mardoyan, H., Tran, P., Pardo, O.B., Verluise, F., Achouche, M., Boutin, A., Blache, F., Dupuy, J.-Y., Bigo, S.: Transmission of 16.4-bit/s capacity over 2550 km using PDM QPSK modulation format and coherent receiver. IEEE/OSA J. Lightwave Tech. **27**, 153–157 (2009)

Cimini, L.J.: Analysis and simulation of a digital mobile channel using orthogonal frequency division multiplexing. IEEE Trans. Commun. **CM-33**, 665–675 (1985)

Cooley, J.W., Tukey, O.W.: An algorithm for the machine calculation of complex Fourier series. Math. Comput. **19**, 297–301 (1965)

Dischler, R., Buchali, F.: Experimental investigation of non-linear threshold of 113Gbit/s O-OFDM signals on DCF free transmission links. In: Proceedings of the Optical Fibre Communication Conference 2009, OWW3, San Diego, California, USA, 22–26 March (2009)

Dischler, R., Buchali, F.: Transmission of 1.2 Tb/s continuous waveband PDM-OFDM-FDM signal with spectral efficiency of 3.3 bit/s/Hz over 400 km of SSMF. In: Proceedings of the Optical Fibre Communication Conference 2009, PDPC2, San Diego, California, USA, 22–26 March (2009b)

Dixon, B.J., Pollard, R.D., Iezekeil, S.: Orthogonal frequency-division multiplexing in wireless communication systems with multimode fibre feeds. IEEE Trans. Microw Theory Techniques **49**, 1404–1409 (2001)

Duhamel, P., Vetterli, M.: Improved Fourier and Hartley transform algorithms: application to cyclic convolution of real data. IEEE Trans. Acoust. Speech Signal Process. **ASSP-35**, 818–824 (1987)

Ellis, A.D., Zhao, J., Cotter, D.: Approaching the non-linear Shannon limit. J Lightwave Technol. **28**, 423–433 (2010)

Essiambre, R.-J., Foschini, G.J., Kramer, G., Winzer, P.J.: Capacity limits of information transport in fibre-optic networks. Phys. Rev. Lett. **101**, 163901 (2008)

Fatadin, I., Ives, D., Savory, S.J.: Blind equalization and carrier phase recovery in a 16-QAM optical coherent system. IEEE/OSA J. Lightwave Technol. **27**, 3042–3049 (2009)

Faure, J.P., Lavigne, B., Bresson, C., Bertran-Pardo, O., Colomer, A.C., Canto, R.: 40 G and 100 G deployment on 10 G infrastructure: market overview and trends, coherent versus conventional technology. In: Proceedings of the Optical Fibre Communication Conference 2010, OThE3, San Diego, California, USA, 21–25 March (2010)

Fludger, C.R.S., Duthel, T., van den Borne, D., Schulien, C., Schmidt, E.-D., Wuth, T., Geyer, J., de Man, E., Khoe, G.-D., de Waardt, H.: Coherent equalization and POLMUX-RZ-DQPSK for robust 100-GE transmission. IEEE/OSA J. Lightwave Technol. **26**, 64–72 (2008)

Fuerst, C., Elbers, J.P., Camera, M., et al.: 43Gbit/s RZ-DQPSK DWDM field trial over 1047 km with mixed 43 Gbit/s and 10.7 Gbit/s channels at 50 and 100 GHz channel spacing. In: Proceedings of the European Conference Optical Communication 2006, Th4.1.4, Anaheim, California, USA, 5–10 March (2006)

Fuerst, C., Wernz, H., Camera, M., Nibbs, P., Pribil, J., Iskra, R., Parsons, G.: 43 Gbit/s RZ-DQPSK field upgrade trial in a 10 Gbit/s DWDM ultra-long-haul live traffic system in Australia.

In: Proceedings of the Optical Fibre Communication Conference 2008, NTuB2, San Diego, California, USA, 24–28 February (2008)

Gnauck, A.H., Winzer, P.J.: Optical phase-shift-keyed transmission. IEEE/OSA J. Lightwave Technol. **23**, 115–130 (2005)

Gordon, J.P., Mollenauer, L.M.: Phase noise in photonic communications systems using linear amplifiers. Opt. Lett. **15**, 1351–1353 (1990)

Govan, D.S., Doran, N.J.: An RZ DPSK receiver design with significantly improved dispersion tolerance. Opt. Express **15**, 16916–16921 (2007)

Griffin, R.A., Carter, A.C.: Optical differential quadrature phase shift keying (oDQPSK) for high-capacity optical transmission. In: Proceedings of the Optical Fibre Communication Conference 2002, WX6, Los Angeles, California, USA, 17–22 March (2002)

Hou, H.S.: The fast Hartley transform algorithm. IEEE Trans. Comput. **C-36**, 147–156 (1987)

Ip, E.M., Kahn, J.M.: Fibre impairment compensation using coherent detection and digital signal processing. IEEE/OSA J. Lightwave Technol. **28**, 502–519 (2010)

Ito, T., et al.: 3.2 Tb/s – 1500 km WDM transmission experiment using 64 nm hybrid repeater amplifiers. In: Proceedings of the Optical Fibre Communication Conference 2000, PDP24, Baltimore, Maryland, USA, 7–10 March (2000)

Jansen, S.L., Morita, I., Takeda, N., Tanaka, H.: 20-Gb/s OFDM transmission over 4,160-km SSMF enabled by RF-Pilot tone phase noise compensation. In: Proceedings of the Optical Fibre Communication Conference 2007, PDP15, Anaheim, California, USA, 25–29 March (2007a)

Jansen, S.L., Morita, I., Tanaka, H.: 16 52.5-Gb/s, 50-GHz spaced, POLMUX-COOFDM transmission over 4,160 km of SSMF enabled by MIMO processing KDDI R&D Laboratories. In: Proceedings of the European Conference Optical Communication 2007, PDP1.3, Berlin, Germany, 16–20 September (2007b)

Jansen, S.L., Morita, I., Schenk, T.C.W., van den Borne, D., Tanaka, H.: Optical OFDM – A candidate for future long-haul optical transmission systems. In: Proceedings of the Optical Fibre Communication Conference 2008, OMU3, San Diego, California, USA, 24–28 February (2008a)

Jansen, S.L., Morita, I., Tanaka, H.: 10x121.9-Gb/s PDM-OFDM transmission with 2-b/s/Hz spectral efficiency over 1,000 km of SSMF. In: Proceedings of the Optical Fibre Communication Conference 2008, PDP2, San Diego, California, USA, 24–28 February (2008b)

Jansen, S.L., Morita, I., Schenk, T.C.W., Takeda, N., Tanaka, H.: Coherent optical 25.8-Gb/s OFDM transmission over 4160-km SSMF. IEEE/OSA J. Lightwave Technol. **26**, 6–15 (2008c)

Jansen, S.L., Spinnler, B., Morita, I., Randel, S., Tanaka, H.: 100GbE: QPSK versus OFDM. Opt. Fibre Technol. **15**, 407–413 (2009a)

Jansen, S.L., Morita, I., Schenk, T.C.W., Tanaka, H.: 121.9-Gb/s PMD-OFDM Transmission with 2-b/s/Hz spectral efficiency over 1000- km of SSMF. IEEE/OSA J. Lightwave Technol. **27**, 177–188 (2009b)

Kao, Y., Leven, A., Baeyens, Y., Chen, Y., Grosz, D.F., Bannon, F.D., Fang, W., Kung, A.P., Maywar, D.N., Lakoba, T., Agarwal, A., Banerjee, S., Wood, T.H.: 10 Gbit/s soliton generation for ULH transmission using a wideband GaAs pHemt amplifier. In: Proceedings of the Optical Fibre Communication Conference 2003, FF6, Baltimore, Maryland, USA, 23–28 March (2003)

Kazovsky, L.: Balanced phase-locked loops for optical homodyne receivers: performance analysis, design considerations, and laser linewidth requirements. IEEE/OSA J. Lightwave Technol. **4**, 182–195 (1986)

Killey, R.I., Watts, P.M., Mikhailov, V., Glick, M., Bayvel, P.: Electronic dispersion compensation by signal predistortion using digital processing and a dual-drive Mach-Zehnder modulator. IEEE Photon Technol. Lett. **17**, 714–716 (2005)

Killey, R.I., Watts, P.M., Glick, M., Bayvel, P.: Electronic dispersion compensation by signal predistortion. In: Proceedings of the Optical Fibre Communication Conference 2006, OWB3, Anaheim, California, USA, 5–10 March (2006)

Laperle, C., Villeneuve, B., Zhang, Z., McGhan, D., Sun, H., O'Sullivan, M.: WDM performance and PMD tolerance of a coherent 40-Gbit/s dual-polarisation QPSK transceiver. IEEE/OSA J. Lightwave Technol. **26**, 168–175 (2008)

Lee, S.C.J., Breyer, F., Randel, S., Schuster, M., Zeng, J., Huijskens, F., van den Boom, H.P.A., Koonen, A.M.J., Hanik, N.: 24-Gb/s transmission over 730 m of multimode fibre by direct modulation of an 850-nm VCSEL using discrete multi-tone modulation. In: Proceedings of the Optical Fibre Communication Conference 2007, PDP6, Anaheim, California, USA, 25–29 March (2007)

Lee, S.C.J., Breyer, F., Randel, S., van den Boom, H.P.A., Koonen, A.M.J.: High-speed transmission over multimode fibre using discrete multitone modulation. J. Opt. Networking **7**, 183–196 (2008)

Lee, S.C.J., Breyer, F., Randel, S., Gaudino, R., Bosco, G., Bluschke, A., Matthews, M., Rietzsch, P., Steglich, R., van den Boom, H.P.A., Koonen, A.M.J.: Discrete multitone modulation for maximizing transmission rate in step-index plastic optical fibres. IEEE/OSA J. Lightwave Technol. **27**, 1503–1513 (2009)

Leven, A., Kaneda, N., Koc, U.-V., Chen, Y.-K.: Frequency estimation in intradyne reception. IEEE Photon Tehnol. Lett. **19**, 366–368 (2007)

Li, G.: Recent advances in coherent optical communication. Adv. Opt. Photon **1**, 279–307 (2009)

Li, X., Chen, X., Goldfard, G., Mateo, E., Kim, I., Yaman, F., Li, G.: Electronic post-compensation of WDM transmission impairments using coherent detection and digital signal processing. Opt. Express **16**, 880–888 (2008)

Liu, X., Chandrasekhar, S., Zhu, B., Winzer, P.J., Gnauck, A.H., Peckham, D.W.: Transmission of a 448-Gb/s Reduced-Guard-Interval CO-OFDM Signal with a 60-GHz Optical Bandwidth over 2000 Km of ULAF and Five 80-GHz-Grid ROADMs. In: Proceedings of the Optical Fibre Communication Conference 2010, PDPC2, San Diego, California, USA, 21–25 March (2010)

Lowery, A.J.: Optical OFDM. In: Proceedings of the Conference Lasers Electro-Optics 2008, CWN1, San Jose, CA, USA, 4–9 May (2008)

Lowery, A.J., Armstrong, J.: Orthogonal-frequency-division multiplexing for dispersion compensation of long haul optical systems. Opt. Express **14**, 2079–2084 (2006)

Lowery, A.J., Du, L.B., Armstrong, J.: Performance of optical OFDM in ultralong-haul WDM lightwave systems. IEEE/OSA J. Lightwave Technol. **25**, 131–138 (2008)

Ly-Gagnon, D.S., Tsukarnoto, S., Katoh, K., Kikuchi, K.: Coherent detection of optical quadrature phase-shift keying signals with carrier phase estimation. IEEE/OSA J. Lightwave Technol. **24**, 12–21 (2006)

Ma, X., Kuo, G.S.: Optical switching technology comparison: optical MEMS vs. other technologies. IEEE Commun. Mag. **41**, S16–S23 (2003)

Ma, Y., Yang, Q., Shieh, W.: 107 Gb/s coherent optical OFDM reception using orthogonal band multiplexing. In: Proceedings of the Optical Fibre Communication Conference 2008, PDP7, San Diego, California, USA, 24–28 February (2008)

Makovejs, S., Millar, D.S., Lavery, D., Behrens, C., Killey, R.I., Savory, S.J., Bayvel, P.: Characterisation of long-haul 112 Gbit/s PDM-QAM-16 transmission with and without digital nonlinearity compensation. Opt. Express **18**, 12939–12947 (2010)

Malouin, C., Bennike, J., Schmidt, T.J.: Differential phase-shift keying receiver design applied to strong optical filtering. IEEE/OSA J. Lightwave Technol. **25**, 3536–3542 (2007)

Mateo, E., Yaman, F., Li, G.: Efficient compensation of inter-channel nonlinear effects via digital backward propagation in WDM optical transmission. Opt. Express **18**, 15144–15154 (2010)

Merched, R.: On OFDM and single-carrier frequency-domain systems based on trigonometric transforms. IEEE Signal Process Lett. **13**, 473–476 (2006)

Mikkelson, B., Rasmussen, P., Mamyshev, P., Liu, F.: Partial DPSK with excellent filter tolerance and OSNR sensitivity. Electron. Lett. **42**, 1363–1364 (2006)

Millar, D.S., Makovejs, S., Mikhailov, V., Killey, R.I., Bayvel, P., Savory, S.J.: Experimental comparison of nonlinear compensation in long-haul PDM-QPSK transmission. In: Proceedings of the European Conference Optical Communication 2009, 9.4.4, Vienna, Austria, 20–24 September (2009)

Moose, P.: A technique for orthogonal frequency division multiplexing frequency offset correction. IEEE Trans. Commun. **42**, 2908–2914 (1994)

Nielsen, T.N., Stentz, A.J., Rottwitt, K., Vengsarkar, D.S., Chen, Z.J., Hansen, P.B., Park, J.H., Feder, K.S., Strasser, T.A., Cabot, S., Stulz, S., Peckham, D.W., Hsu, L., Kan, C.K., Judy, A.F., Sulhoff, J., Park, S.Y., Nelson, L.E., Gruner-Nielsen, L.: 3.28 Tb/s (82 × 40Gb/s) transmission over 3 × 100km nonzero-dispersion fibre using dual C-band and L-band hybrid Raman/Erbium-doped inline amplifiers. In: Proceedings of the Optical Fibre Communication Conference 2000, PD23, Baltimore, Maryland, USA, 6–12 March (2000)

Nijhof, J.H.B., Forysiak, W.: Investigation of optimal dispersion management for terrestrial 40Gbit/s RZ-DQPSK transmission systems. In: Proceedings of the Conference Lasers Electro-Optics 2006, CThE3, Long Beach, California, USA, 21–26 May (2006)

Noe, R.: PLL-free synchronous QPSK polarisation multiplex/diversity receiver concept with digital I & Q baseband processing. IEEE Photon Technol. Lett. **17**, 887–889 (2005)

Okoshi, T., Kikuchi, K.: Coherent Optical Fibre Communications. Springer, Berlin (1988)

Optical Internetworking Forum (2010) Implementation agreement for integrated dual polarisation intradyne coherent receivers. OIF-DPC-RX-01.0, April 2010

Pan, Q., Green, R.J.: Bit-error-rate performance of lightwave hybrid AM/OFDM systems with comparison with AM/QAM systems in presence of clipping impulse noise. IEEE Photon Technol. Lett. **8**, 278–280 (1996)

Peng, W.-R., Wu, X., Arbab, V.R., Feng, K.-M., Shamee, B., Christen, L.C., Yang, J.-Y., Willner, A.E.: Theoretical and experimental investigations of direct-detected RF-tone-assisted optical OFDM systems. IEEE/OSA J. Lightwave Technol. **27**, 1332–1339 (2009)

Pennickx, D., Chbat, M., Pierre, L., Thiery, J.-P.: The phase shaped binary transmission (PSBT): a new technique to transmit far beyond the chromatic dispersion limit. IEEE Photon Technol. Lett. **9**, 259–261 (1997)

Poggiolini, P., Carena, A., Curri, V., Forghieri, F.: Evaluation of the computational effort for chromatic dispersion compensation in coherent optical PM-OFDM and PM-QAM systems. Opt. Express **17**, 1385–1403 (2009)

Roberts, K., Li, C., Strawczynski, L., O'Sullivan, M., Hardcastle, I.: Electronic precompensation of optical nonlinearity. IEEE Photon Technol. Lett. **18**, 403–405 (2006)

Roberts, K., Beckett, D., Boertjes, D., Berthold, J., Laperle, C.: 100 G and beyond with digital coherent signal processing. IEEE Commun. Mag. **48**, 62–69 (2010)

Ruhl, F., Sorbello, L., Evans, D., Diaconescu, L., Doucet, D., Fasken, D., Nimiczeck, M., Sitch, J., O'Sullivan, M., Belanger, M.P.: 2038 km and four 50 GHz OADM/filters transmission field trial of 115.2Gb/s coherent CoDDM modem in the Telstra network. In: Proceedings of the Optical Fibre Communication Conference 2010, NTuC4, San Diego, California, USA, 21–25 March (2010)

Sakamoto, T., Chiba, A., Kawanashi, T.: 50-km SMF transmission of 50-Gb/s 16QAM generated by quad-parallel MZM. In: Proceedings of the European Conference Optical Communication 2008, Tu.1.E.3, Brussels, Belgium, 21–25 September (2008)

Sano, A., Miyamoto, Y.: Performance evaluation of prechirped RZ and CS-RZ formats in high-speed transmission systems with dispersion management. IEEE/OSA J. Lightwave Technol. **19**, 1864–1871 (2010)

Sano, A., Masuda, H., Kobayashi, T., Fujiwara, M., Horikoshi, K., Yoshida, E., Miyamoto, Y., Matsui, M., Mizoguchi, M., Yamazaki, H., Sakamaki, Y., Ishii, H.: 69.1 Tb/s (432x171-Gb/s) C- and extended L-band transmission over 240 km using PDM-16-QAM modulation and digital coherent detection. In: Proceedings of the Optical Fibre Communication Conference 2010, PDPB7, San Diego, California, USA, 21–25 March (2010)

Savory, S.J.: Digital filters for coherent optical receivers. Opt. Express **16**, 804–817 (2008)

Savory, S.J., Gavioli, G., Killey, R.I., Bayvel, P.: Electronic compensation of chromatic dispersion using a digital coherent receiver. Opt. Express **15**, 2120–2126 (2007)

Savory, S.J., Gavioli, G., Torrengo, E., Poggiolini, P.: Impact of interchannel nonlinearities on a split-step intrachannel nonlinear equalizer. IEEE Photon Technol. Lett. **22**, 673–675 (2010)

Schmidt, T., Hong, J.: 40 G DWDM: a case study in market fragmentation. In: Proceedings of the Asia Communication Photon Conference 2009, ThL1, Shangai, China, 2–9 November (2009)

Schmidt, B.J.C., Lowery, A.J., Armstrong, J.: Experimental demonstration of 20 Gbit/s direct-detection optical OFDM and 12 Gbit/s with a colorless transmitter. In: Proceedings of the Optical Fibre Communication Conference 2007, PDP18, Anaheim, California, USA, 25–29 March (2007)

Schmidt, B.J.C., Lowery, A.J., Armstrong, J.: Experimental demonstration of electronic dispersion compensation for long-haul transmission using direct-detection optical OFDM. IEEE/OSA J. Lightwave Technol. **26**, 196–203 (2008)

Sethares, W.A., Rey, G.A., Johnson, C.R.: Approaches to blind equalization of signals with multiple modulus. In: Proceedings of International Conference Acoustics Speech Signal Process 1989, D3.21, Glasgow, Scotland, 23–26 May (1989)

Shieh, W., Djordjevic, I.: OFDM for Optical Communications. Elsevier, Burlington (2010)

Shieh, W., Bao, H., Tang, Y.: Coherent optical OFDM: theory and design. Opt. Express **16**, 841–859 (2008a)

Shieh, W., Yi, X., Ma, Y., Yang, Q.: Coherent optical OFDM: has its time come? J. Opt. Networking **7**, 234–355 (2008b)

Sorensen, H.V., Jones, D.L., Burrus, C.S., Heideman, M.T.: On computing the discrete Hartley transform. IEEE Trans. Acoust. Speech Signal Process **ASSP-33**, 1231–1238 (1985)

Sun, H., Wu, K.T., Roberts, K.: Real-time measurements of a 40 Gb/s coherent system. Opt. Express **16**, 873–879 (2008)

Sun, H., Gaudette, J., Pan, Y., O'Sullivan, M., Roberts, K., Wu, K.-T.: Modulation formats for 100 Gb/s coherent optical systems. In: Proceedings of the Optical Fibre Communication Conference 2009, OTuN1, San Diego, California, USA, 22–26 March (2009)

Svaluto Moreolo, M.: Power efficient and cost-effective solutions for optical OFDM systems using direct detection. In: Proceedings of the International Conference Transparent Optical Networks 2010, Mo.D1.5, Munich, Germany, June 27 – July 1 (2010)

Svaluto Moreolo, M., Muñoz, R., Junyent, G.: Novel power efficient optical OFDM based on Hartley transform for intensity-modulated direct-detection systems. IEEE/OSA J. Lightwave Technol. **28**, 798–805 (2010)

Tang, J.M., Shore, K.L.: 30-Gb/s signal transmission over 40-km directly modulated DFB-laser-based single-mode-fibre links without optical amplification and dispersion compensation. IEEE/OSA J. Lightwave Technol. **24**, 2318–2327 (2006)

Tang, J.M., Shore, K.L.: Maximizing the transmission performance of adaptively modulated optical OFDM signals in multimode-fibre links by optimizing analog-to-digital converters. IEEE/OSA J. Lightwave Technol. **25**, 787–798 (2007)

Taylor, M.G.: Coherent detection method using DSP for demodulation of signal and subsequent equalization of propagation impairments. IEEE Photon Technol. Lett. **16**, 674–676 (2004)

Taylor, M.G.: Phase estimation methods for optical coherent detection using digital signal processing. IEEE/OSA J. Lightwave Technol. **27**, 901–914 (2009)

Tibuleac, S., Filer, M.: Transmission impairments in DWDM networks with reconfigurable optical add-drop multiplexers. IEEE/OSA J. Lightwave Technol. **28**, 557–598 (2010)

van den Borne, D., Jansen, S.L., Gottwald, E., Khoe, G.D., de Waardt, H.: Lumped dispersion management in long-haul 42.8-Gbit/s RZ-DQPSK transmission. In: Proceedings of the European Conference Optical Communication 2006, Mo3.2.2, Cannes, France, 24–28 September (2006)

van den Borne, D., Jansen, S.L., Gottwald, E., Krummrich, P.M., Khoe, G.-D., de Waardt, H.: 1.6-b/s/Hz spectrally efficient transmission over 1700 km of SSMF using 40×85.6-Gb/s POLMUX-RZ-DQPSK. IEEE/OSA J. Lightwave Technol. **25**, 222–232 (2007)

Viterbi, A.J., Viterbi, A.N.: Nonlinear estimation of PSK-modulated carrier phase with application to burst digital transmission. IEEE Trans. Inf. Theory. **29**, 543–551 (1983)

Waegemans, R., Herbst, S., Holbein, L., Watts, P., Bayvel, P., Fürst, C., Killey, R.I.: 10.7 Gb/s electronic predistortion transmitter using commercial FPGAs and D/A converters implementing real-time DSP for chromatic dispersion and SPM compensation. Opt. Express **17**, 8630–8640 (2009)

Wang, D., Liu, D., Liu, F., Yue, G.: A novel DHT-based ultra-wideband system. Proc. Int. Symp. Commun. Inform. Technol. **50**, 172–184 (2005)

Watts, P., Waegemans, R., Glick, M., Bayvel, P., Killey, R.: An FPGA-based optical transmitter design using real-time DSP for advanced signal formats and electronic predistortion. IEEE/OSA J. Lightwave Technol. **25**, 3089–3099 (2007)

Weinstein, S.B., Ebert, P.M.: Data transmission by frequency-division multiplexing using the discrete Fourier transform. IEEE Trans. Commun. **19**, 628–634 (1971)

Winzer, P.J., Essiambre, R.-J.: Advanced optical modulation formats. In: Kaminow, I.P., Li, T., Wilner, A.E. (eds.) Optical Fibre Telecommunications V B: Networks and Systems. Academic, New York (2008). Chapter 2

Winzer, P.J., Gnauck, A.H., Doerr, C.R., Magarini, M., Buhl, L.L.: Spectrally efficient long-haul optical networking using 112-Gb/s polarisation-multiplexed 16-QAM. IEEE/OSA J. Lightwave Technol. **28**, 547–556 (2010)

Yamazaki, E., Inuzuka, F., Yonenaga, K., Takada, A., Koga, A.: Compensation of interchannel crosstalk induced by optical fibre nonlinearity in carrier phase-locked WDM system. IEEE Photon Technol. Lett. **19**, 9–11 (2007)

Yang, Q., Tang, Y., Ma, Y., Shieh, W.: Experimental demonstration and numerical simulation of 107-Gb/s high spectral efficiency coherent optical OFDM. IEEE/OSA J. Lightwave Technol. **27**, 168–176 (2009)

Zhou, X., Yu, J., Huang, M.-F., Shao, Y., Wang, T., Nelson, L., Magill, P., Birk, M., Borel, P.I., Peckham, W., Jr Lingle, R.: 64-Tb/s (640x107-Gb/s) PDM-36QAM transmission over 320 km using both pre- and post-transmission digital equalization. In: Proceedings of the Optical Fibre Communication Conference 2010, PDPB9, San Diego, California, USA, 21–25 March (2010)

Chapter 2
Signal Processing, Management and Monitoring in Transmission Networks

Carmen Vázquez, Julio Montalvo, Jawaad Ahmed, David Bolt, Christophe Caucheteur, Gerald Franzl, Philippe Gravey, David Larrabeiti, Jose A. Lazaro, Tatiana Loukina, Veronique Moeyaert, Josep Prat, Lena Wosinska, and Kivilcim Yüksel

2.1 Introduction

For transparent optical networks, it is important to support the demanded signal quality of physical paths from source to destination nodes, in response to a given call request. Due to the lack of regeneration in current optical networks, optical signal impairments accumulate along paths. With channel bit rates of 10 Gbit/s or

C. Vázquez (✉) • J. Montalvo • D. Larrabeiti
Department of Electronic Engineering, Universidad Carlos III de Madrid, 28911 Madrid, Spain
e-mail: cvazquez@ing.uc3.es; julio.montalvo@uc3m.es; dlarra@ing.uc3.es

J. Ahmed
The Royal Institute of Technology (KTH), Electrum 229, 16440 Kista, Stockholm, Sweden
e-mail: jawwad@kth.se

D. Bolt
College of Engineering, Swansea University, Singleton Park, SA2 8PP Swansea, West Glamorgan, UK
e-mail: 393729@Swansea.ac.uk

C. Caucheteur • V. Moeyaert • K. Yüksel
Faculté Polytechnique, Electromagnetism and Telecommunication Unit, Université de Mons, 31, Boulevard Dolez, Mons 7000, Belgium
e-mail: christophe.caucheteur@umons.ac.be; Veronique.MOEYAERT@umons.ac.be; kivilcim.yuksel@umons.ac.be

G. Franzl
Institute of Telecommunications, Vienna University of Technology, Favoritenstraße 9/389 A-1040, Vienna, Austria
e-mail: Gerald.Franzl@tuwien.ac.at

P. Gravey • T. Loukina
Telecom Bretagne, Technopôle Brest-Iroise, CS, 83818–29238 Brest Cedex 3, France
e-mail: philippe.gravey@telecom-bretagne.eu; Tatiana.loukina@telecom-bretagne.eu

A. Teixeira and G.M.T. Beleffi (eds.), *Optical Transmission: The FP7 BONE Project Experience*, Signals and Communication Technology,
DOI 10.1007/978-94-007-1767-1_2,

higher, linear and non-linear fibre impairments become prominent factors affecting the signal quality. Thus, new techniques in both physical layer and network layer are necessary for mitigating impairments to accommodate high-speed traffic (Tomkos et al. 2002).

On the other hand, the flexibility of modern networks with dynamic and distributed management, where lightpaths can be dynamically and automatically assigned end-to-end, increases the risk of changing path conditions that affect signal quality and consequently quality of service (QoS) over the lifetime of a connection as well as for subsequent connection set-up requests (on-demand routing and wavelength assignment – RWA). For traditional connection provisioning an ideal physical layer, ignoring transmission impairments (Chlamtac et al. 1992) could be assumed because the implicit per hop regeneration (optics-to-electronics-to-optics conversion – O/E/O) compensated signal impairments hop-by-hop in a rather static manner (section commissioning). For end-to-end all-optically switched connections using photonic cross-connect switches (PXC), the conditions change and common practice is not applicable. Anyhow, intelligent connection provisioning is an important traffic engineering problem, and minimising operational cost as well as efficiently utilising network resources is the main driver. In this context, it seems important to have flexible routing protocols that take into consideration the most relevant physical impairments, and are able to exchange messages with their values as part of the route information with others. This condition, if definitely used to calculate routing, will surely assure better success in data delivery over the network (Huang et al. 2005).

Irrespective of the control architecture chosen by an operator of an optical network, the included control plane (in charge of setting up and tearing down optical circuits; also known as lightpaths) needs to have a proper set of tools to satisfyingly deal with physical impairments. The most essential tools are: optical signal monitoring, signal processing and impairment compensation techniques (each all-optically and/or electrically).

With the help of these tools, it is possible to predict the QoT (quality of transmission) attainable for a lightpath at a given time (i.e. the latest network-wide monitoring instant), and the control plane can route requests across the network

J.A. Lazaro
BarcelonaTech. Department of Signal Theory and Communication, Universitat Politècnica de Catalunya, 08034 Barcelona, Spain
e-mail: jose.lazaro@tsc.upc.edu

J. Prat
Department of Signal Theory and Communication, Polytechnic University of Catalonia, 08034 Barcelona, Spain
e-mail: jprat@tsc.upc.edu

L. Wosinska
The Royal Institute of Technology (KTH), Electrum 229, 16440 Kista, Stockholm, Sweden
e-mail: wosinska@kth.se

according to this information, keeping the circuit as much as possible in the optical layer. In this chapter, we introduce and use a specific signalling model, the IETF GMPLS framework, to illustrate how impairment information and impairment compensation tools can be dealt with under a distributed control plane paradigm. The use of GMPLS obeys to simplicity of presentation of concepts and to the fact that ITU-T and OIF have adopted GMPLS signalling and routing protocols as valid implementation alternatives. However, the techniques for impairment management described here can be easily applied to any other optical network control architecture under standardisation. Optical burst and packet switching are thus not considered, as these are currently not under standardisation. They actually demand sophisticated signalling latency independent and thus less accurate, but in response more robust, approaches.

Real-time monitoring of physical impairments related to transmission media properties such as chromatic dispersion, attenuation, and non-linear effects at different wavelengths can be achieved by different optical monitoring techniques, which have greatly evolved in the last years. Even optical compensation techniques that alleviate the effect of impairments are important to predict the signal quality along a lightpath. Having in mind that, in a wavelength-routed optical network spanning a large geographical area, optical signals may traverse several long fibre segments and a number of different compensation devices, we recognise that an integration view of their effects should be taken into account for selecting a path (Teixeira et al. 2009). Impairment-aware routing and wavelength assignment (IA-RWA) algorithms considering TE information together with the locally introduced physical impairments and the availability of compensation devices through the network can be employed to compute a feasible and least-costly (optimal) lightpaths from sources to destinations. Impairment-aware on-demand routing is conceptually superior to shortest-path routing with maximum transparent distance constraint as well as exhaustive offline constraint-based routing algorithms (Pachnicke et al. 2009).

The basic ideas of general multi-protocol label switching (GMPLS), being the protocol framework intended to control dynamic circuit set-up and teardown in the next generation of optical transport networks, are reported in Sect. 2.2, along with the description of the main routing protocols, new tendencies and their operation procedures and flexibility in selecting routes for lightpaths. At the end of the section, some aspects about convergence between transport and access networks, mainly in terms of physical impairment monitoring and compensation, are discussed.

Section 2.3 presents different optical monitoring techniques (including spectral and time domain techniques) and some dispersion compensation techniques. Approaches respecting constraints for end-to-end QoT control (related to changing physical properties along lightpaths) and different impairment-aware routing alternatives are discussed in Sect. 2.4. Finally, summarising conclusions are outlined on Sect. 2.5.

2.2 Dynamic and Distributed Control in Transmission Networks

2.2.1 *IP Networks and Circuits: IP-MPLS*

Nowadays, there are two clearly differentiated overlaid networks making up the backbone of wide area data networks: the packet-switched networks and the underlying electro-optical circuit-switched long distance transport networks. Today, IP is the dominating WAN packet-switching network technology in the world. Given its importance, there is a growing interest to design optical transport networks tailored for IP-based data communication. Conversely, packet networks have tried to incorporate features primarily available in circuit networks. This latter trend has made MPLS become a commodity on IP backbone networks: it enables to extend IP routers with circuit-oriented capabilities. We will start to review the basic ideas that led to MPLS in order to understand GMPLS (generalised MPLS), the protocol framework proposed by IETF intended to control circuit set-up and teardown in the next generation of optical transport networks.

IP is a network layer protocol (layer-3 in the OSI reference model) intended to provide end-to-end packet delivery between two end systems (usually hosts) connected to an IP network. To this end, hosts are assigned IP addresses, and the packet headers carry the source and destination addresses (32 bits in the case of IPv4 and 128 bits in IPv6). The forwarding of packets is performed by packet switches (routers) interconnected by links. Since IP has not a single predefined link layer technology to work upon a logical IP link, connecting two neighbouring routers can be realised by any digital communication service between them adapted by a suitable link layer. Today, Internet core routers are typically interconnected by point-to-point 2.5, 10 and 40 Gbit/s electro-optical channels (direct or circuit-switched), and IP packets are sent in different ways, such as Ethernet link-layer frames. Towards the edge of the network, links between IP nodes consist of one or more link layer sections spanning a greater variety of physical transmission media (usually radio, copper or fibre). Each link technology has its own link and medium access control protocols that are opaque to IP.

In IP, packet forwarding is basically driven by the packet's destination IP address and the router's routing table. The next hop selected for a packet is determined by the longest-matched network prefix in the routing table and usually by additional policy rules. This is a relatively complex operation that, in core routers, is expedited either by route lookup hardware (e.g. TCAM) or by SRAM forwarding caches. Each router constructs and maintains its own routing table and takes the forwarding decision autonomously, in a fully distributed way. The routing table is created on each router's CPU either by static configuration of routes or by a routing protocol. An IP network administrator needs to decide what routing protocol to use within its own domain, i.e. the IGP, and which routing protocol the domain's border routers need to use to exchange routes with other domains (also called autonomous systems – AS). This latter is the EGP which, in the case of the public Internet, is agreed to be the BGP. All routing tables are thus automatically populated by these two routing protocols.

From the two major families of IGP routing protocols, distance-vector and link state-based, derived recent link-state protocols which became predominant for their globally faster recovery from link failures. The most popular standard link-state routing protocols are: OSPF and ISIS. With link-state protocols, each router monitors the state of the link to each neighbour and floods this information over the network either periodically or upon an event (e.g. a link-down event). Whenever a link-state advertisement is received by a router, the router reviews its routing table and updates the next-hop entry for each route's network prefix. The next hop is the IP address of the router on the shortest path to the destination according to the established metrics. If a link or router fails, the neighbouring routers react and convey the news to the rest of the network, causing an update of the network topology as seen by the routing process at each node.

There are a number of intrinsic limitations to the way IP networks operate that can actually be summarised into one: IP does not provide enough flexibility to support TE.

Firstly, since, as stated, the IP forwarding decision is distributed and only based on the IP address, all packet flows tend to take the shortest path, which is not always the right choice: Alternative longer paths may be underutilised and could be used to transport part of the total traffic. Available routing-based traffic balancing mechanisms (such as OSPF's equal-cost multipath or the use of a RTT metric) are not flexible enough and have side effects difficult to foresee and handle.

Secondly, it is not possible to implement fast-recovery protection mechanisms for traffic trunks. Due to the distributed nature of IP routing, recovery times are by far longer than the ones of optical networks, such as SDH or SONET, assuring a convergence time of less than 50 ms (Valenti et al. 2009). Depending on the size and complexity of the network topology, the recovery process of IP networks may take up to seconds before all routers converge and create new loop-free paths. This effect, which could be regarded unimportant in the past, today is an obstacle to realise the *multi-service* network concept, where the IP backbone is expected to carry voice and TV traffic trunks in a very reliable way (substituting legacy telecommunication services), as well as regular TCP/IP traffic. Hence, it is essential to have in IP networks the same capability known from SDH/SONET networks, i.e. the ability to set up primary and back-up paths for fast switchover in the event of a failure, as nowadays readily provided by Carrier Ethernet. To be effective, this should be made in such a way that the backup path is predetermined, insensitive to transitory routing instabilities, and its spare capacity needs to be reserved in advance in order to carry instantly all the diverted traffic.

Consequently, a mechanism to set up virtual circuits in an IP network for traffic balancing and path protection was deemed very useful. This functionality is enabled by MPLS. MPLS provides a way to create unidirectional virtual circuits, named LSP. A router enhanced with MPLS capability is known as a LSR. In MPLS, IP packets (or any other protocol data unit, PDU) are encapsulated into MPLS

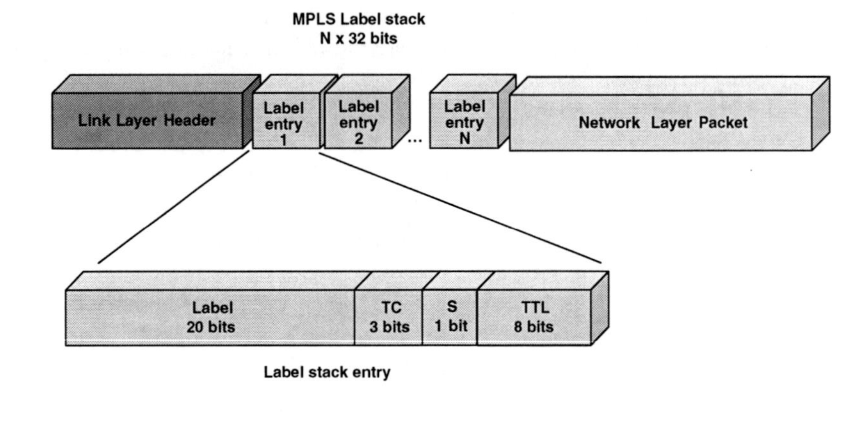

Label:	Label #	Label space can be per-interface or per-platform
TC:	Traffic Class of frame	Usually maps QoS bits (802.1p, IP ToS)
S:	Bottom of Stack.	Value 1 for the innermost label stack entry
TTL:	Time to live.	Used to detect loops

Fig. 2.1 MPLS generic label encapsulation

frames at the ingress LSR and label-switched all the way through a pre-established LSP to the egress LSR, where the MPLS header is removed. MPLS headers are very simple (see Fig. 2.1) and so is the forwarding mechanism: given an input port and input label, a single lookup into the LFIB provides the output port, next hop and the operation to perform on the frame (label swap, pop, push or a combination). In fact, the original purpose of label switching was speeding up packet forwarding. This forwarding mechanism based on stackable 32-bit labels is a fixed aspect of MPLS. What is variable or open is the control plane, i.e. the protocols that determine the route and signal the set-up and teardown of LSPs.

The signalling protocol must create a binding between an FEC and a label. An FEC is an identifier for a specific set of packet flows that need to be forwarded in exactly the same way by the LSRs along the LSP created for that FEC; in other words, packets of the same FEC follow the same path(s) and receive identical QoS treatment. An LSP may be created:

- either by following the forwarding path determined by the routing protocol. In this case, the most commonly used protocol is the LDP. In this setting, LDP creates automatically LSPs from each node to every other network node along the shortest paths.
- Or over an arbitrarily specified path. This can be made either by manual configuration of all LSRs along a path or by signalling initiated by the LSP's ingress router using RSVP-TE, RFC3209. This is the signalling protocol commonly used

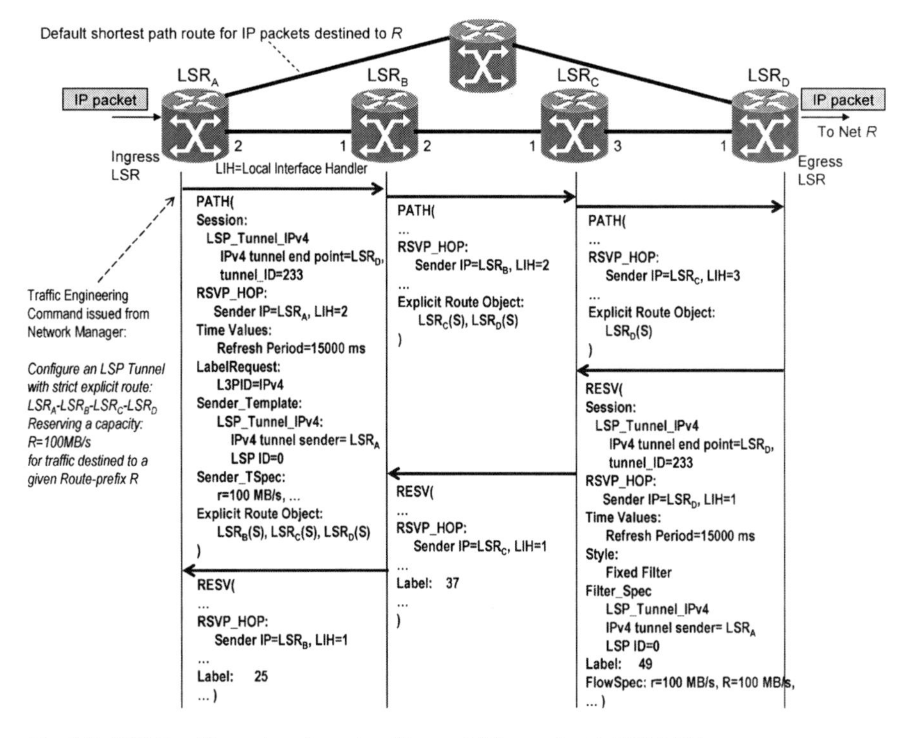

Fig. 2.2 MPLS traffic engineering: signalling an LSP tunnel with RSVP-TE

to set up traffic engineering LSPs, also known as MPLS LSP Tunnels[1], and with GMPLS circuits, as described below.

We focus on this latter type of LSPs for its significance to optical networks. The idea is to be able to command an ingress or head-end LSR to set up an LSP to an egress or tail-end LSR that fulfils a set of constraints (e.g. have a residual capacity of 200 Mbit/s available). Given such command, the head-end router would compute the path and initiate the required RSVP-TE signalling (Fig. 2.2). For this to happen, the head-end router must have considerably extended link information about the network, such as maximum link bandwidth, maximum reservable bandwidth, currently unreserved bandwidth, the administrative group, etc. This is accomplished by means of TE extensions to an IGP protocol, such as OSPF-TE (RFC 3630 – traffic engineering extensions to OSPF).

[1] IP tunnels have also been used for simple traffic engineering operations. LSP tunnels have several advantages over IP tunnels, namely efficiency (label switching vs. route lookup forwarding), explicit route specification and forwarding (IP source routing could be used, but the processing of this IP option is less efficient than label swapping), immunity to routing instabilities and bandwidth reservation capability.

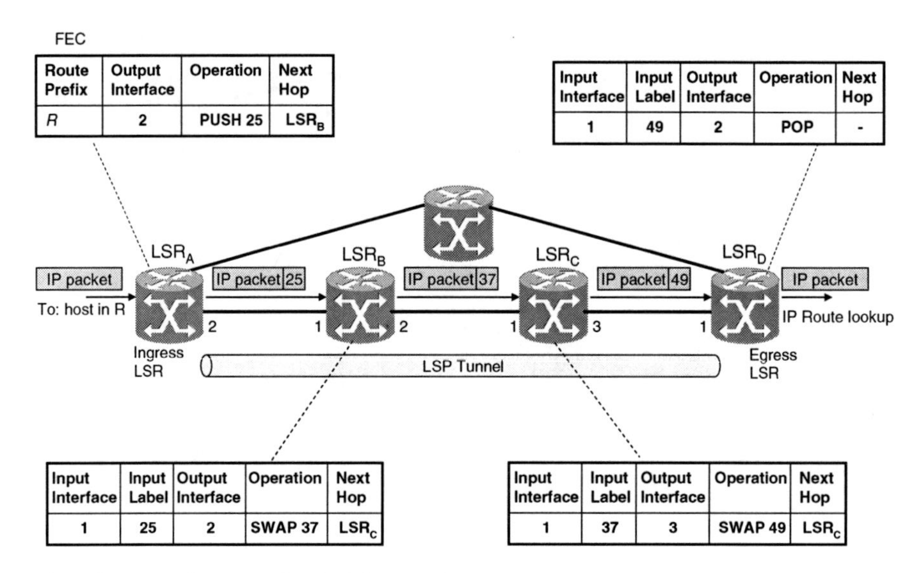

Fig. 2.3 Forwarding state after the LSP tunnel set-up

Once all the extended link information is known at the head-end and the LSP path is computed, the router starts the RSVP-TE signalling to set up the LSP Tunnel. RSVP-TE is an extension of the IETF resource reservation protocol RSVP, designed to deal with FEC-label assignments and reservations for traffic trunks (aggregation of flows of the same class to be placed in an LSP (see RFC2702)), rather than for individual IP flows. Figure 2.2 shows an example of LSP tunnel set-up and Fig. 2.3 depicts the resultant forwarding state along the new path for the diverted traffic. The additional route configuration required to reroute the traffic over the new LSP is not shown. Figure 2.3 also illustrates the forwarding of an IP packet over the LSP. Similarly to RSVP, RSVP-TE is a soft-state signalling protocol that implies periodic refreshing of signalled LSPs. RSVP-TE allows to specify the path from the source node in the ERO and, if a reservation is desired, a traffic descriptor, usually represented by a token bucket specification, denoting the average rate, peak rate and maximum burst size of the traffic trunk. In addition to improved traffic balancing across network resources, the LSPs created can also be used to divert part of the traffic from the shortest path, for example to deal with congestion, or as a predetermined backup route to fast reroute traffic traversing a failing network segment.

An interesting issue is path computation for the LSP Tunnel. This is out of the scope of RSVP-TE. In principle, within a single unstructured domain, OSPF-TE can supply the extended link-state information required for the head-end router to compute the shortest explicit route fulfilling the requirement (this is called *constraint-based routing*). However, this autonomous decision by the head-end LSRs can have the negative side effect on racing for network resources that induces potential blocking and degradation[2] of LSPs requests. Therefore, the idea of having global sparsely distributed path computation elements (RFC4655) to calculate paths considering demands from all ingress routers in a correlated way is gaining interest.

Especial mention deserves the ITU-T activity on MPLS. The multi-protocol properties of MPLS and its dual ability to build both IP-driven and IP-agnostic circuits are making MPLS one of the predominant technologies in today's packet-switched transport networks. This trend impelled ITU-T SG-15 to enhance its transport network model to embrace MPLS circuits and to try to provide service levels comparable to SDH with them. To prevent non-interoperable implementations of transport-specific MPLS architectures, IETF is working on a set of RFCs to support the requirements of ITU-T transport networks in terms of QoS, end-to-end OAM and protection. This version of MPLS, convergent with ITU-T T-MPLS specifications, is known as MPLS-TP (MPLS-transport profile) (Bocci et al. 2010). This transport profile is expected to become the basis for carrier-grade implementation of services, such as VPLS in the near future. Other alternatives under study by network operators, such as IEEE PBB-TE (provider backbone bridging–traffic engineering) (IEE8021), are only applicable to Ethernet-based backbones.

Before closing this short overview of MPLS concepts, we shall recall once more that the switching technology implied by MPLS is packet-based store-and-forward switching, not circuit-based cut-through switching. An attempt to integrate both worlds under the same framework came with GMPLS.

2.2.2 Distributed Control of Optical Networks: GMPLS

GMPLS (Mannie et al. 2004) is an extension of the MPLS Traffic Engineering scheme to support circuit switching networks, such as TDM and WDM Networks. In our context, GMPLS signalling is intended to drive the dynamic set-up of circuits in transparent or translucent optical networks by using the same signalling protocols as MPLS-TE and extensions.

The technological shift to dynamically switched circuits has still a long way to go before multivendor interoperability of optical equipment is in place. The reason for this is the inherent complexity of analogue optical interfaces in WDM, hindering consensus between vendors. Therefore, nowadays, circuit set-up and management in optical networks is single-vendor and mainly performed manually by operators from a central network operation centre. Provisioning is facilitated by proprietary applications that connect to every node to configure cross-connections and carry out other OAM tasks. It is common to find vendors offering web-based interfaces for circuit provisioning and management of their equipment. GMPLS tries to change

[2]This is especially the case of GMPLS in DWDM, as explained later in the chapter, where distributed and concurrent requests add extra burden to the already complex process of optical circuit provisioning.

this centralised approach to a more distributed one (Banerjee et al. 2001). In GMPLS, all nodes exchange control information over a signalling channel, and connection set-up can be initiated by the head-end of a connection (following a TE request from the NMS) and is realised by the exchange of messages between adjacent nodes using RSVP-TE. This requires a control network based on IP, i.e. optical switch's control electronics need to have a TCP/IP stack, an IP address, run RSVP-TE and OSPF-TE or ISIS-TE, alongside with a LMP. Thus, a single control protocol could eventually provide end-to-end circuit provisioning across routers, SDH multiplexers and optical switches, as well as a unified framework for protection, restoration, monitoring, management, etc., including all layers and switching devices involved in a connection.

In the beginning, GMPLS was regarded as a competing alternative to ITU-T ASON signaling (G.8080/Y.1304), but the open nature of these signalling specifications (G.7713 (G7713)) made it possible to include IETF GMPLS signalling protocols with specific extensions (RFC4974 (RFC4974)) as an ITU-T-recommended alternative to ITU-T PNNI/Q.2931 (G.7713.1/Y.1704.1 (G77131) and G.7713.2/Y.1704.2 (G77132), respectively). Furthermore, the OIF has tried to adopt the best of ASON and GMPLS concepts to define and impulse the take-up of specific signalling protocols. These recommendations include both signalling protocols too and include guidelines for interworking of PNNI and RSVP-TE/OSPF-TE subnetworks.

Next, we briefly outline the main changes introduced by GMPLS with respect to MPLS. A recommended, more detailed overview can be found in Halabi (2003).

- **Circuit switching network support**. Compared to MPLS virtual circuits over packet-switched networks, GMPLS makes true circuits over a circuit-switched network; these circuits are still called LSPs. In GMPLS, the data units do not carry an explicit label; the label is only managed by the control plane to identify the timeslot, wavelength or port through which the connection data must be switched. To deal with each specific technology, GMPLS defines five types of LSR interfaces: **PSC**, **L2SC** (e.g. ATM), **TDM** (e.g. over SONET/SDH channels; in this case, forwarding is based on the incoming timeslot of data, i.e. the implicit label is the timeslot), **LSC** (e.g. optical cross connect with wavelength switching granularity; the implicit label is the lambda) and **FSC** (Fibre-switch capable, e.g. photonic cross-connect with light switching granularity; the label is the port number). The interface switching capability descriptors are advertised by means of the routing protocol.
- **Link management**. Unlike MPLS LSRs, GMPLS nodes may have many links with the neighbouring nodes. In order to drive bundles of WDM channels in a suitable and efficient way, instead of overloading the IGP with new functions, a specific LMP (RFC4204) has been proposed to manage TE links between every two neighbouring nodes over a common control channel. Link provisioning, bundling, protection, fault isolation, signalling control channel monitoring, connectivity verification, configuration and verification are some of the functions carried out by this protocol.

- **Routing: topology and resource discovery**. In GMPLS, network nodes run IP routing protocols with Traffic Engineering extensions, usually OSPF-TE, to convey not just topology information but also available link capacity, termination capabilities and protection properties. To deal with multiple links between adjacent nodes in an efficient way, the routing allows to bundle these links into a *TE link* and to announce them as such, summarising the information of its component links. Component links can be uniquely identified by tuples <node ID, link bundle, link identifier>. This is called "unnumbered link" support because no explicit subnet is built per individual link or link bundle. Other information conveyed by routing protocols for GMPLS is the *SRLG* and the *link protection type*. A set of links may constitute a "shared risk link group" (RFC4202) (RFC4202) if they share a resource (e.g. a fibre conduit) whose failure may affect all links in the set. The SRLG information is a list made up with all the SRLGs that the link belongs to. An SRLG is identified by a 32-bit number that is unique within an IGP domain. The SRLG of a LSP is the union of the SRLGs of the links in the LSP. Disjoint SRLG LSP path computation is important to build protection circuits. The *link protection type* information is used by TE algorithms to search for paths with a target protection level. The protection types defined are *extra traffic, unprotected, shared, dedicated 1:1, dedicated 1 + 1 and enhanced.*
- **Signalling**. In order to support the new interface classes, a number of changes on signalling are required. The most relevant follow:
  - Generalised labels are introduced to identify time slots, wavelengths range in a waveband being switched, fibres in a bundle and the MPLS label. Generalised Label requests determine the technology-specific LSP type being requested. Unlike MPLS, GMPLS permits an upstream node to suggest a label to the downstream node before the label mapping is received from downstream, and it is possible to constrain the label range to be used to set up LSPs between peers. This is especially useful to deal with wavelength switching and conversion limitations of optical cross-connects. Furthermore, it is possible to determine a common end-to-end label for all-optical LSPs without wavelength conversion.
  - Bidirectional LSPs. Unlike MPLS-TE, GMPLS enables the specification of both unidirectional and bidirectional LSPs to improve set-up latency and path consistency, especially for networks with bidirectional circuits (e.g. SDH/SONET).
  - Control/data plane separation. In MPLS-TE, signalling normally uses the same interfaces used for data, and hence data and control links share their fate. In GMPLS, control and data interfaces are usually different. This implies the need for mechanisms for proper identification of the control and data channels being controlled, and a differentiated fault handling of data, control channel and nodes. Consequently, the application of the soft-state model used in MPLS RSVP-TE is diluted in GMPLS since control plane and data plane failures may not be correlated, and hence LSPs are not released whenever a

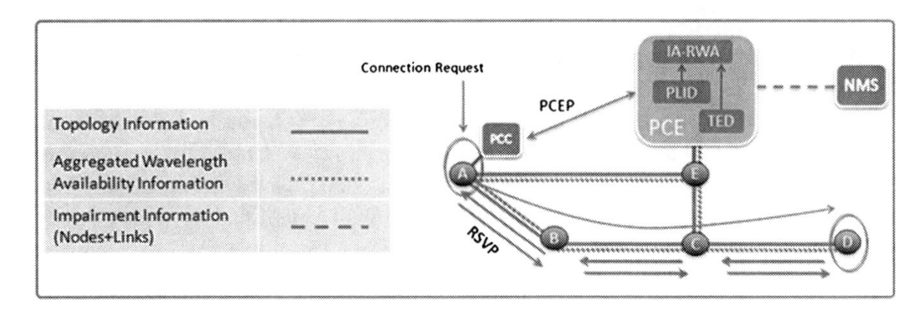

Fig. 2.4 Centralised approach – topology and aggregated wavelength availability information are collected through OSPF-TE, and impairment information through the NMS

sequence of path/resv messages are missing. Control plane restart processes (RFC3473, RFC5063) allow adjacent nodes to resynchronise their control plane state to reinstate information about LSPs that have persisted in the data plane after a node or link failure. GMPLS also proposes a new RSVP-TE Notify message to notify errors to non-adjacent nodes along routes that are not tied to the connection path.

– Hierarchical LSPs. Signalling must support LSP nesting even with unnumbered links in order to improve the scalability of GMPLS (Kompella 2005). An LSP may be advertised as a TE link for use within the same instance of the control plane as was used to set up the LSP.

In transparent optical networks, it is important to assess the signal quality of a path that needs to be established from a source to a destination node. Due to the lack of electrical regeneration, optical signal impairments accumulate along the path. Consequently, it is not sufficient anymore to just find free resources in the network to establish a feasible path. However, there is not yet a standard to support physical impairments in the GMPLS framework. In order to cater for this problem, there are a number of solutions that have emerged over the last couple of years. There are two fundamental ways to incorporate physical impairments in GMPLS-based optical networks: a centralised approach and a distributed approach. They are outlined and discussed next.

1. Centralised approach
   In the centralised approach (Fig. 2.4), an entity referred to as the PCE (RFC4655) is employed in the network and is responsible for computing the routes for the connection requests within its domain. The PCE maintains a centralised TED (e.g. populated using an OSPF-TE routing protocol) for the network topology and aggregated wavelength availability information for all the links in the network. In addition, the PCE maintains a PLID, e.g. populated by the NMS. PLID can either contain information related only to static impairments or include also dynamic impairments information, which is updated at regular intervals. Alternatively, the PCE may fetch this information directly from the ingress nodes.

There are different ways to employ a PCE in this centralised approach. One possibility is to have all connection requests sent directly to the PCE from the ingress nodes, also known as path computation clients in the PCE terminology, for the path computation. The PCE, in turn, responds to the path computation clients with the computed paths. A signalling protocol (e.g. RSVP-TE) is then used to establish the path (Castoldi et al. 2007). Another possibility is to have the NMS responsible for coordinating with the PCE for the path computation and for establishing the computed paths, manually or through a signalling protocol (Tsuritani 2006).

On the positive side, this approach does not require any modification to either the signalling or the routing protocol, while impairments can be catered for, with minimal effort, during the path computation process in the GMPLS framework. Furthermore, since both the TED and PLID are maintained at the PCE, an IA RWA algorithm (see Chap. 2.4) considering the TE information together with the physical impairments can be employed at the PCE to compute an optimal path from source to destination.

However, there are a number of issues associated with this approach. First, this solution does not scale well, particularly when there is only one PCE available per domain. This is mainly because of the intensive computation associated with constraint-based routing, in general, and with IA-RWA, in particular. Moreover, a single PCE approach is vulnerable to TED and PLID errors and failures, aside from the inevitably delayed response to dynamic changes (e.g. link failures) in the data plane, which likely demand recalculation of many paths at once. Finally, there is a certain control overhead associated with the PCEP for every exchange of PCEP related messages. However, this overhead can be mitigated by enabling the bundling of connection requests (Ahmed et al. 2010), which provides also an opportunity for concurrent path optimisation, an efficient way of reducing the connection blocking probability.

2. Distributed approaches
   Distributed solutions can be further categorised in two groups: routing-based approaches and signalling-based approaches, which are outlined and discussed next.

   - Routing-based approach
     In the routing-based approach (Strand et al. 2001), OSPF-TE is extended to convey impairment parameters between nodes in the network. For this purpose, some new top-level type, length and value (TLV) objects are incorporated. In this approach, the impairment parameters for the network links are disseminated via OSPF-TE in the same way as network topology, QoS and TE information. At each routing node in the network, a global PLID database is maintained in addition to the TED (Martinez et al. 2006), with the up-to-date physical-impairment-related information for all links in the network. For each incoming connection request, the ingress node runs an impairment-aware routing algorithm (Chap. 2.4.2 presents some candidates) to determine a feasible path from source to destination (see Fig. 2.5). Next,

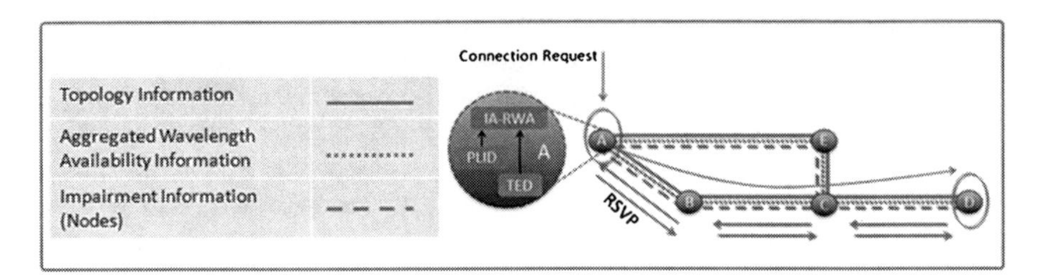

Fig. 2.5 Distributed routing approach – topology aggregated wavelength availability information and impairments information are all collected through OSPF-TE

similarly to the centralised approach, RSVP-TE-based signalling is used to establish the path. Normally, a set of K-shortest paths is computed using the IA-RWA algorithm in order to provide alternative routes in case the path establishment on the first computed path fails during the signalling phase.

This approach has the advantage that a global TED and PLID are maintained locally at all the nodes with path computation capabilities. In addition, path set-up time is shorter, assuming that there are no significant convergence time issues for OSPF-TE. However, there are also several problems associated with this approach. First of all, it does not scale well. In particular, the control overhead can grow very fast if the network domain is large and if there are frequent changes in the TE- and PLID-related information. An approach to mitigate the control overhead has been presented in (Halabi 2003), where the authors propose a TI-LSA. According to the TI-LSA idea, the OSPF-TE LSA flooding principle is modified to take advantage of the node's TI. TI-LSA limits the number of flooded entries by constraining the scope of advertisements to only the nodes that may require that information, i.e. within the boundaries of the node's TI. As a result, the TI-LSA protocol may cope with increasingly large size networks and represents a scalable solution to the problem of topology discovery and update when the TI size is small relative to the whole network. A second problem with this centralised approach is related to inconsistencies in the TED and PLID, mainly due to information propagation latencies. These inconsistencies may lead to suboptimal routing and may ultimately increase the average LSP request blocking probability and/or the resource consumption. The applicability of this approach also depends upon the processing power of each NE. If there are NEs with low computation power, the processing of the LSA may take longer, hence an increased risk of database inconsistency among different nodes. This, in turn, has a direct effect on the LSP request blocking probability, as demonstrated by the work in (Salvadori et al. 2007a). It can be concluded that this approach is not appropriate for large and highly dynamic network scenarios. Another drawback is the significant modifications that need to be incorporated in OSPF-TE.

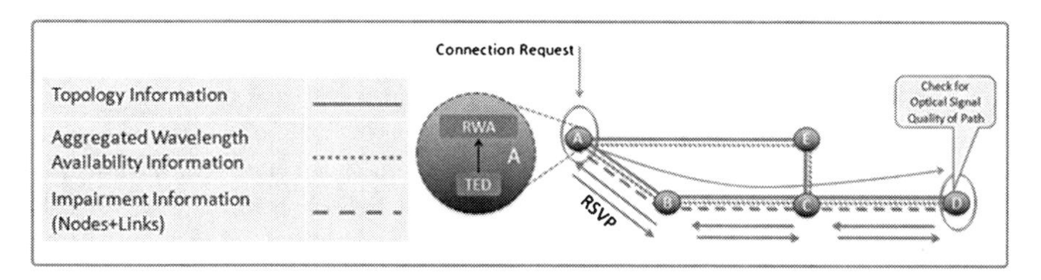

Fig. 2.6 Distributed signalling approach – topology and aggregated wavelength availability information collected through OSPF-TE, and physical impairments information collected during signalling process

In (Salvadori et al. 2007a), a different solution for the distributed routing approach is specified. Here, detailed impairment-related models are uploaded to all the nodes in the network, while only consolidated wavelength availability information, provided by OSPF-TE, is utilised by the nodes to estimate the effect of static link impairments. However, it is assumed that all NEs are of the same kind. In fact, when different optical devices (e.g. from different vendors) are incorporated in a network domain, the estimation of end-to-end LSP impairments becomes much more difficult, given that different models need to be considered hop-by-hop.

– Signalling-based approach
A lot of recent work addresses a distributed solution referred to as signalling-based approach (Cugini et al. 2004). As opposed to routing-based solutions where OSPF-TE was extended to distribute physical layer information, in a signalling-based approach, RSVP-TE is extended with more objects to incorporate and communicate impairment-related information. Similarly to the routing-based approach, a local PLID is maintained at each node, but this time, only impairment information for the current node and its associated links is stored. In a signalling-based solution, the RSVP-TE Path message collects the impairment-related information along all the traversed links and nodes from the source to destination. At the destination node, an evaluation is then made, based on the collected impairment information, to get an estimate of the optical signal quality and to decide whether a path should be accepted (i.e. an RSVP-TE Resv message is sent on the reverse path from destination to source) or it should be rejected because of unsatisfactory signal quality (i.e. a RSVP-TE PathErr message is sent back to the ingress node). This is illustrated in Fig. 2.6 and compares to the IA-RWA approach outlined in Sect. 2.4.2.6. Note that in this case the ingress node computes a set of K-shortest paths (possibly link-disjoint) using an RWA algorithm based only on resource availability information and tries to establish a path using the modified RSVP-TE protocol. If the first path cannot satisfy the signal quality requirement, then the next one is tried and so on. The decision on whether to use a computed path or not is made at the destination (egress) node. Furthermore, in (Sambo et al.

2006), a modification for the PathErr message is proposed. If the egress node fails to establish the path, it may also send back some additional information about the reason of the failure (e.g. insufficient signal quality on a link along the route). In this way, ingress node can get an estimate of the signal quality of a number of links and nodes in the network. It can store this information in the local PLID, populate it and use it for the new connection requests. This kind of the signalling approach is referred to as lightpath provisioning using signalling feedback.

A possible drawback of the signalling-based approach is that it uses RWA based only on the resource availability information, and path feasibility in terms of optical signal quality is only checked at the egress node. Therefore, the computed path may not be optimal in terms of signal quality. Due to this reason, it may increase the average path set-up delay because it may take several attempts before a path is finally established, e.g. a route with an acceptable signal quality is found. A possible solution to this problem is presented in Salvadori et al. (2007b) where, rather than trying to establish the path using a set of computed K-routes in a sequential manner, a Path message is sent along all the computed K-routes in parallel to try to establish a path with an acceptable signal quality, although in this case, there will be a higher signalling control overhead for each connection request. Another variant of this approach is also proposed in (Salvadori et al. 2007b) where the route is computed on a hop-by-hop basis so that each node only considers information related to its directly connected links.

On the positive side, this approach requires minimal changes to RSVP-TE and no changes at all to OSPF-TE (Cugini et al. 2005). Furthermore, the computational load exerted on the ingress nodes is much less intensive (i.e. absence of IA-RWA). Also, the performance of this approach is less prone to the dynamism in the network since local PLID is not populated through the OSPF-TE and hence not dependent on the significant LSA processing delays in larger and more dynamic network scenarios. For the above-mentioned reasons, this approach is much more scalable. Finally, note that this approach can also cater for the intra-node impairment parameters in addition to the link-based impairments as in the routing-based approach.

2.2.3 Convergence Between Metro and Access

The development of transmission techniques is increasing the maximum distance and the number of users served by the access networks. On the one hand, the provided data rate is significantly increasing, from 155 Mbit/s to currently under deployment 1 Gbit/s and 2.5 Gbit/s of IEEE EPON and ITU-T GPON, while also the next recommendations are already focusing in 10 Gbit/s as the 10 Gbit/s Ethernet Passive Optical Network standard (10 G-EPON) recently approved in September

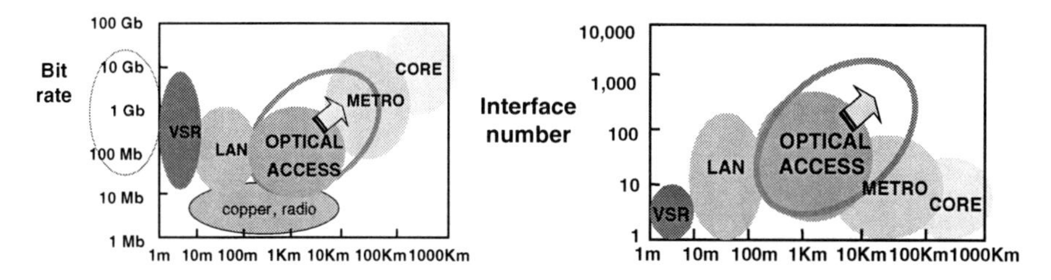

Fig. 2.7 Network segment definition in terms of bit rate, interface number and distance served

2009 as IEEE 802.3av (IEEE8023). On the other hand, ITU-T and FSAN are also working on the standardisation of the XG-PON (gigabit passive optical network).

In parallel to the progress in standardisation of higher bit rates, also the maximum reach distance of access networks and the maximum number of users have been significantly enhanced, and new recommendation proposals are including signal regeneration techniques as the "extender boxes" (EB) over PONs (IEEE8023) which can offer significant extra optical budget (up to 30 dB). The EB is inserted between the OLT and ONUs to provide extra budget to the system while being transparent to protocols and ranging process. In February 2008, the ITU GPON standard (Huang et al. 2005) for optical reach extension, also known as G984.re, was finalised. This standard defines 60 km reach with an optical budget higher than 27.5 dB achieved in both spans adjacent to the EB. Moreover, compatibility with existing equipment is maintained. Both optoelectronic regeneration and optical amplification are considered as solutions for the EB (ITUG984).

The evolution of the demand in date rata and power budget in access networks, leading to extended reach and higher number of users, is forcing the convergence between access and metro transport networks, as shown in Fig. 2.7.

Several research projects are focused on the convergence process as ISTMUSE, Photonic Integrated Extended Metro and Access Network ISTPIEMAN, Fibre-Optic Networks for Distributed Extendible Heterogeneous Radio Architectures and Service Provisioning ISTFUTON and Scalable Advanced Ring-based Passive Dense Access Network Architecture ISTSARDANA. While the projects, MUSE, PIEMAN and FUTON, are considering the use of elements requiring electrical supply in the outside fibre plant of PON, the SARDANA project focuses on maintaining a fully passive outside plant. Obviously, some kind of optical regeneration is required for providing the extra budget of converged transport and access networks. This is achieved in SARDANA by means of remotely pumped optical amplification as Raman and/or hybrid Raman and EDF. We describe the main characteristic of SARDANA approach and the physical monitoring and compensation challenges for CD, Rayleigh Back-scattering RBS, OSNR and physical link protection.

The SARDANA network implements an alternative architecture of the conventional tree WDM/TDM-PON. It consists in the organisation of the optical distribution network as a WDM bidirectional ring and TDM access trees, interconnected by means of scalable optical passive add & drop RN, as shown in Fig. 2.8.

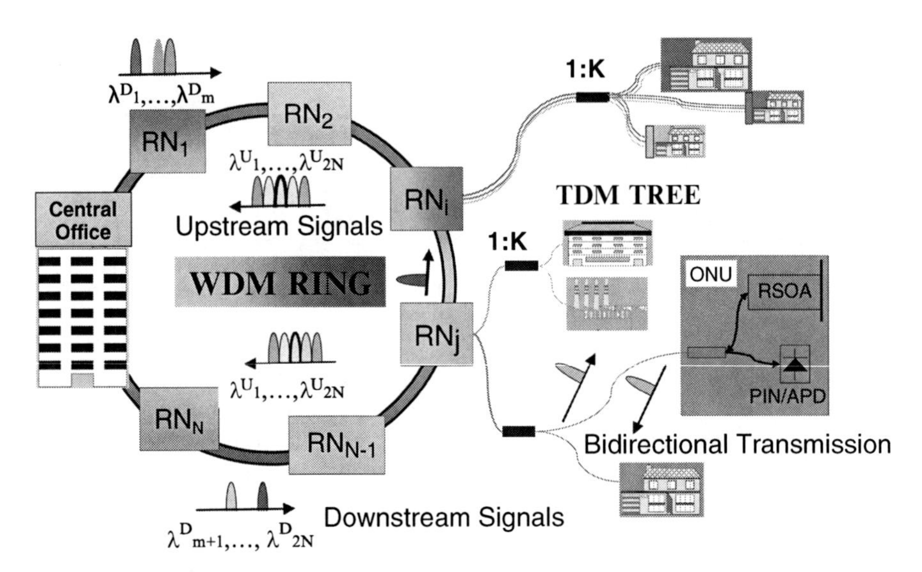

Fig. 2.8 Network architecture and implemented test bed comprising a remote node and reflective SOA as colourless ONU

The proposed network aims serving more than 1,000 users spread along distances up to 100 km, at 10 Gbit/s, with 100 Mbit/s to 10 Gbit/s per user in a flexible way (Bock et al. 2007; Lazaro et al. 2007a; Lazaro et al. 2008).

The ring and tree topology can be considered as a natural evolution from the conventional situation where Metro and Access networks are connected by heterogeneous O/E/O equipment at the interfaces between the FTTH OLTs and the Metro network nodes towards an integrated Metro-Access network. In this case, the network is covering similar geographical area, users and services, but the equipment, previously scattered, will now sit at a unique CO, and implementing an all-optical passive alternative, operating as a resilient TDM over WDM overlay. Depending on the scenario, the ring and tree mixed topology optimises the usage of the fibre infrastructure in the ODN and also offers enhanced scalability and flexible distribution, as new RNs can be installed.

This network transparently merges TDM single-fibre tree sections with a main WDM double-fibre-ring by means of the passive RNs. The 100 km WDM ring transports, e.g. 32 wavelengths for >1,000 users if the TDM trees implement a splitting factor of 1: K = 32, and only 1 wavelength per TDM tree is required. Protection and traffic balancing properties of the network are provided by the ring configuration and the design of the RNs (Prat et al. 2009), providing always a connection between each RN and the CO even in the case of fibre cut.

Physical monitoring is proposed by using several techniques. Implementation of OTDR techniques can be used by adapting these techniques to a WDM network, such as the SARDANA network (Teixeira et al. 2009).

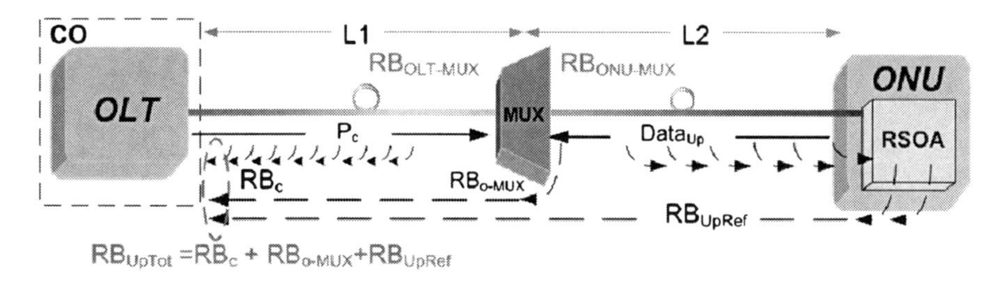

Fig. 2.9 Schematic of a plain WDM-PON with MUX at the fibre plant

In addition to the problem of physical layer monitoring, several transmission challenges are faced as the network evolves towards a more transparent, passive and extended network. Some of the transmission impairments are not specific from SARDANA network but are common for extended WDM PONs.

For example, Rayleigh Backscattering is a common challenge for single-fibre networks with centralised optical signal distribution. As shown in Fig. 2.9, the CW carrier signal provided by the CO is modulated with the user data at the ONU and back reflected in the upstream direction on the same wavelength. In this full-duplex single-fibre bidirectional transmission context, RBS is a dominant impairment. For compensating this impairment, several approaches can be followed: frequency dithering of the optical source at the CO (Lazaro et al. 2007b); wavelength shifting or the centralised distributed signal by the ONU (Omella et al. 2009a); the analysis the best network design and ONU gain for reducing its impact (Lopez et al. 2010). Following this approach, it is possible to analyse the most adequate location of the MUX in the PONs and to determine the optimum ONU gain on each case.

Figure 2.10 shows the simulated and experimental results of the BER values obtained as a function of the MUX position and the different ONU gain values.

Regarding the extension of the access networks, the incorporation of the transparent interfaces between the transport and the access networks makes the accumulation of OSNR as one of the more challenging impairments. As in the case of RBS, this is not a challenge which is unique to SARDANA network but also for other transport plus access converged networks.

Also, the WDM PONs can be considered a transport plus access network, as a trunk fibre is used for transporting downstream and upstream signals through longer distances, as shown in Fig. 2.11.

In this case, a ROPA is analysed as box extender for increasing the transport (trunk section) power budget while maintaining the access budget. Figure 2.12a shows as a reference, the typical access budget (G.984.2, 2.4/1.2 Gbit/s) for range from 13 to 31 dB at 0 dB of trunk budget, providing 3 dB margin to the 28 dB of maximum access losses specified by B+. The small trunk budget margin (shown by the vertical line) from 0 to 4 dB is compatible with the 15 dB ODN loss variations and maximum loss of B + recommendation (marked by the box). Using the same RX and TX and including a ROPA, Fig. 2.12b shows that the ROPA can provide

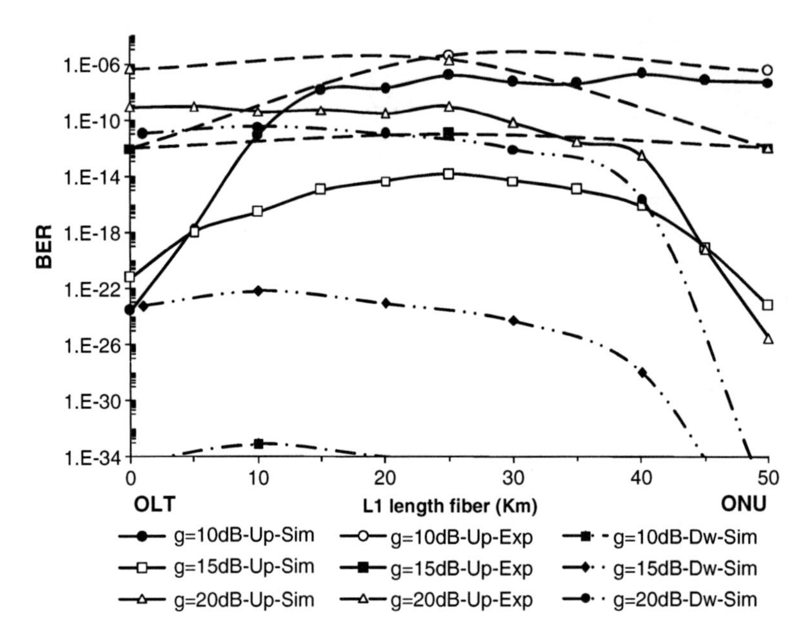

Fig. 2.10 BER as a function of the MUX position to different ONU gain values (simulations and experimental results)

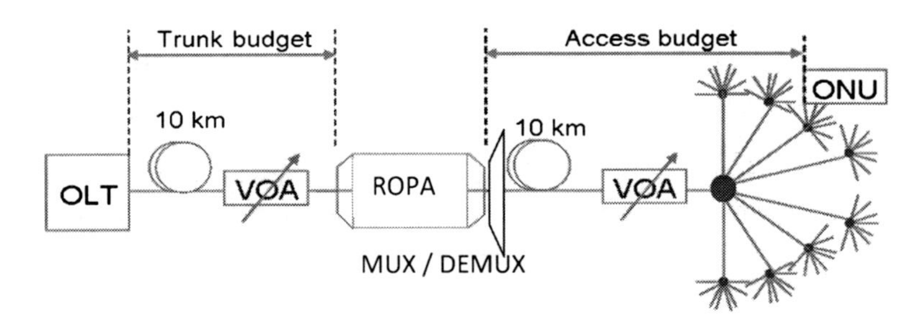

Fig. 2.11 Architecture analysed for the ROPA as extender box, including a Mux/Demux for the overlaying of several G-PON architectures in a WDM/TDM network

enough gain to provide an extra trunk budget from 0 to 14 dB compatible with B + access budget from 13 to 28 dB by implementing a ROPA with 15 m of EDF and providing 20dBm of pump, at 1,480 nm, at the ROPA. Thus, a total budget of 42 dB is achieved.

In SARDANA architecture, the power budget extension is performed by hybrid Raman and EDF amplification. Remote pump power is transmitted through both upstream and, if required, downstream fibres.

A similar analysis to the WDM–PON can be done, while in this case, due to the higher complexity of the network, the BER values for the furthest ONU in resilience mode (a fibre failure has been produced and all the traffic has to be directed through one side of the WDM ring architecture) are shown in Fig. 2.13. It is shown that the

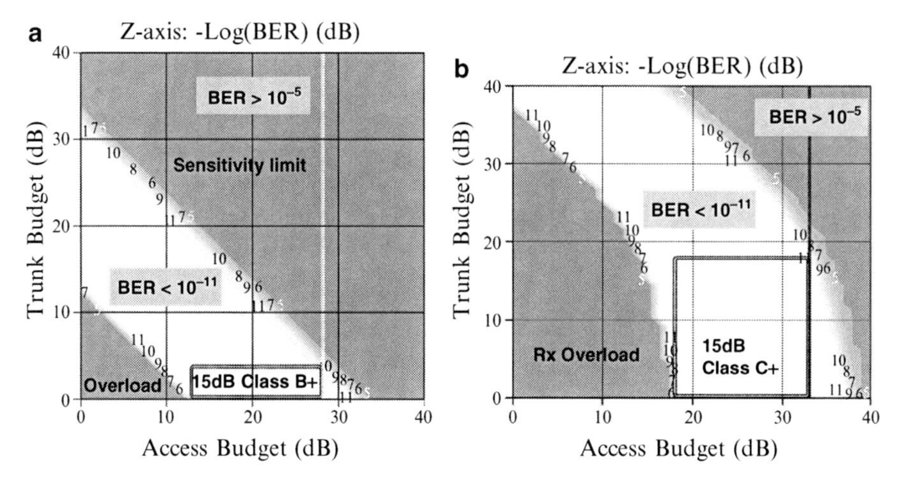

Fig. 2.12 (**a**) BER values for a G-PON (2.4/1.2 Gbit/s) using PIN detectors with minimum sensitivity and overload values –28 dBm and −8 dBm, respectively, and TXs providing the maximum mean launched power (+1.5 dBm); (**b**) BER values with a ROPA for 8 WDM wavelength channels for a ROPA with 15 m of HE980 EDF and 20 dBm of pump power, the *vertical line* indicating the maximum access budget for B + (28 dB); both for downstream transmission at the central channel at 1,550 nm (*Max* BER: 10–11 and *min* BER: 10–5)

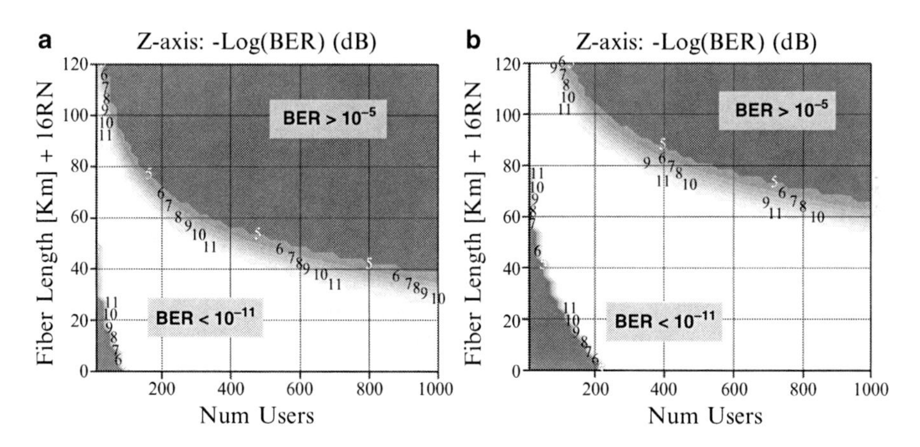

Fig. 2.13 (**a**) DS BER values for the furthest ONU in resilience mode of a Raman + In-line remotely pumped SARDANA network of 32 channels sending 3dBm per channel from the OLT, 50% splitting signal at the ONU, with 1.2 W of pump at 1,480 nm for both US/DS fibres and an ONU RX showing a sensitivity of –24dBm at 10Gbit/s.; (**b**) DS BER values 2.4 W of pump at 1,480 nm both US/DS fibres

combination of Raman and in-line EDF amplification can provide adequate signal quality for up to 100 km reach and 128 users (under the assumption of first 20–25 km of the ring free of RNs) or 20 km reach and 1,024 users by remotely pumping with

1.2 W of pump at 1,480 nm for both US/DS fibres and a higher number of users and distances, by rising the available pump power per fibre to 2.4 W, as shown in Fig. 2.13b.

The SARDANA project focuses on providing services up to 10 Gbit/s. The transmission of signals at this data rates through distances in the range of 100 km results in the CD impairment. In this project, two approaches are analysed: compensation of fibre CD by dispersion-compensating fibres located at the CO and by electronic equalisation techniques (Omella et al. 2009b).

2.3 Monitoring and Signal Processing in Optical Networks

2.3.1 Monitoring

The engineering of high-bit-rate WDM optical transmission systems requires a careful control of each channel characteristic in order to limit the detrimental effects of the different types of physical impairments taking place in single-mode optical fibres. The initial values of the parameters, set at the system installation, may need further adjustment due to many reasons:

- Evolution of the characteristics of optoelectronic devices
- Fluctuation of the fibre characteristics
- Deployment of additional wavelength channels
- Upgrade of the line rate of the channel

The required flexibility tends to be increasingly important because optical transport networks become more dynamic and transparent. For instance, as ROADMs are now implemented in long haul transmission systems and metropolitan rings, the different wavelength channels may experiment a new transmission path according to the actual ROADM configuration. This issue will become even more complex in the case of meshed networks using transparent (i.e. without optoelectronic regeneration) or partly transparent networks based on optical cross-connects.

Considering all these possible changes in the network, it seems quite impossible to base the control of the signal characteristics only on initial tests performed on a new deployed channel. It is clear that some amount of real-time monitoring of the characteristics is mandatory to provide information to the system and/or network controllers. This fact is mandatory in networks using IA-RWA algorithms.

On the other hand, it is an important requirement for an optical network, comprised of multiple point-to-point links, that the signals propagating throughout being of sufficient quality to detect. Historically, this has been achieved by the use of electrical repeaters. These convert the incoming optical signal into an electrical signal from which the base data is recovered before being used to transmit a new optical signal. This OEO conversion is undesirable when striving for high-bit-rate systems in which the conversion becomes a limiting factor.

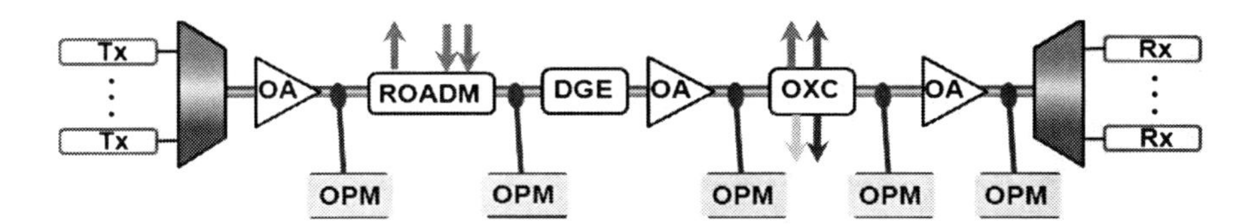

Fig. 2.14 Place of optical performance monitors (*OPM*) in reconfigurable and dynamic all-optical network with optical amplifiers (*OA*), reconfigurable add/drop multiplexer (*ROADM*), dynamic gain equaliser (*DGE*) and optical cross-connect switches (*OXC*)

Optical amplifiers have removed the OEO conversion but added ASE noise to the optical signal. On the other hand, in future optical transport technologies of 100 Gbit/s transmission over around 1,000 km, CD has a strong effect in limiting transmission bandwidth. Signal regeneration and CD compensation in the optical domain are two approaches to solve the problem. In the case of optical regenerator, additional processing functions such as amplitude equalisation (reshaping) and temporal repositioning (retiming) of the optical pulses are developed.

Some, not exhaustive, description of monitoring, compensation and signal processing techniques are presented in the next sections, not pretending to be an exhaustive description of state of the art; but some examples to show their potential, with some specific contributions from the authors in them.

2.3.1.1 Optical Performance Monitoring

The term OPM (Chung 2008) generally refers to monitoring techniques operating at a lower level than the data protocol monitoring, which measures protocol performance information. OPM techniques include spectral (optical or electrical) and time (optical or electrical) domain techniques. Some examples of the spectral domain techniques are presented in Sect. 2.3.1.1. Section 2.3.1.2 focuses on asynchronous time-domain sampling of the photo-detected signal and Sect. 2.3.1.3 presents optical time domain reflectometry applications to OPM.

Figure 2.14 shows different strategic places for transport signal quality monitoring (Kilper et al. 2009; Bendelli et al. 2000).

Key requirements regarding OPM are: (1) small size; (2) fast and flexible measurements; (3) operation at low input power; (4) multichannel operation: monitoring of several channels in parallel or consecutively; (5) bit rate and modulation format transparency (mixed traffic can be present on the line or signal formats may change during the lifetime of the OPM) (Bendelli et al. 2000). Moreover, the OPM should be: passive, remotely configurable and low cost compared to conventional test equipment.

Depending on the type of physical parameters which are used to perform OPM, one can distinguish basic OCM and advanced signal quality monitoring (Kilper et al. 2004a). Nowadays, OCM becomes very common in WDM systems. Key parameters

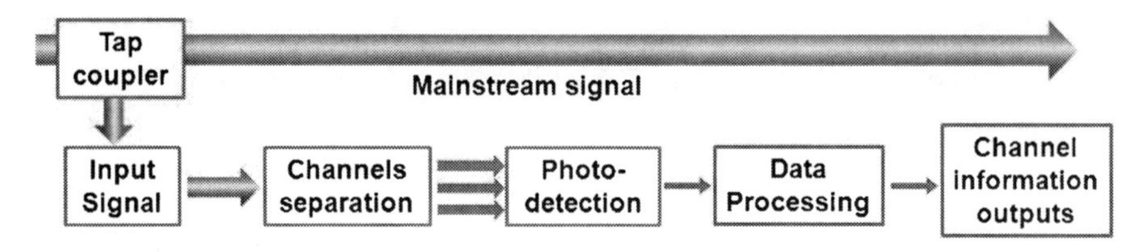

Fig. 2.15 Optical channel monitoring functional blocks

to be monitored are: channel wavelength, channel power, OSNR and their respective drifts. According to (Kirstaedter et al. 2005), the values have to be obtained every 10 ms for power and wavelength and 100 ms for optical OSNR. On the other hand, such signal distortions as in-band OSNR, accumulated CD and PMD are considered as advanced parameters which need more complex monitoring techniques.

Optical Channel Monitoring

Optical power at a given wavelength is the basic parameter for any WDM network. For monitoring purposes, a fraction (typically 1%) of the light power is tapped from the mainstream optical signal. Then, this tapped weak signal is optically demultiplexed or filtered, in order to separate the channels, and then directed to the photodetector. Optical signal is converted to electrical signal for processing and finally channel information is transmitted to the network manager (Fig. 2.15).

A simple way to accomplish this can be using a convenient diffraction grating, such as a free space VPHG, a FBG or an AWG with a photodiode array (Pinart et al. 2005) (ENABLENCE). However, this approach is still quite expensive as it requires a large number of photodetectors to cover a wide spectral span at high resolution.

Another way to monitor the WDM channels consists of using a single detector combined with one of various types of tunable filters, such as a thin-film filter, an MEMs tunable filter, a PZT-tuned Fabry–Perot filter, an acousto-optic tunable filter and a temperature-tuned etalon filter (Cahill et al. 2006). But these techniques require complex tuning mechanisms and sometimes have insufficient resolution.

Nowadays, both approaches have been commercialised, and current standard OPM technology with OSA approach ensures standardised measurements according to ITU-T G.697 (ITU-T G.697). Table 2.1 presents typical specifications for this category of monitors.

The main difference between these devices is the response time, determined as the sum of scan, data processing and report times. Depending on the measurement resolution and parameters to be monitored, full scanning can take from about 10 ms to few hundreds of ms to complete a measurement across the entire C-band. Nevertheless, some of equipment manufactures add the OCM module to their products, such as DGE, ROADM, optical switch, etc. (LIGHTWAVE) (JSDUNPH).

Table 2.1 Typical specifications of commercial optical channel monitors

Parameters	Value	Units
Channel spacing	50 100	GHz
Wavelength range	C-, L- or C + L-band	nm
Channel number (for C-band)	>80 >40	
Absolute wavelength accuracy	±50	pm
Relative wavelength accuracy	±30	pm
Dynamic range	>30 (typically 50)	dB
Maximum input channel power	From −10 to +5	dBm
Absolute channel power resolution	±0.5	dB
Relative channel power resolution	±0.3	dB
PDL	<0.3	dB
OSNR out-of-band	>25 >28	dB
OSNR accuracy	±0.75 (typically ±1.5)	dB
Scan and report time	From 10 to 1,000	ms

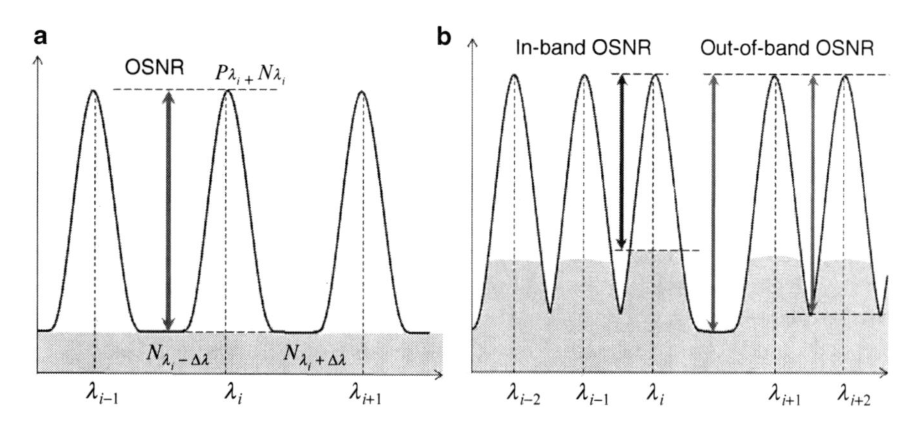

Fig. 2.16 (**a**) Linear interpolation method for OSNR measurements; (**b**) Comparison between out-of-band OSNR method and in-band OSNR method

But the real limitation of this OSA-based OPM is the optical noise measurement. For calculating OSNR, the most appropriate noise power value is that at the channel wavelength. However, with a direct spectral measurement, the noise power at the channel wavelength is included in signal power and is difficult to extract. An estimation of the channel noise power can be made by interpolating between the noise power values on both sides of the channel (Fig. 2.16a).

This assumption becomes invalid for current DWDM networks due to signal overlap from neighbouring channels, in-line filtering, spectrum broadening from non-linear effects, four wave mixing introduced noise, etc. With higher modulation rates and narrower channel spacing, the modulation sidebands from adjacent channels interfere and limit the ability to measure the noise level between channels (Fig. 2.16b). Increasing the resolution of the optical spectrum analyser does not remove this limitation.

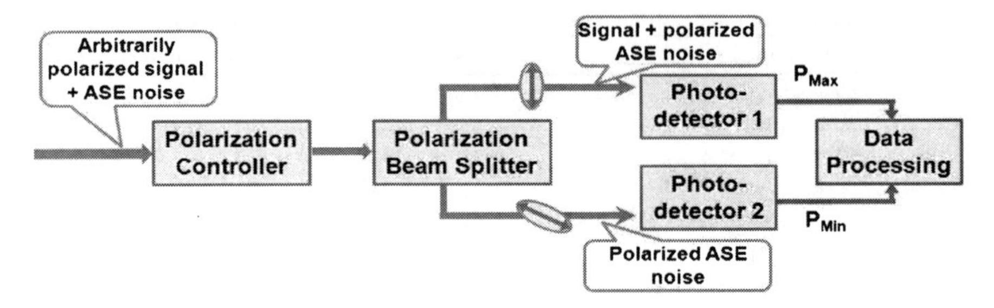

Fig. 2.17 Schematic diagram of the polarisation nulling method

Modulation tone techniques have also been used as a low-cost alternative to spectral measurements. But the principal limitation is the same: optical noise power is extrapolated from the power level adjacent to the channel (Pan et al. 2010).

As the OSNR is the key performance parameter in optical networks that predicts the bit error rate of the system, the in-band OSNR becomes essential in reconfigurable networks.

In-Band OSNR Monitoring

The challenge in this case is to discriminate the noise and the signal in the same spectral band. The polarisation nulling method overcomes some of the limitations of conventional OSA for OSNR measurement. This approach is based upon the hypothesis that an optical signal has a well-defined polarisation, while the ASE noise component is unpolarised, which allows using the polarisation extinction ratio as a measure of the OSNR (Pan et al. 2010; Kirstaedter et al. 2005; Lee et al. 2006).

As shown in Fig. 2.17, a high extinction ratio polarisation beam splitter is used to split the input signal into two arms, both being polarised in orthogonal linear states, and then detected simultaneously (P_1 and P_2, respectively). An adjustable PC is used to find the maximum extinction of the signal when one component consists of signal and polarised noise, while the other contains only polarised noise. A measurement of the in-band OSNR will need multiple scans with different settings of the PC. The sum of $P_{1\min}$ and $P_{2\min}$ indicates the non-polarised in-band noise (P_{Noise}), whereas for a given polarisation state of the signal, the sum of P_1 and P_2 corresponds to ($P_{\text{Signal}} + P_{\text{Noise}}$). At the end of the measurement, the in-band OSNR values for each channel are calculated with the following equation:

$$\text{OSNR} = \frac{P_1 + P_2 - (P_{1,\min} + P_{2,\min})}{(P_{1,\min} + P_{2,\min})} \tag{2.1}$$

Unfortunately, the performance of this technique could be affected by various polarisation effects in the transmission link. For example, it could be seriously deteriorated if the signal is depolarised by PMD and non-linear birefringence or the ASE noise is partially polarised due to polarisation-dependent loss.

This method has been successfully implemented in dual port optical spectrum analysers, which became recently commercially available (EXFO; JDSU).

Another method is the optical subcarrier monitoring in which each WDM channel is associated with a subcarrier (small amplitude-modulated RF frequency pilot tone) (Rossi et al. 2000). Because the tone is at a single, low frequency, it can be easily generated and processed using conventional electronics. The average power in these tones will be proportional to the average optical power in the channel, and the aggregate WDM optical signal on the line can be detected; the tones of all the channels will appear in the RF power spectrum in much the same way they would appear in the optical spectrum. Thus, optical parameters can be monitored without using the expensive optical devices, such as tunable optical filter or diffraction grating. The electrical CNR of the subcarrier will be determined and the OSNR is obtained through a mathematical relationship with CNR. This method has an advantage in that it involves monitoring on the actual data signal as it has propagated along the impairment path of the signal itself and can be implemented with narrowband electronics. Moreover, the monitoring of RF tones can be used for measuring the accumulation of CD and PMD on a digital signal (Rossi et al. 2000).

The major drawbacks of this technique are that the AM tone and data could interfere with each other and cause deleterious effects. Thus, the amplitude of the pilot tone should be large enough to discern the tone signals from the noise-like random data, but small enough not to induce a significant degradation in the receiver sensitivity for data.

MZI method is based on the difference of behaviour between a coherent signal, which is able to interfere at the output of the interferometer, and non-coherent ASE noise. By adjusting the path difference between the two arms, the maximum (constructive interference) and minimum (destructive interference) output powers are obtained, and OSNR can be derived while it is proportional to the ratio P_{const}/P_{dest}. With increasing ASE power (i.e. decreasing ONSR), P_{dest} increases faster than P_{const} because of the random phase of the noise (Liu et al. 2006).

The most promising results was obtain with a 1/4-bit delay method. Since the phase relationship between successive bits is not important, the method is applicable to multiple modulation formats (Lizé et al. 2007).

Uncorrelated beat noise can also be used for OSNR monitoring (Chen et al. 2005). This method is compatible with different modulation formats, independent of the pattern length and insensitive to PMD.

Chromatic Dispersion and Polarisation-Mode Dispersion Monitoring

We give here a short description of existent technologies for real-time CD and PMD monitoring which are summarised in Pan et al. 2010.

Firstly, monitoring techniques based on RF tone (conversion of a phase modulated signal into an amplitude-modulated one by inserting a subcarrier at the transmitter) are relatively simple and quite fast but may require transmitter modification.

The RF clock techniques are based upon the same concepts as the RF pilot tones techniques, with a monitored frequency corresponding to the bit rate. The clock power detection technique has been used as CD and PMD monitors, whereas the technique based on phase detection is used for CD monitoring only. The main advantage of the clock techniques is the absence of modification of the transmitter; however, they are potentially expensive (single channel operation).

Impact of New Modulation Formats

The techniques presented in this section have been first introduced to monitor OOK (mostly NRZ) 10 Gbit/s signals. Most of them can be applied to more advanced modulation formats that are envisioned for 40 or 100 Gbit/s transmission. This trend towards more complex modulation schemes could, however, have an impact on the deployment of OPM functions. There will still remain a need for the monitoring of the basic parameters (power, OSNR) of multiplexed channels. On the other hand, the high spectral efficiency and related robustness against DC and PMD of these modulations could reduce the need of in-line monitoring of DC and PMD. For instance, it has been shown that CO-OFDM signal is robust against PMD and tolerates a chromatic dispersion equivalent to 3,000 km standard single-mode fibre. Moreover, these modulation formats involve advanced signal processing algorithms in the receivers which can provide information about the impairments experienced by the incoming signals. In (Shieh et al. 2007), OCE through receiver signal processing is proposed as one approach to optical performance monitoring. Most importantly, performance monitoring by OCE is basically free because it is embedded as a part of the intrinsic receiver signal processing. Such a monitoring device could also be placed anywhere in the network without concern about the large residual chromatic dispersion of the monitored signal. Cost and standardisation issues will be determinant to select among the different per-channel monitoring techniques: optical and/or RF spectrum analysis, digital signal processing (which implies clock recovery) and asynchronous sampling which will be discussed in the next section.

2.3.1.2 Asynchronous Performance Monitoring

In the previous section, we introduced several techniques for the monitoring of a WDM channel. These techniques are based on the analysis of the optical or electrical spectrum of a group of channels or of a single channel, where some extra monitoring signals (e.g. pilot tones) have been possibly added. The present section is dedicated to time-domain monitoring techniques, which involve the sampling of the channel

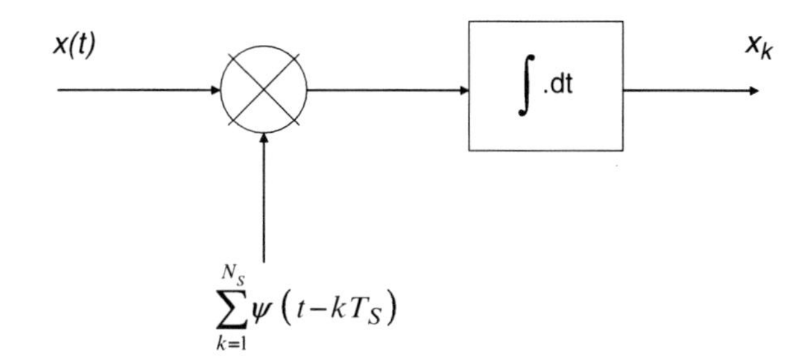

Fig. 2.18 Representation of the sampling of a signal $x(t)$ with a sampling period T_S and a sampling window defined by the function ψ

to be monitored and a statistical analysis of the acquired samples. For all these techniques, it is assumed that the channel to be monitored has been isolated from the rest of the optical comb.

We will first consider an amplitude-modulated binary digital signal $x(t)$, with a bit duration TB and bit frequency $f_B = 1/T_B$. Figure 2.18 provides a diagram of a simplified sampling system where $x(t)$ is multiplied by a train of periodic sampling pulses centred in the sampling instants $k \cdot T_S$, where T_S is the sampling period (and f_S the sampling frequency). Each sampling pulse $\psi(t)$ has duration T_{res} and is generally assumed to be a gate function. The multiplication can be performed either in the optical domain (for instance, by using sum-frequency generation in a non-linear crystal (Shake et al. 1998)) or, more commonly, in the electrical domain by gating the photo-detected signal (e.g. in (Mueller 1998)). A set of N_S samples is acquired. N_S should be high enough to contain the entire statistics of the signal.

Let us assume, that $f_S = \frac{n}{m} f_B + f_{off}$, where n and m are two natural numbers which minimise $\left| f_S - \frac{n}{m} f_B \right|$ and f_{off}, is the offset frequency. In the conventional synchronous sampling technique, f_S is synchronised with f_S in order to satisfy (Shake et al. 2004):

$$T_{step} = \frac{1}{f_S} - \frac{1}{\left(\frac{n}{m}\right) f_B} = \frac{1}{p f_B} \tag{2.2}$$

where T_{step} is the interval between the p sampling time positions inside the bit duration. This implies a clock recovery of f_B. The above relationship determines the offset frequency for synchronous sampling.

In the case of asynchronous sampling, the offset frequency does not satisfy (2); thus, if N_S is high enough, the sampling instants will be uniformly spread across the entire bit period. Figure 2.19 shows an example of both synchronous and asynchronous histograms and the corresponding eye diagram.

An example of typical asynchronous sampling parameters for a 10 Gbit/s NRZ or RZ signal, taken from (Shake et al. 1998), is $f_S = (f_B/1{,}024) - 10\,\text{kHz} \approx 9.7\,\text{MHz}$, $T_{res} = 1$ ps and NS $\approx$ 1.5 104. The T_{res} value is generally fixed, much shorter than the bit period, in order to avoid loss of information due to averaging effects. However, by noting that the averaging effect mostly concerns the noise, it is possible to relax this constraint and use sampling durations close to the half bit period. This value

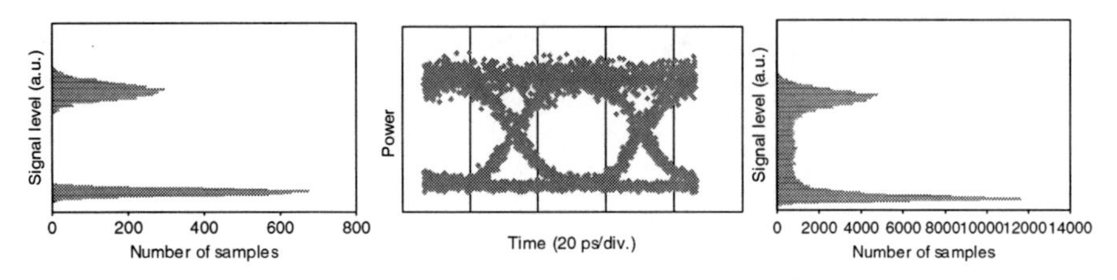

Fig. 2.19 Eye diagram example (*centre*) of a 10 Gbit/s NRZ signal with associated synchronous (*left*) and asynchronous (*right*) histograms

may even nearly reach T_B if the sample processing takes into account inter-symbol interference (Luis et al. 2004) to the expense of an increased processing complexity.

The main motivation for asynchronous sampling is the absence of clock recovery which makes it less expensive than synchronous sampling and enables it to work at a wide variety of bit rates. This is a clear advantage in the context of transparent optical networks, but several issues need to be solved in order to apply it as a monitoring technique. In particular, it should allow identifying the strength and the origin of signal perturbation.

Since the early proposals of asynchronous performance monitoring (Shake et al. 1998; Mueller 1998), different studies have been carried to address this issue, especially by deducing the Q-factor from the asynchronous histogram. A simple analysis can be provided for NRZ coding and negligible inter-symbol interference (Luis et al. 2004). At a fixed timing phase t_0, $Q(t_0)$ is defined by:

$$Q(t_0) = |\mu_1(t_0) - \mu_0(t_0)| \;/\; |\sigma_1(t_0) + \sigma_0(t_0)| \tag{2.3}$$

where $\mu_i(t_0)$ and $\sigma_i(t_0)$ are the mean and standard deviation of the mark(1) and space(0) levels at t_0, respectively. If the choice of t_0 corresponds to the optimum decision time, $Q(t_0)$ reduces to the usual Q-factor, which (assuming Gaussian distributions of mark and space amplitudes) is linked to the BER by:

$$\text{BER} = \frac{1}{2} erfc\left(\frac{Q}{\sqrt{2}}\right) \tag{2.4}$$

When performing asynchronous sampling, it is only possible to measure the average Q-factor (Q_{ave}), defined by:

$$Q_{\text{ave}} = |\mu_{1,\text{ave}} - \mu_{0,\text{ave}}| \;/\; |\sigma_{1,\text{ave}} + \sigma_{0,\text{ave}}| \tag{2.5}$$

where $\mu_{i,\text{ave}}$ and $\sigma_{I,\text{ave}}$ are the mean and standard deviation of the mark(1) and space(0) of all sampled data, respectively.

To get useful information from asynchronous sampling, one needs to derive a relationship between Q_{ave} and Q. It is quite intuitive that Q_{ave} will be smaller than Q because the data obtained by asynchronous sampling include unwanted cross-point

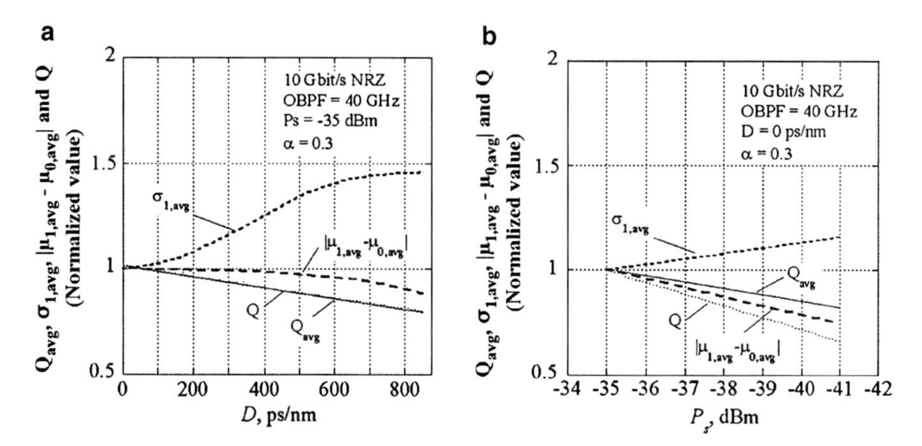

Fig. 2.20 Chromatic dispersion dependence (**a**) and OSNR (**b**) dependence of $\sigma_{1,\text{ave}}$, $|\mu_{1,\text{ave}} - \mu_{0,\text{ave}}|$, and Q_{ave} for a 10 Gbit/s NRZ optical signal with a 25 ps rise time and $T_{\text{res}} = 1/64$ T_{B}, $N_{\text{S}} = 16{,}384$, $B_{\text{opt}} = 40$ GHz, $B_{\text{el}} = 7.5$ GHz and $\alpha = 0.3$ (© 2003 IEEE, after Ref (Shake 2003))

data in the eye diagram. The effect of these cross-points can be limited by using a threshold α ($0<\alpha<0.5$) in order to define $\mu_{\text{th0}} = \mu_{0,\text{ave}} + \alpha\ |\mu_{1,\text{ave}} - \mu_{0,\text{ave}}|$ and $\mu_{\text{th1}} = \mu_{1,\text{ave}} - \alpha\ |\mu_{1,\text{ave}} - \mu_{0,\text{ave}}|$ and discarding the values comprised between μth_0 and μth_1. In practice, values of α between 0.1 and 0.5 have been used and result in linear relationship between Q_{ave} and Q, when 13 dB $< Q <$ 20 dB (for the sake of simplicity, the same notation is used for the linear and decibel expressions of Q). The actual slope of this curve depends mainly on the $B_{\text{opt}}/f_{\text{B}}$ and $B_{\text{el}}/f_{\text{B}}$ ratios where B_{opt} and B_{el} are the optical and electrical bandwidths of the sampling system, respectively (Shake and Takara 2002). Reference (Luis et al. 2004) provides a model based on an equivalent filter $h_{\text{eq}}(t) = \psi\ (-t)$ (where ψ has been introduced in Fig. 2.18), which confirms this linear dependence for IM/DD signals without inter-symbol interference, when the spontaneous-spontaneous beat noise can be neglected.

Beyond the estimation of the Q-factor, the analysis of asynchronous histograms enables to detect signal degradations due to noise, crosstalk or pulse distortion. Indeed, these effects have different impacts on the various μ and σ values. ASE noise will reduce $|\mu_{1,\text{ave}} - \mu_{0,\text{ave}}|$and increase both $\sigma_{0,\text{ave}}$ and $\sigma_{1,\text{ave}}$, while crosstalk will mainly increase σ1,ave; finally, chromatic dispersion will have a more pronounced impact on σ1,ave than on $|\mu_{1,\text{ave}} - \mu_{0,\text{ave}}|$. All these perturbations yield noticeable modifications of the asynchronous histograms for B.E.R. > 10^{-12}. A detailed analysis of the evolution of the histograms in the presence of both noise and dispersion can be found in Shake (2003). Figure 2.20, reproduced from (Shake 2003), shows an example of the evolution of $\sigma_{1,\text{ave}}$ and $|\mu_{1,\text{ave}} - \mu_{0,\text{ave}}|$as a function of received power and line chromatic dispersion.

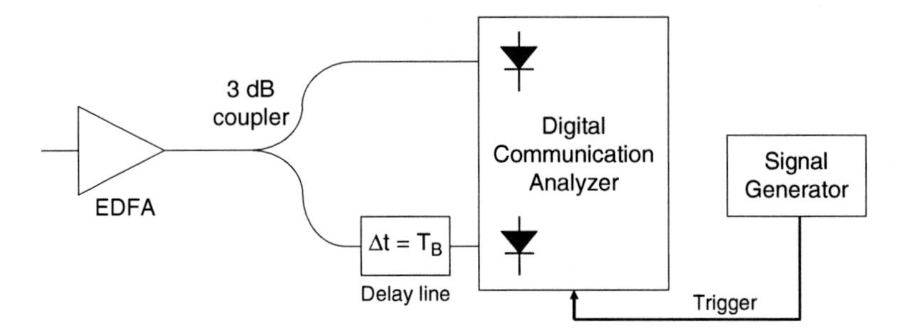

Fig. 2.21 Two-tap asynchronous sampling set-up with a separation between sample pairs $= T_{\mathrm{B}}$

Nevertheless, the interpretation of conventional asynchronous histograms remains difficult in the presence of simultaneous transmission impairments. In order to solve this issue, a two-tap sampling technique has been proposed (Anderson et al. 2006; Dods 2006).

In this technique, the asynchronous samples are collected by pairs (x_k, y_k), with $x_k + 1 = x_k + T_{\mathrm{S}}$ and $y_k = x_k + (T_{\mathrm{B}}/p)$ where usually $p = 1$, 2 or 4 (Fig. 2.21). The two-tap histogram is obtained by plotting the N_{S} points with coordinates (x_k, y_k).

For an NRZ modulation with $p = 1$, the ideal figure (without impairments) represents the edges of a square with its secondary diagonal: the edges correspond to 001, 011, 110 and 100 bit sequences, while the diagonal corresponds to the 010 and 101 sequences. In contrast to conventional asynchronous or synchronous sampling techniques, this method captures information about the distribution of closely spaced samples, which makes it particularly suited to the monitoring of pulse distortion.

The ability of two-tap sampling to provide information about simultaneous impairments is illustrated in Fig. 2.22 which shows the impact on the histograms of up to four combined impairments, for delay values of T_{B}, $T_{\mathrm{B}}/2$ and $T_{\mathrm{B}}/4$. The different figures can generally be well distinguished, except for the limited effect of PMD on the histogram with T_{B} delay.

The analysis of these two-tap histograms can be performed automatically with estimation techniques used in pattern recognition (Anderson et al. 2006), including three-layer artificial neural networks (Jargon et al. 2009).

Until recently, most of the work on asynchronous performance monitoring was carried on OOK data. New generations of high capacity optical transmission systems will likely involve more complex modulation schemes and coherent detection. This could increase the complexity of the sampled patterns and make more difficult a quantitative analysis of the different impairments. However, recent results suggest that the two-tap asynchronous sampling technique can be applied to various PSK modulation schemes. An example is given on Fig. 2.23, which shows the results of the processing by a three-layer artificial neural network of 40 Gbit/s RZ-BPSK data after two-tap asynchronous sampling, with $T_{\mathrm{B}}/2$ delay. Both direct and balanced

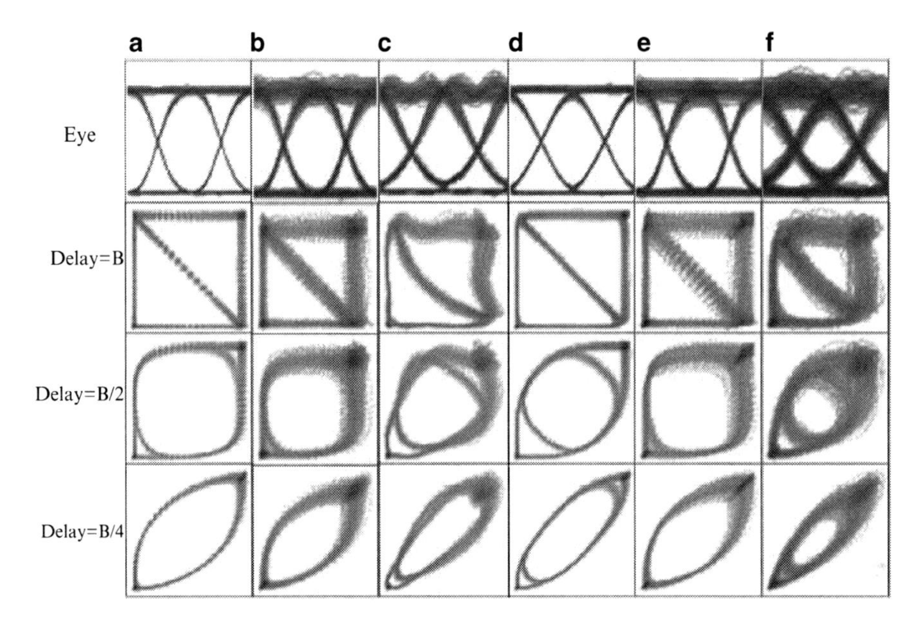

Fig. 2.22 Eye diagram and two-tap plots for delay = T_B, $T_B/2$ and T_B /4 for (**a**) OSNR = 35 dB and no impairment, (**b**) OSNR = 25 dB and no impairment, (**c**) OSNR = 35 dB and $D = 800$ ps/nm, (**d**) OSNR = 35 dB and PMD = 40 ps, (**e**) OSNR = 35 dB and interferometric crosstalk = −25 dB, (**f**) OSNR = 25 dB, $D = 800$ ps/nm, PMD = 40 ps and crosstalk = −25 dB (© 2006 IEEE, after Ref (Anderson et al. 2006))

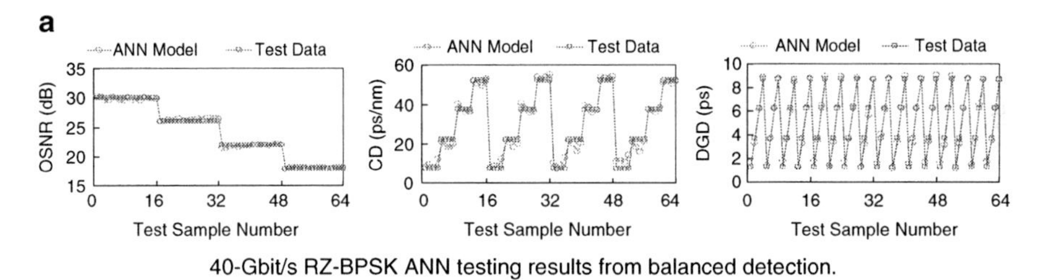

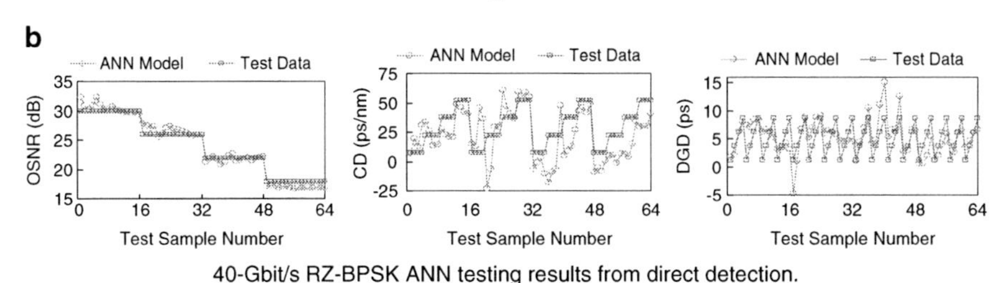

Fig. 2.23 Comparison of testing and artificial neural networks – model data for a 40 Gbit/s RZ-BPSK channel (© 2009 VDE Verlag after Ref. (Wu et al. 2009))

detection schemes are compared. The later provide a much more accurate prediction of the impairments, especially for CD and PMD.

This approach can be extended to more complex modulation schemes: asynchronously generated constellation patterns have been used to estimate the amount of CD and DGD of DQPSK signals (Arbab et al. 2009).

In conclusion, asynchronous sampling monitoring can provide as much useful information on the impairments experienced by an optical signal as the conventional eye-diagram monitoring, but without necessity of a clock recovery. This method is also quite sensitive as it can detect impairment levels responsible of a rough (before FEC) 10^{-12} BER in a few milliseconds, much faster than direct BER measurement. The simple one-tap technique allows measuring an averaged Q-factor and is well suited to NRZ or RZ modulations, when one type of impairment is likely to predominate. It is, to a large extent, bit-rate independent, even though the accuracy of its prediction can be improved when the sampling parameters are adapted to the bit rate. The two-tap technique is, by nature, bit-rate dependent, but has proven to be much more powerful to provide reliable results by using, e.g. artificial neural network-based algorithm, to process the samples, in the presence of simultaneous impairments, and it offers a good potential for operation with advanced modulation schemes.

2.3.1.3 Infrastructure Monitoring (OTDR)

Installation of PONs as a main choice of the operators has been accelerating mainly due to their future proof nature and lower OPEX.

Even though PON outside plant (OSP) should be sustainable over the expected lifetime of the system, there is still room for the operators to save significant amount of OPEX using effective and easy preventive maintenance of the physical infrastructure. In today's PON systems, the physical infrastructure is usually not entirely visible to the NMS. As a direct consequence, a physical failure cannot be detected before creating service outage in upper layers which in turn may lead to tremendous loss in business for the operators. We can mention not only direct financial losses due to service interruption but also indirect financial losses due to bad reputation. These arguments have been gaining importance as the warranty on the quality of the infrastructure becomes a deciding factor in the strongly competitive marketplace.

The convergence between transport and access networks with more complex hybrid topologies and high bit rates (10 Gbit/s) over longer reach/higher split options gives rise to capacity increase and make a network failure more disastrous.

The aim of preventive maintenance is to detect any kind of deterioration in the network that can cause suspended services and to localise these faults in order to avoid specially trained people deployed with dedicated and often expensive equipments for troubleshooting. PON infrastructure does not only suffer from accidental damages and environmental effects (e.g. water penetration in splice closures) but are also subject to a lot of changes after the network is installed and activated. As an example, the optical access network may not be initially fully loaded; subscribers would be turned up, possibly over an extended period of time (Frigo et al. 2004). Hence, network operators should continuously be aware if a change noticed by its monitoring system is service-oriented or indeed a fault. In addition to that, it is crucial to discriminate the faults (accidental interruptions) from attacks (intentional interruptions) results in a strengthening of relations between optical maintenance functions and the security management. All arguments mentioned above mean that the existing maintenance methods need to be updated making the monitoring in PONs an active research area.

The number of scientific publications has significantly increased in the last few years. Authors propose different approaches which are addressing some of the challenges of PON monitoring. Still, there are no standardised and mature monitoring methods. Ideal optical monitoring framework in PONs has the following general requirements (Yüksel et al. 2008):

- It should provide continuous, remote, automatic, and cost-effective supervision of the physical layer.
- It should provide rapid and accurate detection of performance degradation as well as service disruption.
- It should unambiguously provide failure source location.
- The testing should not affect normal data transmission (non-intrusive testing).
- It should distinguish between a failure in the end users' own equipment and a failure in the operator's network. Monitoring results should be conveyed to the NMS and evaluated here in detail enabling preventive countermeasures (like protection and restoration, isolation of attacking port...).
- It should be interoperable with many network variants (bit rate, protocol ...).

The most common maintenance tool employed for troubleshooting in long-haul, point-to-point fibre-optic links is an optical time domain reflectometer (OTDR). However, implementation of OTDRs into PONs brings some testing challenges which are: the lack of dynamic range to monitor the infrastructure after the splitter, a long measurement time due to averaging necessary to obtain an OTDR trace and repetition of the measurement on large number of ONTs, and the reflection dead-zone that makes it impossible to distinguish the monitoring reflection peaks from two nearly located ONTs. Beside these general considerations, one of the main technical issues on maintenance in PON system is known as point-to-multipoint problem. In the PONs, the OTDR pulses launched into the fibre are passively split and propagate simultaneously in every branch after the splitting point. As a result, the backreflected and backscattered light signals from each branch add up together to form a "global" or composite trace that makes the interpretation of the OTDR trace from the CO a difficult task. Looking at the global trace without any additional information, one cannot identify the faulty branch. The fault quantisation is another problem; in order to determine the "real loss" or "real return loss" of a fault, some further calculations are needed based on the network's parameters, such as power levels measured during installation, splitting ratio and splitting insertion losses (Wuilmart et al. 1996).

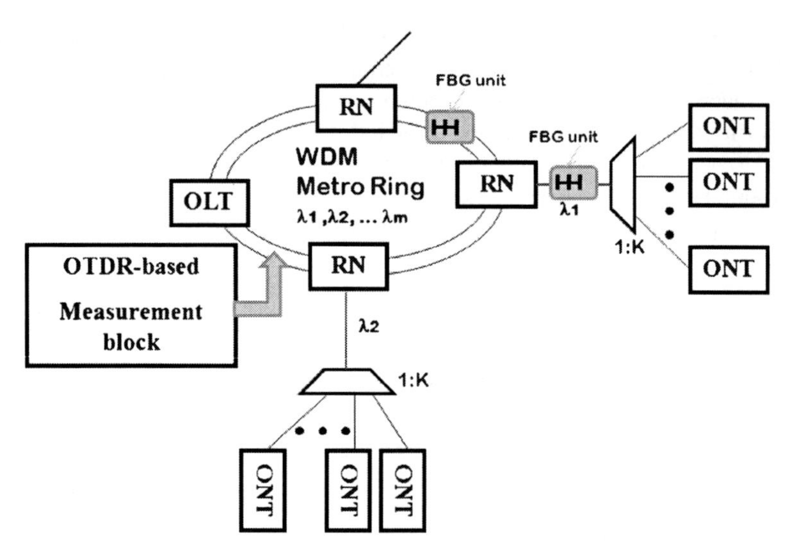

Fig. 2.24 Ring-tree PON scheme with the proposed monitoring method

When converged metro/access network topologies are considered, guarantying the reliability in these networks might be even more crucial for the operators as the architecture becomes more complex, and the first expected aim is to transport high capacity services to business customers.

In particular, the combination of WDM bidirectional rings and TDM access trees as proposed in the framework of ISTSARDANA requires an elaborated monitoring approach. It is obvious that diagnostic requirements for such a converged network are different from that of classical TDM-PONs. A conventional OTDR which operates at a single wavelength is not suitable to probe branches beyond the wavelength selective component at the RN which realises the add/drop functionality to assign a fixed wavelength to each TDM access tree. Hence, OTDR must have wavelength tuning capability.

A new method for monitoring such a complex hybrid topology was studied in the framework of SARDANA project (Militello et al. 2009). This method is based on a tunable OTDR and reference reflections created by FBGs.

Figure 2.24 shows the schematic of the combined ring/PON topology implementing the proposed monitoring system. A tunable OTDR is used at the OLT side and interrogates the FBGs at different central wavelengths located in the ring or the access parts of the network. For each OTDR measurement, wavelength is adjusted so that one or a group of FBGs creates reference reflection peaks on the OTDR trace which are used to check the integrity of the network.

As represented in Fig. 2.25, tunable OTDR can be implemented by using a commercially available OTDR and a wavelength conversion system (WCS). WCS includes two optical circulators (C1 and C2 in Fig. 2.25), a TLS and an optical/electrical (O/E) converter. The optical pulses emitted by the OTDR are

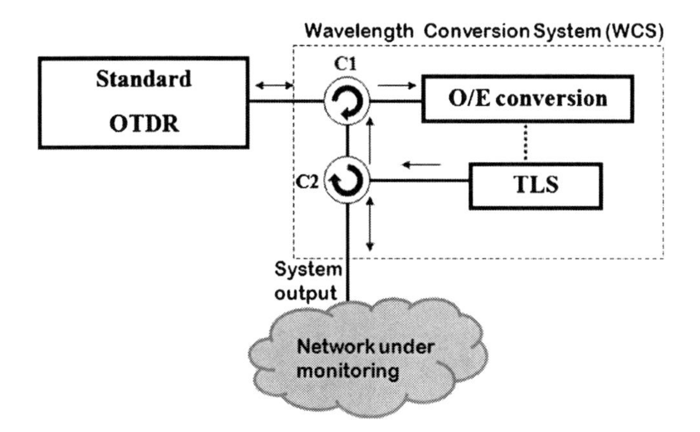

Fig. 2.25 Implementation of tunable OTDR

directed onto O/E converter via a first circulator (C1). The electrical pulses obtained at the output of O/E converter are amplified and modulate the optical power emitted by the TLS. As a consequence, optical pulses at a desired wavelength are produced at the OTDR repetition rate (with a certain pulse delay) and directed into the network. The standard OTDR receives the reflected and backscattered signals via the two circulators C2 and C1 and stores the associated trace.

Monitoring results for an example scenario are shown in Fig. 2.26. In this example case, two FBGs with central wavelengths of 1557.36 nm and 1560.61 nm are placed respectively into the ring and one PON section of the network. The OTDR trace when TLS generates pulses at 1,550 nm (Fig. 2.26a) shows reflection peaks initially present in the network (e.g. connectors). FBGs are not involved in this trace as 1,550 nm is out of the reflection bands of the FBGs. In Fig. 2.26b, the wavelength of the TLS is set to 1557.3 nm where the FBG in the ring (placed about 34 km from the OTDR) creates a high intensity peak. A non-assigned wavelength in the WDM-ring should be chosen to test this part of the network. Then, TLS wavelength is set to 1560.6 nm to detect the reflection peak due to the second FBG placed in the feeder line of a tree PON connected to the ring through the RN. The optical pulse at the TLS output is shown in Fig. 2.26d.

Apart from the solution based on OTDR, other monitoring solutions based on OFDR recently appeared in the literature as an alternative approach (Zou et al. 2007; Effenberger 2008). A new monitoring method for PONs using an OFDR at the OLT and interferometer units (IF-unit) at the ONTs/ONUs is under study. Each IF-unit includes a uniform FBG and creates a beat term (a peak) on the OFDR trace which is used to check the integrity of the corresponding branch. Analysing the beat terms of all branches allows an easy distinction of the faulty branch after the splitter. In addition to the easy determination of the faulty branch, the system directly measures the temperature variation in the network terminals, such as ONU/ONT, fibre distribution hub or network access terminals. Temperature measurement is realised by using the temperature sensitivity of the FBG's spectrum

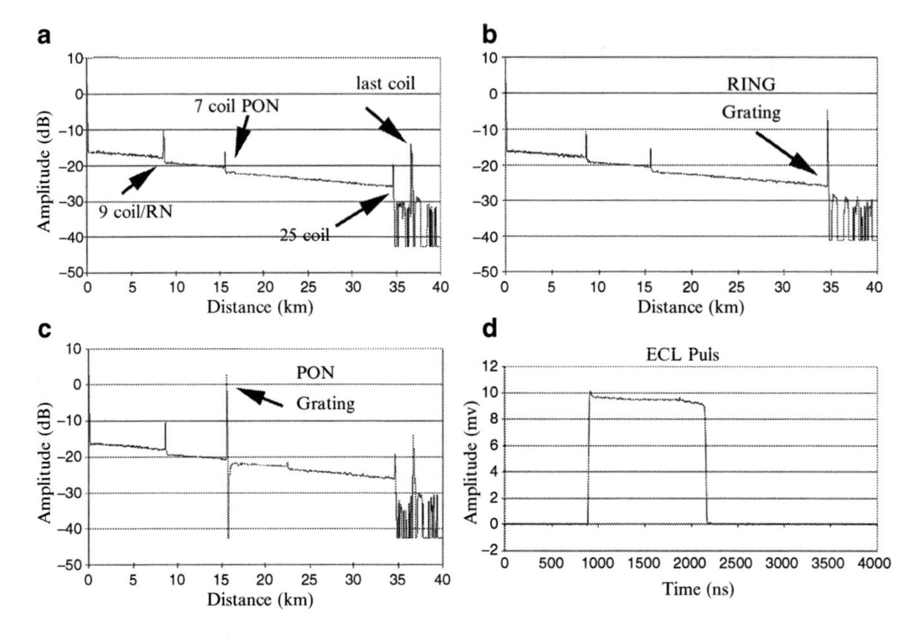

Fig. 2.26 (**a**) Monitoring trace at 1,550 nm, (**b**) monitoring trace at 1,557.3 nm, (**c**) monitoring trace at 1,560.6 nm, (**d**) optical pulse at the TLS output

inside the IF-units. In this method, simple signal processing steps are applied on the OFDR trace to deduce the Bragg wavelength shift of each FBG located in each IF-unit. This information in turn gives the temperature evolution of interferometer device's position (Yuksel et al. 2010).

2.3.2 Signal Processing and Compensation

2.3.2.1 Optical CD Compensation

CD is an intra-channel degradation effect that takes place as a short light pulse propagates along an optical fibre, and it can be one of the main restrictions for optical digital transmission systems. CD is due to the dispersion property of the fibre, that is, the fact that its index of refraction, and thus the light velocity, varies as a function of wavelength. Short pulses are not monochromatic but rather have a certain spectral bandwidth; the shorter the pulse, the wider its spectrum and thus the stronger the temporal broadening due to CD. This temporal broadening is also proportional to the length of the fibre span, so the CD becomes larger as the length of the fibre increases. In the near future, the performance of optical transport technologies such as optical Ethernet, SONET/SDH, CWDM, DWDM and OTN/ASON is expected to reach the range of 100 Gbit/s transmission over

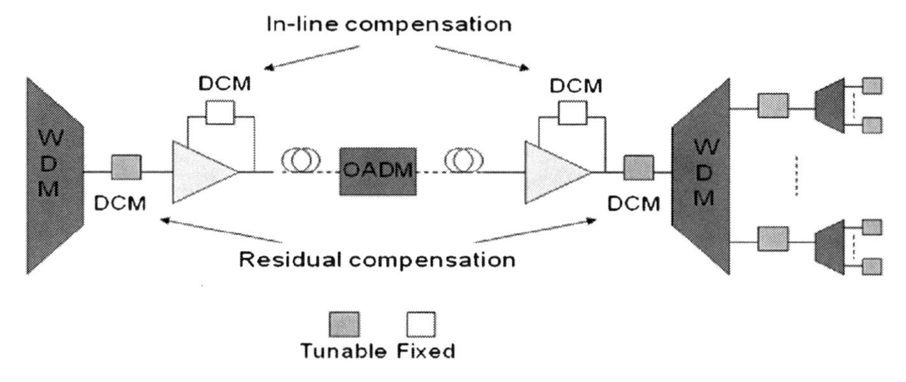

Fig. 2.27 General scenario of a WDM transmission link with different dispersion compensation modules (DCM). *OADM*, optical add-drop multiplexer

around 1,000 km without need for optical regeneration of the digital signals. In this scenario, it is absolutely mandatory to keep the effects of fibre dispersion under control to guarantee the required quality of transmission.

A general scenario of a DWDM amplified fibre link with chromatic dispersion compensation is shown in Fig. 2.27.

In order to exploit the existing SSMF links under the emerging transmission standards, the CD effect is one of the most important issues to take into account. Signal regeneration and CD compensation in the optical domain are two approaches to solve the problem. Monitoring techniques to estimate the accumulated CD in reconfigurable and scalable optical networks are also very desirable.

Cost-effective CD monitoring techniques have been developed in order to manage reconfigurable and scalable optical networks, in which the accumulated CD may change. Online CD monitoring with no need for tunable filters has been demonstrated by adding small sinusoidal components (pilot tones) to the WDM optical signal, either employing amplitude modulation (AM) (Petersen et al. 2002) or phase modulation (PM) (Park et al. 2003).

Figure 2.28 illustrates the RF-fading effect of an AM pilot tone, showing the baseband magnitude response of an SMF link with a length of 10 ~ km. The figure has been obtained using the software VPItransmissionMaker™ Cable Access v.7.0.1.

The magnitude of the received AM pilot tone at a fixed frequency near those resonance values also change with the accumulated CD of the fibre link, so the measurement of this magnitude can be employed as the basis of a CD monitoring technique (Petersen et al. 2002).

A different approach for avoiding the effects of signal degradation after transmission is optical regeneration, which consists of three steps (3R) of signal processing,: firstly, an optical amplification and amplitude equalisation of the signal (1R); secondly, a reshaping of the previously amplified signal (2R); and finally, a transmission of the reshaped signal following a retiming obtained with a clock

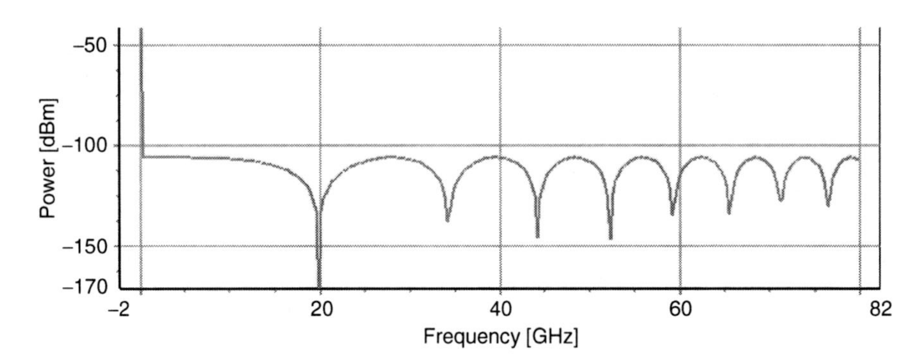

Fig. 2.28 Magnitude response of an AM fibre-optic link with a length of 10 km and ideal components except for fibre dispersion ($D = 16$ ps/nm·km) and attenuation ($A = 0.2$ dB/km) at 193.1 THz

recovery stage (3R). Simple amplification and reshaping of the signal (2R) are usually sufficient to avoid problems derived from the amplitude noise and extinction ratio degradation. Nevertheless, random time deviations of optical pulses can, in some cases, arise from the interaction between ASE and signal, caused by fibre dispersion. In this case, a synchronous pulse stream clock recovery is required to keep timing jitter below acceptable levels.

Other approach to manage signal degradation in high capacity long-haul transmission systems is the dispersion compensation approach. Equalised optical amplification and filtering to reduce noise can manage, in some cases, the power attenuation in optical fibre, but chromatic dispersion must also be taken into account. The effect of CD can be addressed either by DCF with negative dispersion coefficient, using advanced fibre Bragg gratings or employing optical filters.

In the following, we will focus on the optical filter approaches for dispersion compensation.

All-pass optical filters are excellent candidates to perform dispersion management in long-haul WDM networks because they can be designed following a desired periodical phase response without inducing any amplitude distortion and with low insertion losses. With the current integrated-optics fabrication technologies, compact and lightweight devices can be obtained for all-optical CD managing.

Since more than two decades, fibre-optic interferometers such as the FP and the RR have been taken into account as possible laser chirp and chromatic dispersion-compensating filters in high-bit-rate digital transmission systems (Gnauck et al. 1990) (Dilwali 1992) PLC in silica waveguides were demonstrated shortly afterwards as optical dispersion equalisers (Takiguchi et al. 1994) by synthesising a FIR lattice filter with five MZI.

Novel techniques based on RR lattice architectures were proposed as general optical filter design and synthesis algorithms (Orta et al. 1995; Madsen 1996), and chips designed with these methods were fabricated using Ge-doped silica planar waveguides on silicon substrate (Madsen 1996). Following the general design and

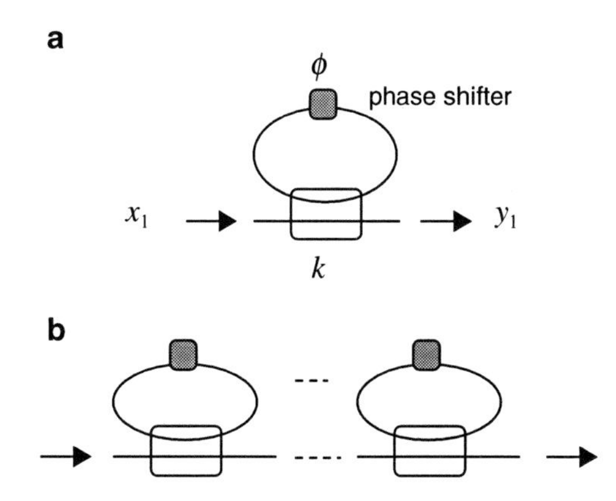

Fig. 2.29 (**a**) Single-stage optical all-pass filter based on the RR and (**b**) multistage all-pass filter architecture using RRs in cascade (Madsen 1998; Vargas et al. 2010)

synthesis framework of all-pass filters proposed for dispersion control applications in (Madsen 1998; Lenz 1999), a two RR in series filter (Madsen et al. 1999) and a four stages Fabry–Perot tunable micro-electro-mechanical (MEM) actuated all-pass filter (Madsen et al. 2000) were fabricated (Fig. 2.29).

The ring resonator with lossless coupler and lossless feedback path is a single-stage all-pass filter (Madsen 1998). In Fig. 2.30, the group delay and the quadratic dispersion of a 6th order RR-based all-pass filter with spectral periodicity of 25 GHz is simulated versus optical frequency using VPIphotonics^{TM} software, using ideal couplers and phase-shifters inside the RR with 0.1 dB of loop attenuation. It can be seen how the left sideband shows a negative quadratic dispersion around −2,400 ps/nm, which is enough, in principle, for compensating the chromatic dispersion of a fibre span of standard SMF with a length of 150 km.

The analysis of a Sagnac (SG) loop in RR has been reported as a tunable optical filter (Vargas et al. 2001; Vargas 2007) with ultra-narrow bandwidth for use in Dense WDM systems; see Fig. 2.31.

This configuration is a second-order all-pass filter in a single stage and can be cascaded, as in the case of the single ring resonator in order to form a multistage photonic architecture for digital IIR filter synthesis. The configuration offers the advantages of avoiding the use of phase-shifters and a simple mechanism through the coupling factor κ_2 for fixing the frequency of the zeros and poles, which appear as two complex conjugated pairs. On the other hand, it also offers a considerable immunity to variations in the ring length due to the fact that the clockwise and counterclockwise recirculations propagate along the same optical length. The magnitude distortion induced by a single stage is only caused by the excess loss of the optical coupler, as in the case of the RR all-pass filter, so it is not a critical restriction.

A modified RR-SG configuration with different transmission functions in the clockwise and counterclockwise recirculations has been proposed as novel all-pass filter photonic structure and studied following the Z-transform formalism in (Montalvo et al. 2008). Under certain conditions, it is demonstrated that a modified

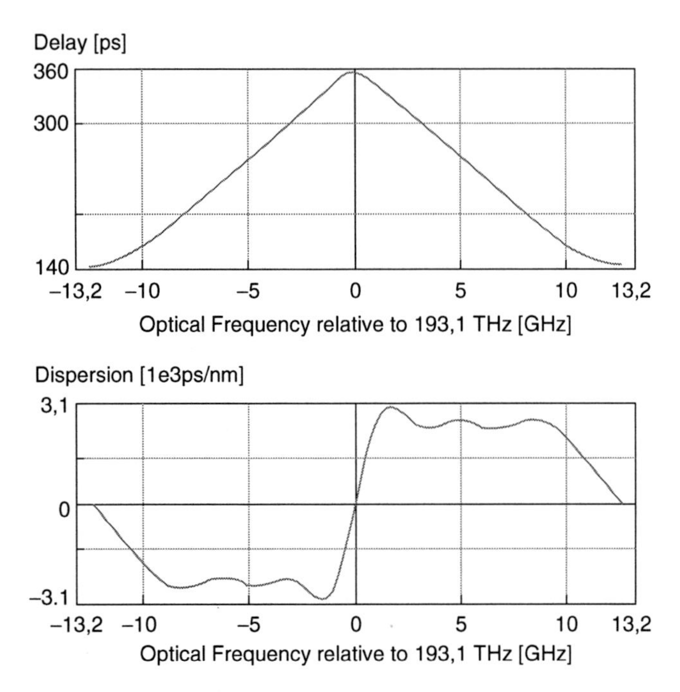

Fig. 2.30 Group delay and quadratic dispersion of a 6th-order RR in series design for chromatic dispersion compensation

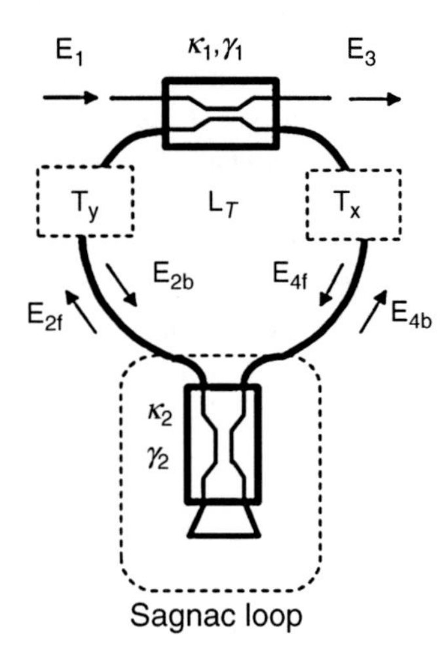

Fig. 2.31 General filter architecture consisting of a ring resonator with internal transmission transfer functions (*Tx*, *Ty*) and a Sagnac loop as transmission-reflection function (TRF). LT is the total length of the feedback path

RR-SG configuration using integrated optical circulators and amplifiers can perform as an all-pass filter response with no magnitude distortion and arbitrarily high quadratic dispersion peaks.

Simulations of optical components and networks using PDA software are the next step after the design and photonic synthesis procedures of an optical filter, being very important before the fabrication of an integrated chip, which can be extremely expensive.

Simulations are very important to study the effect of employing an optical filter in an optical transmission link for in-line chromatic dispersion compensation because the impact of the filter in the link cannot be completely derived by the shapes of its quadratic dispersion and its magnitude response but rather depends on several other aspects, such as the bit rate and the optical signal to noise ratio at the receiver's end.

On the other hand, the value of the quadratic dispersion can be related, in principle, with a specific length of a fibre link. Quick and flexible simulations are a cost-effective and useful tool to consider all these aspects, thus providing information related to the performance of an optical filter which is not evident from the observation of its transfer function or its quadratic dispersion.

2.3.2.2 FBG Telecommunication Applications in Compensation and Signal Processing

During the past couple of years, FBGs-based dispersion compensators have become a real alternative to the incumbent DCFs. This results from the fact that the FBG solution brings several advantages. Its low insertion loss provides significant cost saving and OSNR improvement through the decrease in amplification requirements. For instance, when used in-line, FBG-based dispersion compensators associated to single-stage EDFAs can replace DCF-based compensators in dual-stage EDFAs, providing a more cost-effective solution. Moreover, FBGs are compact devices that are easy to manufacture and that provide negligible distortion.

In parallel to this important development, FBGs have also been used for optical pulse shaping and manipulation purposes dedicated to ultrafast optics. They have indeed the potential to replace the bulk or micro-optics configurations that are bulky, costly and that cannot be easily integrated in fibre optics.

This section focuses on the major achievements obtained by the use of linearly CFBGs in the fields of chromatic dispersion compensation, optical pulse shaping and tunable delay line.

An FBG consists in a permanent and periodic (or quasi-periodic) modulation of the core refractive index over a certain fibre length. This modulation is often created by exposure to an ultraviolet interference light pattern incident transversely along the fibre. The periodic nature of the index modulation yields a resonant spectral response. Indeed, the FBG reflects light preferentially at the Bragg wavelength defined by $\lambda B = 2{\cdot}n_{\mathrm{eff}}{\cdot}\Lambda$ where n_{eff} is the average refractive index of the fibre core and Λ is the grating period (Othonos 1999).

Contrary to uniform FBGs, the period of chirped FBGs varies along the fibre axis at a rate (also called chirp coefficient) that can reach several nm/cm. A location z along a CFBG reflects light at the wavelength $\lambda\ (z) = 2{\cdot}n_{\mathrm{eff}}{\cdot}\Lambda\ (z)$ where $\Lambda\ (z)$ stands for the local index modulation period at the position z (Othonos 1999).

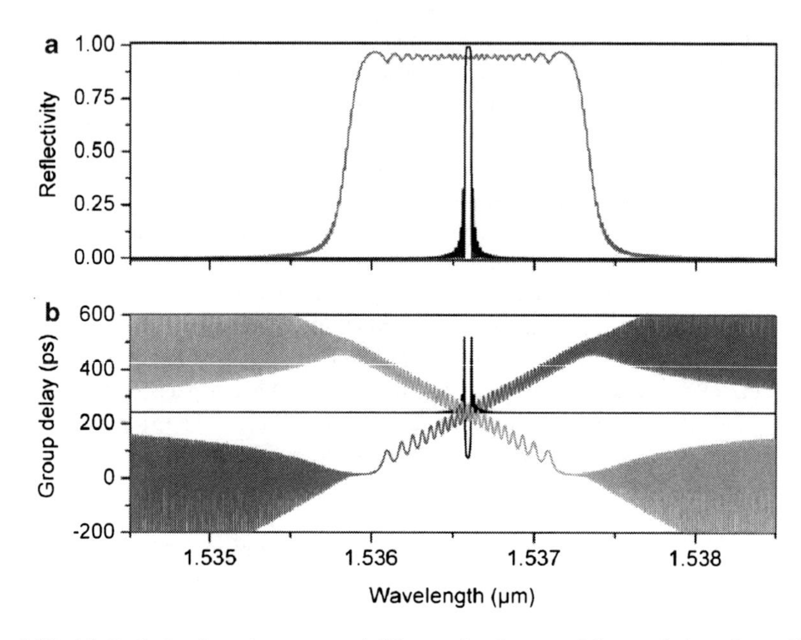

Fig. 2.32 (**a**) Typical reflected spectra and (**b**) associated group delay evolutions for uniform (*black curve*) and chirped (*grey curves* – light launched through the short-wavelengths port (*dark grey*), light launched through the long-wavelengths port (*light grey*)) FBGs (parameters used for the simulation: grating length $L = 5$ cm, nominal grating period $\Lambda_0 = 530.0$ nm, chirp coefficient $C = 0.1$ nm/cm and refractive index modulation $\delta n = 2.5\ 10^{-4}$)

Figure 2.32a presents the typical reflected amplitude spectra of both a uniform and linearly chirped FBGs obtained by means of the coupled mode equations. A total chirp of 0.5 nm is defined along the grating length, yielding a CFBG reflection bandwidth (grey line) of about 1.50 nm, nearly one order of magnitude higher than the one of the uniform FBG (0.20 nm, dark line). Figure 2.32b depicts the group delay evolutions in reflection of both gratings.

The group delay of uniform FBGs strongly evolves in narrow wavelength regions (a few picometers) matching the edges of the reflection band. The evolution with wavelength of the group delay of chirped FBGs is rather different: it monotonously evolves on the whole reflection band, starting from zero (a maximum value) at the beginning and reaching a maximum value (zero) at the end of the grating when the light is injected through the short (long) wavelengths port. The maximum value of the delay (in ps) is equal to twice the grating length divided by the light velocity in silica. The slope (dispersion in ps/nm) of the group delay evolution is positive (negative) when the light is injected via the short (long) wavelengths side.

The CFBG dispersion can readily reach several hundreds of ps/nm. Deviations of the group delay spectrum from a straight line are known as group delay ripple. Depending on the grating reflectivity, this ripple can reach several percents of the mean group delay value. As it can impair the system performances, the ripple should be practically as small as possible. It is the reason why apodisation profiles of the

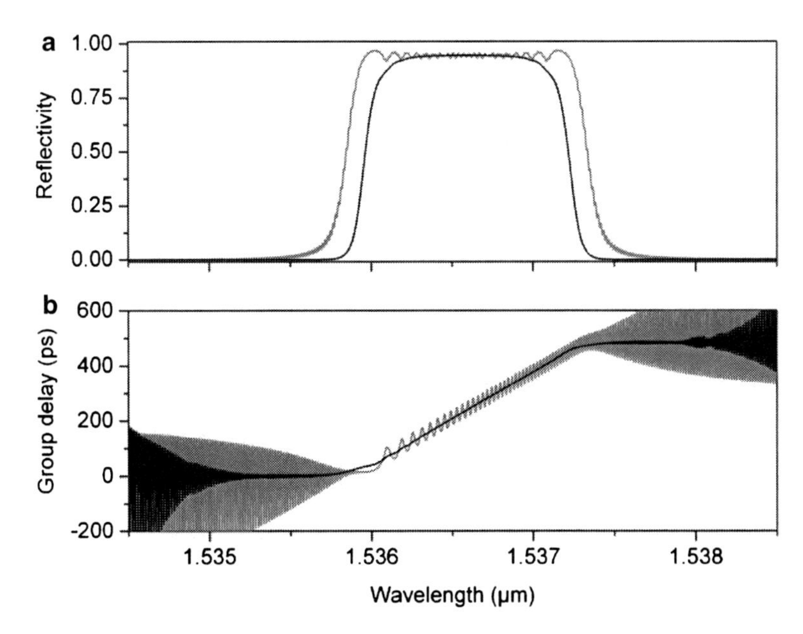

Fig. 2.33 (**a**) Reflected spectra and (**b**) associated group delay evolutions for a CFBG without apodisation profile (*grey curve*) and with a hyperbolic tangent apodisation profile (*black curve*). The parameters used for the simulations are similar to those used in Fig. 2.32

refractive index modulation are privileged for CFBGs used in telecom applications. They indeed reduce the internal interference effects, which in turn decreases the phase ripple. To illustrate this point, Fig. 2.33 shows a comparison between a CFBG without apodisation and a CFBG characterised by a hyperbolic tangent refractive index modulation profile. For the apodised grating, the reflected bandwidth is slightly reduced (a few tens of picometers), while the ripple is strongly decreased. Hence, a careful design of the CFBGs physical parameters is fundamental for a correct operation in practical applications.

CFBGs have been fabricated by using several different methods combined with the use of ultraviolet lasers emitting around 240 nm (continuous-wave frequency-doubled argon-ion laser or pulsed excimer laser). In practice, the chirp is often induced by varying the physical grating period along z. In the commonly used dual-beam holographic technique, the fringe spacing of the interference pattern is made non-uniform by using dissimilar curvatures for the interfering wavefronts, resulting in variations of the period (Othonos 1999). Many other inscription techniques have been reported. For instance, CFBGs have been fabricated by tilting or stretching the fibre, by using strain or temperature gradients and by stitching together multiple uniform sections. The most straightforward inscription technique remains the use of specific phase masks that contain the desired chirp profile.

Fibre chromatic dispersion, i.e. the dependence of the refractive index value as a function of the wavelength, leads to temporal distortion of optical pulses as they

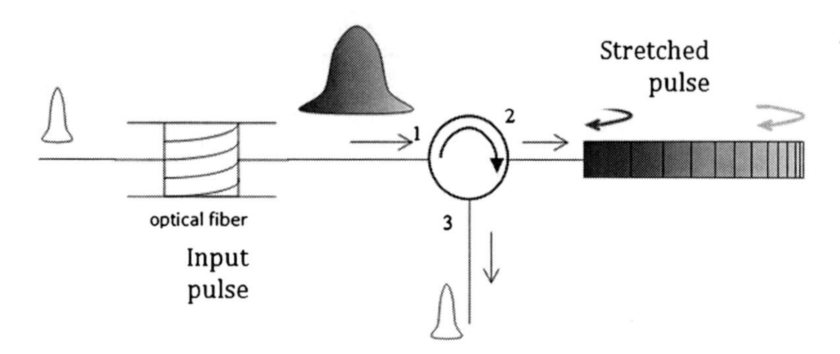

Fig. 2.34 Operating principle of a CFBG pulse stretcher

traverse an optical fibre. As a consequence, this distortion of the signal leads to inter-symbol interferences and consequently, must be compensated. One traditional mean used to overcome the issue of dispersion is to incorporate bundles of DCFs throughout the network. This is a quite straightforward technique that is based on optical fibres having a dispersion coefficient of opposite sign in comparison to standard single-mode fibres used in the network. DCFs have a typical dispersion coefficient four to eight times that of standard single-mode fibres. This level of dispersion is achieved at the expense of a fibre core diameter reduction, which in turn increases the optical loss and limits the levels of optical power that can be sustained without inducing non-linear effects.

Fibre chromatic dispersion compensation using highly reflective CFBGs is based on the introduction of wavelength-specific time delays through the use of precisely designed CFBGs (Ouelette 1987). By combining such a CFBG with a three-port optical circulator, a compact and effective dispersion compensation module can be readily realised. Figure 2.34 illustrates pulse compression by a CFBG with a period that decreases away from the entrance point. Longer wavelengths (dark grey) are reflected early along the grating, while shorter wavelengths (light grey) are reflected later near the back. The optical pulses launched into the CFBG are dispersed temporally after propagation in a bundle of optical fibre. The CFBG placed in the set-up is designed to recompress them to their original shape.

In practice, more elaborate period profiles can be realised to generate complex spectral dispersions. Moreover, the modulation amplitude of the CFBG can be tailored longitudinally in order to shape spectrally the grating reflectance.

The potential of CFBGs for dispersion compensation was demonstrated during the 1990s in several transmission experiments. In 1995, chromatic dispersion compensation over 160 km of standard single-mode fibre at 10 and 20 Gbit/s was realised (Kashyap et al. 1995). One year later, a 12-cm-long CFBG was used to compensate the dispersion accumulated over 270 km of fibre at 10 Gbit/s (Laming et al. 1996). Since then, the transmission distances have been increased up to a few hundreds of kilometres using a few centimetres long apodised CFBGs, which is really impressive with so compact devices.

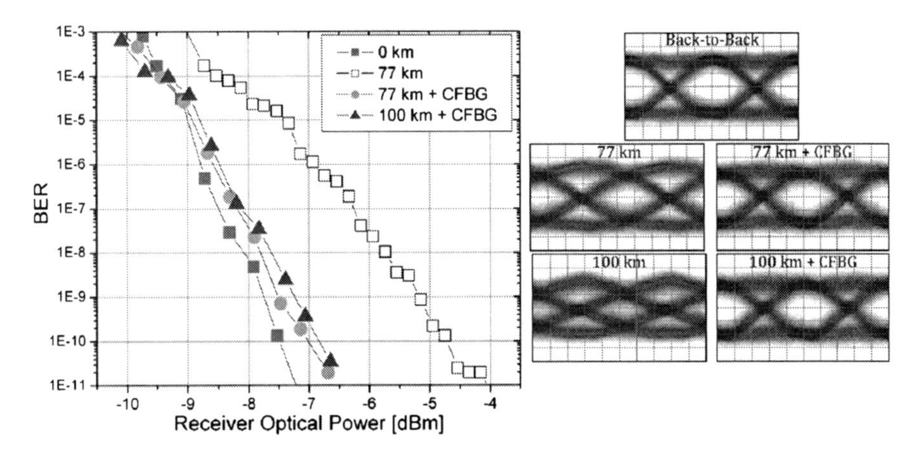

Fig. 2.35 BER measurement as a function of the received optical power and corresponding eye diagrams (scale: time 20 ps/div and amplitude 25 mV/div – power level at the receiver: –5 dBm) illustrating the dispersion compensation by a CFBG

To illustrate this behaviour, measurements were done at 10 Gbit/s using an ~8-cm-long linearly chirped FBG characterised by a chromatic dispersion of about 1,450 ps/nm. An optical attenuator was placed in the set-up to record the performances as a function of the received optical power. Figure 2.35 presents the evolution of the BER for different tested configurations (77 and 100 km of optical fibre). One can see the effect of the CFBG dispersion compensation that decreases the BER by several decades for a given received power. This effect can also be evaluated on the corresponding eye diagram measurements. Due to the chromatic dispersion induced by the bundle of optical fibre, the eye aperture strongly decreases. With the CFBG compensator, the eye retrieves its original shape, as in the back-to-back measurement.

Two main types of FBG-based dispersion compensators are commercially available nowadays: multichannel (or channelised) and broadband. The channelised version provides specific compensation for the different wavelength channels used in an optical network. The broadband type provides, in much of the same manner as a DCF, continuous compensation through the C or L band. In addition, the dynamic compensation of the chromatic dispersion has also been demonstrated through thermal or mechanical actuation of CFBGs.

As CFBGs are dispersive elements in which the group delay is a function of wavelength, they can also be used to manipulate the amplitude and phase of light and consequently, to shape pulses as the amplitude interacts with the reflectivity and the phase with the dispersion. In the Born approximation regime, this interaction maybe separated by manipulating the amplitude separately from the phase. Pulse shaping by CFBGs was first investigated by Rotwitt et al. in 1994 (Rotwitt et al. 1991). Since then, a number of different schemes have been reported, including the use of cascaded gratings with different dispersion characteristics. In such a scheme, a first CFBG is used as a pulse shaper, while the second acts as a dispersion compensator,

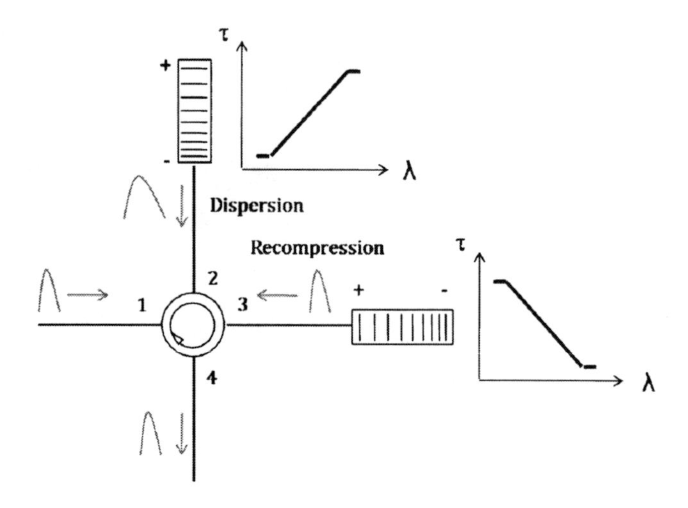

Fig. 2.36 Schematic of a twin CFBG delay line

in a similar way as the scheme presented in Fig. 2.34. With this principle, it is possible to generate triangular or square pulses. CFBGs can also be used to multiply the repetition rate of a stream of pulses, via the application of the temporal Talbot effect.

CFBGs have also been widely used for the realisation of tunable optical delay lines. When a CFBG is stretched, the point at which the incoming pulses are reflected changes while the pulses are simultaneously dispersed (Choi et al. 2005). This dispersion can be practically cancelled by using a second identical grating placed in the opposite direction. In other words, the pulse is recompressed providing that two gratings are used in succession with the sign of the chirp reversed for the second grating, as depicted in Fig. 2.36. By stretching one of the gratings, the physical delay between them is altered, and a time delay is introduced. In practice, the time delay is generally tuned by shifting the CFBG resonance band through thermal or mechanical actuation. Because wavelength shifts are induced, optical pulses with a linewidth comparable to that of the CFBG reflection bandwidth cannot be variously delayed without undergoing important distortions. Moreover, there is an increasing mismatch between the overlap of the reflection spectra of the two gratings with increasing strain. This can be detrimental, especially in the framework of high-speed transmissions with bit rates higher than a few tens of Gbit/s.

To alleviate these drawbacks, a novel set-up has been proposed to generate a tunable delay with a single CFBG (Caucheteur et al. 2010). Instead of using directly the CFBG group delay curve, this solution exploits the DGD evolution. In practice, orthogonally polarised pulses are sent through both ports of the CFBG while its local birefringence is controlled. This leads to a dynamic evolution of the DGD inside the CFBG reflected bandwidth so that tunable delays can be achieved. Two significant advantages are therefore obtained. This set-up does not require any wavelength shift.

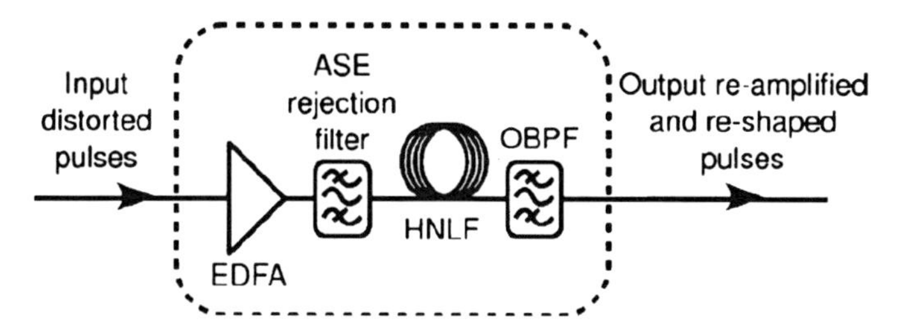

Fig. 2.37 Schematic of the single-cell SBF regenerator. *EDFA* erbium-doped fibre amplifier; *ASE* amplified spontaneous emission; *HNLF* highly non-linear fibre; *OBPF* optical band-pass filter

Consequently, the full CFBG bandwidth can be used to delay optical pulses nearly as broad as the grating reflection band. The double pass of light inside the same CFBG (in both directions) allows to compensate the chromatic dispersion, without the need for a second CFBG perfectly matched with the first one.

In conclusion, CFBGs present spectral characteristics (amplitude and phase) well suited for nowadays optical telecommunication applications, not only for chromatic dispersion compensation but also for pulse shaping and the realisation of tunable optical delay lines.

2.3.2.3 All-Optical Regeneration/Buffer

It is an important requirement for an optical network, comprised of multiple point-to-point links, that the signals propagating throughout be of sufficient quality to detect. Historically, this has been achieved by the use of electrical repeaters. These convert the incoming optical signal into an electrical signal from which the base data is recovered before being used to transmit a new optical signal. This OEO conversion is undesirable when striving for high-bit-rate systems in which the conversion becomes a limiting factor.

Optical amplifiers have removed the OEO conversion but added ASE noise to the optical signal. An optical regenerator adds additional processing functions, such as amplitude equalisation (reshaping) and temporal repositioning (retiming) of the optical pulses.

The purpose of an optical regenerator is to process input distorted pulses, minimising signal degradation through the impact of noise, and contribute an output in which a receiver is better able to extract the original data.

Although there are a selection of optical regeneration schemes, one particularly interesting version, proposed by Mamyshev et al. (1998) is based on the principle of self-phase-modulation-induced spectral broadening followed by offset filtering, which will henceforth be referred to as a SBF regenerator.

Figure 2.37 shows a single SBF cell which consists of a high-power optical amplifier (EDFA), a HNLF and an OBPF. It is useful to include an additional filter after the EDFA to suppress out-of-band ASE noise.

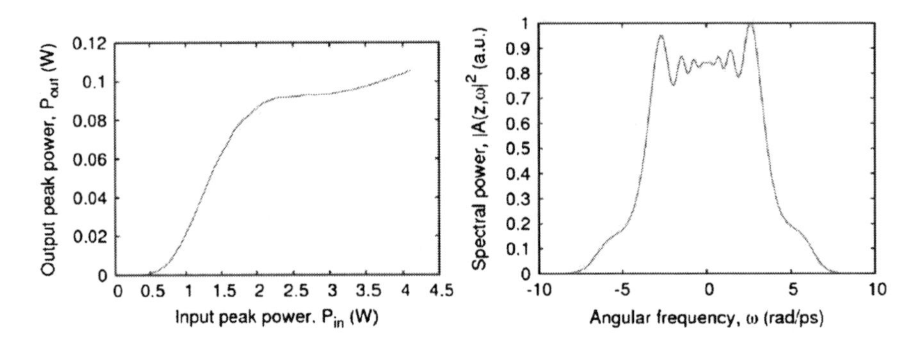

Fig. 2.38 Power transfer function for single pulse passing through one SBF cell (*left*). Spectral plot at the output of HNLF for an input peak power of 2.6 W (*right*); spectral power is normalised (*arbitrary units*)

The EDFA amplifies the peak power of input pulses to a level where considerable spectral broadening can take place within the HNLF due to self-phase modulation. After propagating through the fibre, the spectrum is sliced by the OBPF at a frequency offset to that of the input centre frequency of the pulses.

It is possible to add a second SBF cell after the OBPF of the first cell, allowing the recovery of the original centre frequency of the pulses. This is achieved by using an equal and opposite offset frequency for the OBPF in the second cell.

A plot of PTF for a regenerator shows the peak power exiting the SBF cell (after the offset filter) as a function of the peak power entering the SBF cell (before the amplifier). The input data to the regenerator are represented by RZ pulses, for which zero bits correspond to pulses with (ideally) no power and one bit correspond to pulses with power greater than a fixed threshold level. An ideal PTF for a regenerator is a step function, in which the output peak power is zero for input pulses with peak power below the threshold level and a constant output peak power for input pulses above the threshold.

Generation of the PTF is the first step for regenerator parameter optimisation. The operating region is located at the plateau after the first maxima of the function, and it is here that amplitude jitter (the variation in peak power from that of the average for the pulses) may be reduced. Amplifier gain is adjusted so that input pulses have the required peak power to enter this region. The PTF also provides information on the suppression of ghost pulses, which are low-power pulses located within a "zero" bit.

Figure 2.38 shows a typical PTF in which input pulses with peak power less than 0.5 W are output with almost zero output power, while input pulses with peak power in the range 2.2 to 3.3 W are output at the almost constant peak power of 0.09 W. The SBF regenerator parameters are given in Table 2.2 for a carrier wavelength of 1.550 nm.

Figure 2.38 also shows a spectral plot at the exit of the HNLF for a peak input power of 2.6 W.

Table 2.2 SBF regenerator parameters

Attenuation coefficient, α	2.13 dB/km
Dispersion, D	−1.7 ps/(nm·km)
Dispersion slope, S	0.023 ps/(nm2·km)
Fibre length, L	1 km
Non-linear parameter, γ	18 (W·km)–1
Filter offset	375 GHz
Filter bandwidth	71.25 GHz

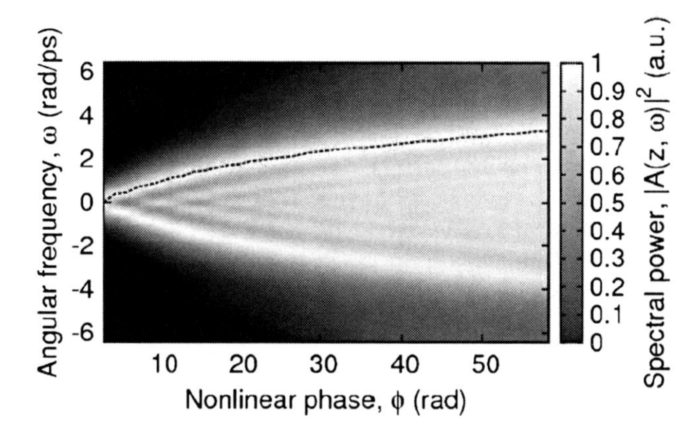

Fig. 2.39 Spectral power map showing broadening of pulse with increasing non-linear phase. *Dotted line* marks the position of the boundary spectral maximum

It is useful to view the spectral evolution of the input pulses as they propagate through the HNLF. As the pulse propagates through the HNLF, it acquires a non-linear phase shift and ultimately starts to broaden. The broadened spectra will consist of multiple peaks, though it is the rightmost or leftmost peak which is of interest. Locating these spectral peaks is useful since the results of (Striegler 2006) indicate that these provide the optimised frequency for the offset filter, while the filter bandwidth should be adjusted to the spectral width of the peak.

Spectral broadening is symmetrical when there is no dispersion or second-order dispersion, while third- or higher-order dispersion causes the spectrum to become asymmetrical. When such asymmetry occurs, the higher-frequency spectral peak should be tracked in preference to the lower-frequency peak. Figure 2.39 shows an example of spectral broadening when 2nd- and 3rd-order dispersion is included with high non-linearity. The spectral peak is traced by the dotted line and indicates the ideal filter frequency in terms of non-linear phase.

Design scaling rules have been proposed (Provost et al. 2007a; Provost et al. 2007b) in which the regenerator is classified according to the shape of its PTF, then optimised according to soliton number. An optimal fibre length and filter offset frequency to bandwidth ratio may then be calculated.

An alternative design rule considers the ratio of dispersion to non-linearity (Baveja et al. 2009). This is useful in categorising a range of SBF regenerators since those that share the same ratio are likely to display the same regeneration capabilities.

When considering noise introduced by amplifiers in an optical link, it has been shown that the inclusion of the ASE-rejection filter within the SBF regenerator is vital (Nguyen et al. 2006), with no improvement in Q-factor unless a filter is included after the amplifier. The filter serves to reduce the noise power over the pass-band of the offset filter, leading to an increase in the optical signal to noise ratio.

In the general case, where higher-order dispersion and non-linear terms are included, there is no analytical solution to the NLSE. Instead, the solution is approximated using numerical techniques.

A common strategy is to use the symmetrised split-step Fourier method (Agrawal 2007). This proceeds by splitting the NLSE into linear and non-linear terms, then applying them over a short segment of the total fibre length in each step. Initially, the field is propagated half a step using linear terms only, followed by including the non-linear terms at the midpoint of the segment, then completing the step using the linear terms over the remaining half step.

Numerical integration is required for an accurate approximation of the fibre segment non-linearity to be applied at the segment midpoint. A popular choice is the (globally) 4th-order Runge–Kutta method.

The NLSE is generally stiff, which can lead to the numerical calculation diverging rapidly to infinity, even when a high precision is used during the calculation. This stiffness may be reduced by transforming to the interaction picture (Hult 2007).

Additional improvements may be included through adaptive step-size control. A particular implementation of this is step-size doubling. A fine estimate using two sub-steps of half the step size is compared to a course estimate which covers the fibre segment in one step. The absolute difference in the two estimates is compared to a user-defined tolerance. If the error estimate is less than or equal to the tolerance, then the simulation continues with the next step (and may increase the step size). Conversely, if the value is greater than the tolerance, then the current step is repeated with a reduced step size.

Alternatives to the explicit step-size doubling method include Fehlberg, Cash–Karp and Dormand–Prince methods. These are implicit methods in which each step is taken once only; the error estimation is calculated internally by the algorithm (normally by using a 4th- and 5th-order comparison).

These high-precision methods are essential when simulating pulse propagation through highly non-linear fibre. Chalcogenide glass fibres have been proposed as the HNLF within the SBF regenerator (Fu et al. 2005), which may have a non-linear parameter of greater than 1,000 $(\mathrm{W{\cdot}km})^{-1}$, which is greater than that of standard silica fibres by more than a factor of 500.

The SBF regenerator has received a relatively broad investigation into its range of operation and optimisation. Most of this research has been applied to a single cell version, usually in the absence of noise and attenuation. Additionally, the regenerator is of 2R classification, that of reamplifying and reshaping of input pulses. A retiming stage is therefore an interesting area of additional investigation.

Since the amplifier within each SBF cell adds additional noise to the incoming bit stream, it would be useful to investigate whether such amplifiers can be removed or be reduced in number. One option is to use a fibre with such high non-linear parameter that the peak power of input pulses is adequate for regeneration without need of additional gain.

Other arrangements allow the reuse of components, typically bidirectional propagation in fibre (Provost et al. 2008).

2.4 Impairment Control and QoT-Constrained Routing

The trend towards service dependent QoS, the demand for guaranteed capacity to integrate telecom services in data communications, the generalisation towards more flexible meshed network topologies and the availability of highly efficient dynamic optical switching architectures bypassing digital signal regeneration leads to novel constraints on routing. The implementation of lightpath routing in a meshed optically switched WDM network is not straightforward as each new request accepted can affect the quality of other previously established circuits. To achieve a requested end-to-end QoS for a specific service, the underlying end-to-end lightpath, sequence of lightpaths in the multi-hop transmission network case, needs to offer a certain minimum quality of transmission (QoT).

The interoperability among network layers based on the introduction of autonomous control planes per layer that share information vertically must cope with the increasing complexity inherent to the deployment of reliable multilayer transport networks. The need to achieve differentiated QoS and to preserve or even enhance network reconfiguration and protection capability and autonomy is thus spreading from network layers towards the physical layer potentially comprising of a multitude of technologies (Saleh 2006).

Following a short introduction outlining the problem, we discuss schemes and approaches for QoT control related to the physical (analogue) layer zero, considering abstract optical signals and the electrical counterparts (prior E/O and post O/E conversion), before we return to the related impairment constraint routing problem.

2.4.1 *Impairment Control*

In contrast to QoS, QoT refers to the physical properties of lightpaths. QoT is a complex metric embracing the physical parameters of the different network components along a lightpath as well as the optical signal transmitted (modulation format, pulse rate, optical filtering). In addition, some parameters depend on the multiplex of signals sharing a resource and, therefore, are load-dependent. This

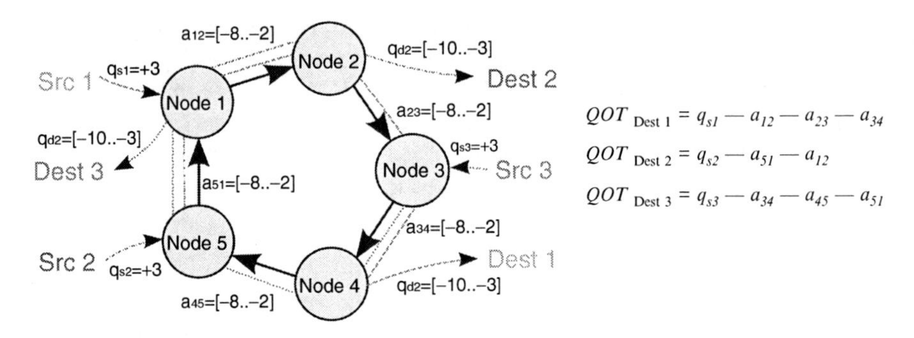

Fig. 2.40 Example illustrating interdependent adjustments

increases the difficulty in maintaining a uniform and acceptable quality for any lightpath across a transparent optical transmission network comprising a multitude of different network components and signals, being generally not feasible (Chen et al. 2004).

Proposed physical impairment compensation and mitigation possibilities yield increased optical reach; however, in contrast to regeneration of bits and bytes, it does not generally provide perfect signal regeneration. It is therefore mandatory to maintain and provide the remaining level of impairment per network element via the responsible control plane. Note that here, the term network element is used for any logically identifiable network resource from any layer, while the term network component shall be used if referring to the subset of physically identifiable network resources.

Physical impairment mitigation schemes based on processing of electrical signals after O/E conversion (and/or before) are applicable at the ends of lightpaths only. Thus, to achieve scalable transparent transmission networks, the multi-hop transmission scenario is essential to integrate electrical impairment compensation and mitigation at intermediate nodes. Optimal (meaning cost-efficient) placement of impairment mitigation features is a hard challenge for network design. However perfectly done, it still relies on impairment-aware routing and monitoring of the remaining physical impairments to grant a QoT that supports the QoS required for offering certain transport services.

Any adjustment that influences QoT must comply with all demands raised by other lightpaths sharing a resource. For instance, optimally adjusting the gain of all-optical amplifiers being part of links carrying many lightpaths may already pose a fundamental mathematical problem. This is illustrated in Fig. 2.40 where an abstract artificial dimensionless additive QoT is used to show the mathematical problem. Demand on received QoT is -10 to -3, degradation per link is -5, input QoT is $+3$ and the compensation can be adjusted between -3 and $+3$, i.e. degradation cannot be completely compensated but deliberately worsened, as it is typical for physical impairments.

To solve the simple dimensioning problem depicted in Fig. 2.40, we have three equations for five variables, a multidimensional continuum of feasible solutions and thus it might be expected that an optimum solution providing equal connection quality among all three connections could be found. However, there is a cyclic dependence and solving the equation system for feasible real numbers yields no such solution. Further, as cycles like the one discussed are likely to exist in every meshed topology, we derive that, in general, solutions that grant equal quality per connection do not exist if impairments cannot be entirely removed hop-by-hop.

We must consequently drop the equal performance requirement and instead aim at obtaining stable performance with least effort, meaning minimal added equipment cost. Therefore, parameters which can be efficiently stabilised per link shall be managed by per link control loops. All other parameters should be controlled per lightpath, utilising impairment control mechanisms integrated in sender and receiver circuitry. This division between link and end-to-end functions is actually not new; it equals the link layer management and the transport layer management defined by the OSI model (ITUX200). This clearly does not grant global optimisation. Nonetheless, stable and reliable physical impairments are more important than squeezing out the least quantum of optical span, especially in the multi-hop transmission network case.

2.4.1.1 Physical Impairments, QoT and Impact on QoS

Customers demand QoS derived from SLAs, which typically consist of statistical metrics on BER, blocking/loss rate and latency among other non-QoS-related issues. Loss-rate and latency are partly related to BER, i.e. loss and repetitions of payload in case of bit errors. Response to buffer overflow and control-plane-related issues are in general not related to physical impairments. Therefore, we need to consider BER as the only QoS requirement dependent on the control of physical impairments. BER is directly related to QoT, although, the relation depends on modulation format and receiver technology. The contribution of the considered physical impairments to QoT degradation is expressed in Eq. 2.6 according to (Cugini et al. 2008).

$$QoT_{[\mathrm{dB}]} = Q_{\mathrm{OSNR}} - Q_{\mathrm{GVD}} - Q_{\mathrm{DGD}} - Q_{\mathrm{FWM}} - \ldots \tag{2.6}$$

where $Q_{\mathrm{OSNR}} = P_{\mathrm{signal}}/P_{\mathrm{noise}}$ reflects the eye-opening at the receiver after consideration of ASE noise. The other Q-factors are eye-closures caused by other linear and non-linear, potentially signal or technology dependent, effects (i.e. GVD, DGD, FWM).

To establish a common parameter for control loops, it is initially assumed that the average QoT per network element is met by design. Consequently, every network element contributes a certain amount of impairments. Considering that BER is related to QoT, a virtual BER contributed per network element can be defined and, comparable to the noise factor F assigned to components along RF connections, used to calculate end-to-end BER. Reducing and stabilising this virtual BER per network element is the target of impairments control, and the virtual BER can be used to identify feasible lightpaths. Next, impairment compensation control loops

Table 2.3 Assumed parameter contribution to QoT (Teixeira et al. 2009)

Parameter	Sensitivity	Impact	Speed
Q_{OSNR}	Low	Linear	Moderate
Q_{GVD}	Low	Exponential	Slow
Q_{DGD}	Moderate	Exponential	Fast
Q_{FWM}	High	Non-linear	n/a

and their influence on actual QoT are discussed before returning to the routing problem and sketching common approaches.

2.4.1.2 Impairment Control and Stabilising of Network Elements

Traditionally, stable transmission performance is gained by controlling the environment, e.g. adding dummy channels to keep optical power constant when not all wavelengths are in use. More dynamic approaches are required since traditional ones are efficient only for entirely opaque network architectures. Reduction and stabilisation of physical parameters is the prime demand to realise reliable dynamically switched optical transmission networks. For OBS and OPS, the problem of keeping the performance of physical components constant, meaning independent of dynamically changing traffic assignments, needs to be addressed.

For stabilisation, we need to consider the dependence of the contributing components on changes of traffic load (sensitivity), their relative impact on QoT (impact) and how fast they possibly could be compensated (speed), alike the assumptions shown in Table 2.3.

Not all contributions can be considered hop-by-hop. Either can they not be monitored all-optically or there are no means to mitigate their impact all-optically. In addition, relevance of several components is dependent on the traffic load. Therefore, for dynamic environments, efficient control strategies need to be deployed.

Assuming that a virtual target BER per network element is achieved by design, the dynamic end-to-end BER is then the sum of these plus the deviations from the targets. The end-to-end deviation thus depends on the number of hops n where adjustments, originated temporarily, increased BER ($\mathrm{BER}_{\mathrm{transient}}$) as follows:

$$
\begin{aligned}
BER_{transient}\left(\mathrm{link_i}\right) &= p_{adj}(i)\; E\left[\mathrm{biterror}|\mathrm{QoT}(t) < QoT_{target}\right] \\
BER_{transient}\left(\mathrm{path}\right) &< \sum_{i=0}^{n} BER_{transient}\left(\mathrm{link_i}\right) + p_{adj}\left(lp\right) \\
&\quad \times E\left[\mathrm{biterror}|QoT(t) < QoT_{target}\right]
\end{aligned}
\tag{2.7}
$$

where $p_{\mathrm{adj}}(i)$ is the probability to be in an adjustment state on the i-th link in the path, $p_{\mathrm{adj}}(lp)$ the probability for lightpath adjustment and $\mathrm{E}[\mathrm{biterror}|\mathrm{QoT}(t) < \mathrm{QoT}_{\mathrm{target}}]$ is the likelihood of bit errors in the case that current QoT is below target QoT, meaning current BER contribution is above the planned.

To get an upper bound, let us assume a worst case: No end-to-end headroom and complete detector malfunction for QoT$(lp) <$ QoT$_{\text{target}}$, i.e. E[biterror| QoT$(t) <$ QoT$_{\text{target}}$] = ½. If we additionally simplify and assume the probability for adjustments to be equal for all links and lightpaths, Eq. 2.7 reduces to the following:

$$BER_{transient} < \frac{n+1}{2} p_{adj} \tag{2.8}$$

To minimise the transient impact, we need to minimise p_{adj}. As p_{adj} is calculated by occurring number of adjustments per time unit multiplied by the time units, these adjustments cause QoT$(t) <$ QoT$_{\text{target}}$; only the latter can be addressed by design of compensation mechanisms. The common problem of control loop design, precision vs. stabilisation time, arises. Along with the acknowledgement of cascaded potentially interdependent control loops, this yields the compensation performance and optimization demand.

Commonly asymptotic control loop behaviour is favoured. However, if there is headroom for sufficient detection, slight overcompensation is acceptable and thus faster operation and reduced number of biterrors per adjustment is possible. To cope with the cascades and interdependencies, a problem, as depicted in Fig. 2.40, arises. This can be handled offline using common methods since control loop performance characteristics are independent of instantaneous traffic load. Nevertheless, the results of such optimisation will depend on actual traffic statistics (distribution and characteristic) and the actual network topology that defines the cascades and interdependencies to consider.

2.4.2 Current IA-RWA Approaches

Routing, a term from graph theory, in communication networks generally performs the identification of a path (route), per connection request, between a source and a destination node across the network. In optical networks, the particular wavelength(s) along the path also need to be determined. The combined problem is in literature often called RWA problem (Zang et al. 2000). Most RWA proposals can be classified into two main categories: (a) considering the effects of physical layer impairments via iterative path testing until a feasible lightpath is found and (b) integrating physical impairments in local routing decisions. Both approaches provide IA-RWA. The latter also represents a specific variant of CBR problems with NP-hard computational complexity, in principle not scalable. However, the limited size of today's transparent networks, conditioned by the limited optical reach, alleviates this scalability issue.

Prior to discussing exemplary approaches, a principal decision is addressed: whether to implement routing as centralised or distributed. Any routing scheme can be implemented centralised, but only a few fit the demands for an efficient distributed implementation. The key demand for decentralised routing is that a correct decision can be made locally based on local information. If every node needs to know about everything going on in the entire network to do routing decisions, this is not economical, and a centralised approach may be considered. However, routing algorithms can be of polynomial complexity if, and only if, no prior knowledge about the reminder of a path is required for a correct routing decision. This is the case if, and only if, the Bellman (2003) or the Hamilton–Jacobi–Bellman (Bertsekas 2005) equations for dynamic programming hold.

2.4.2.1 Centralised Vs. Distributed Routing

Lightpath routing may be performed distributed, on-demand or continuously, end-to-end or hop-by-hop, within the physical layer's control plane, or might be outsourced to be performed in a more centralised way by employing dedicated PCEs that provide explicit routes on request. The latter demands timely flooding of the physical parameters to all PCEs, while the first relies on sufficient information available at each node. Whichever is more efficient depends on the network size, the meshing degree and load variability. However, centralised routing may consider QoT correlation among lightpaths more easily than distributed routing. In any case, considering QoT correlation raises complexity considerably. Thus, if we may assume that the optical routing changes rarely, meaning requests for not already routed all-optical connections and therefore resource assignment changes occur from time to time only and can be scheduled, the centralised approach is, for the optical layer, expedient. This is commonly the case with multilayer networks where lower layers are optimised to statically provide maximal capacity to the layer above independent of the current capacity usage. For such situations, offline routing using ILP-based optimisation and rearranging of assignments scheduled night time is commonly applied. We need to note that (a) static virtual topologies evidently cannot utilise the full potential of statistical cross-layer multiplexing and (b) the intention to reduce the number of layers towards IP over WDM annuls the essential precondition that resource assignment changes occur only rarely.

2.4.2.2 Common Sequential IA-RWA

Most IA-RWA approaches recently proposed consider the QoT problem separately from the RWA problem (Tomkos et al. 2004a; Markidis et al. 2007). If the QoT-independently found route is feasible, this is the fastest approach; however, if it does not fit, it is necessary to find a different path. As every single resource along a path per se commonly provides sufficient QoT, it is not clear which link to exclude for subsequent path searches. Common approaches to this problem are to (a) leave out one by one or (b) find several paths in the first step (k-shortest paths) and then check one candidate after the other. The second approach is more efficient only if the time to find k paths in one step is less than the sum of k subsequent path searches, weighted each with the likelihood $p(k)$ of finding in k subsequent routings only unfeasible paths. Consequently, for well-designed physical layers and not too degradation sensitive signal formats, the single-path routing with occasional subsequent path searches is economic. To decrease the probability of not finding a feasible path in the first routing attempt, we might adjust the weights used by the routing algorithm to reflect the physical layer's state. This is discussed next.

2.4.2.3 Q-Factor-Based Integration of Impairments in Routing Decisions

A straightforward strategy employed to include physical layer impairments in the routing decisions is to incorporate impairments into the cost function. However, a cost function correctly considering linear and non-linear impairments is still an open question. Different analytical models have been developed to describe reference links (Markidis et al. 2007; Brandt-Pearce et al. 2007). Only few studies consider the simultaneous impact of GVD, DGD, ASE and non-linear phase shift (Ezzahdi et al. 2006). Other more universal metrics have been proposed, including the average measured Q (Deng et al. 2005) and noise variance (Brandt-Pearce et al. 2007). In any case, accurate Q estimation is a heavy computational task requiring offline calculation. An approach to derive Q estimates for any path from some end-to-end monitored QoT metrics is outlined next.

2.4.2.4 Network Kriging

In (Chua et al. 2006), a method to predict the performance of paths based on collected information on other paths is presented. It relates to the principle known from applied probability and statistics (Cressie 1993) and is correspondingly called network kriging. In (Sambo et al. 2009a); (b), it is shown that the approach can be applied to predict end-to-end QoT based on monitored end-to-end QoT of other paths. This is achieved by setting up and solving an equation system considering how different paths share resources. The more already monitored paths share each resource of the path to be evaluated, the better the prediction approximates the unknown end-to-end QoT. As only paths that share the same resources contribute to the prediction, it is sufficient to inform only nodes participating in a path on the monitored end-to-end performance via the control plane. Flooding of monitored QoT per path is not required, making the approach scalable. However, to predict the QoT the entire path needs to be known in advance. Therefore, the approach fits best to a scheme where a limited set of candidate paths per ingress–egress node pair is a priori defined. Other than that, the candidate paths are not considered. Seemingly a restriction, this complies with the trend to reduce routing freedom, which can be widely observed in recent proposals and standardisation attempts targeting at more connection oriented IP networks for better traffic engineering options.

2.4.2.5 Ant Colony Optimisation

Over many years now, ant colony has proven to be a heuristic approach that can be applied successfully for network control and other issues that need to consider a multitude of obviously heterogeneous criteria. Ant colony optimisation is based on ant-like mobile agents that cooperate with each other while randomly exploring possible paths. Based on information collected by each agent and the information left behind at every node by all agents, iteratively decentralised routing information is created. In (Lee 2005; Pavani et al. 2008), a distributed IA-RWA scheme based on ant colony optimisation paradigm is presented. The next hop (outgoing optical channel) is determined based on a preference value (pheromone-level) derived from the information left behind by ants that previously passed. To not end up in a static routing, the preference values fade over time (alike pheromones evaporate) and therefore need to be refreshed from time to time. The rate at which preferences fade determines how many agents are required to keep a consistent routing information available and how fast the routing information adopts to abrupt changes like fibre breaks. This approach is rock solid, but being heuristic, it is difficult to optimise and hard to trust in, even though ants are one of the most successful species on earth.

A similar heuristic exploiting routing history instead of agents is presented in (Marin et al. 2007). Other heuristics applicable to perform IA-RWA include the genetic algorithm as proposed in (Ali et al. 1999; Lima et al. 2003), stimulated annealing (Mukherjee et al. 1996) and taboo search (Yan et al. 2001; Yang et al. 2005). These cannot be implemented or distributed and are computationally exhaustive, but are capable to accurately approach the global optimum with increasing computation effort.

2.4.2.6 Multi-Constrained Wavelength Routing (mCBR)

The major disadvantage of most integrated IA-RWA approaches is that the physical layer impairments need to be reduced to a single scalar cost value. Thereby, the information on individual parameters is lost, and it is necessary to define different cost functions that consider specific sensitivities of different signal formats correctly (also known as reference link models). Considering each physical layer impairment/constraint and how it changes hop-by-hop individually would be more convenient. Firstly, this allows finding paths that fit constraint by constraint, and secondly, this allows identification of specific weaknesses of certain paths. Especially in case of multi-hop paths, the latter would enable to profitably combine segments that equalise each other.

DWP provides exactly that by replacing the cost value by a vector of constraints and applying lattice algebra (Jukan 2004). Being a multi-constraint approach (mCBR), dynamic programming rules do not fit and therefore, the approach is not scalable and thus is limited to small networks in terms of node, link and wavelength-conversion counts. The DWP scheme suggests maximal spreading of the computing effort to (a) split the computation effort to as many processing units as possible and (b) to have real-time access to locally monitored parameters. With DWP, all relevant parameters are summed-up along potential paths by distributing path messages; dropping messages where constraints are not met (branch and bound). Being vector-based, any number and type of constraint can be considered in parallel; for example, QoT/BER, delay and reliability, as typically specified within SLAs. The fulfilment of constraints is guaranteed by DWP, as long as the parameters applied during path evaluation do not change. To cope with parameter changes, the selection strategy should be to select the path with sufficient headroom for constraints dependent on dynamic parameters. A strategy how to do so, however, is not presented, neither are the implications evaluated.

2.4.2.7 ILP-Based IA-RWA

Today, ILP is commonly applied to perform offline IA-RWA because of its heavy computational complexity. The challenges with ILP are to (a) define an adequate linear optimisation function and (b) formulate all other demands as side constraints. Once done, all these equations are handed over to a software tool and the result pops out. Seemingly, this is simple; however, results also pop out if some side constraints were not considered or incorrectly defined. There is no way to formally check problem sanity and completeness; that needs to be done by the engineer manually. Many proposals to solve IA-RWA using ILP have been published (Yang et al. 2005; Tomkos et al. 2004b; Cugini et al. 2007). They differ in the type and number of constraints considered, the optimisation target, considered routing demands and network scenario. However, ILP appears inadequate for dynamic environments and is unsuited for real-time control. The inevitable offline operation demands synchronised and, more restrictive, considerably delayed changes in compensation adjustments and traffic assignments, a clear contradiction to any dynamic network/traffic management.

Note that to calculate for a given topology, the control loop specifications and the QoT/BER target per network element should be as outlined in Sect. 2.4.1.2. With impairment control and stabilising of network elements, ILP perfectly fits. The Table 2.4 summarizes typical properties of different approaches.

The properties considered in the table are: complexity – meaning the effort required to determine a feasible path; optimality – is the found path optimal in terms of QoT; providence – does the routing metric consider potential future changes of parameters (respectively history); and robustness – how good does the approach per se respond to spontaneous events (e.g. a link failure). Note that some properties depend on implementation and that some shortcomings can be addressed via add-on mechanisms (i.e. link failures can be signalled if they do not affect the routing metric) (Table 2.4).

Table 2.4 Summary of IA-RWA approach's typical properties

Approach	Complexity	Optimality	Providence	Robustness
Sequ. IA-RWA	Moderate	Poor (first-fit)	No	High
Static Q-estim.	Low	Limited	Yes	Low
On-line Q-estim.	Moderate	Good	No	High
Krieging Q-estim.	High	Limited	Partly	Low
Ant colony	Low	Adjustable	Partly	Adjustable
mCBR	High	Perfect	No	High
ILP	Exhaustive	Perfect	Possibly	High

2.4.3 Other Demands and Considerations Related to IA-RWA

For an IA-RWA strategy to be actually implemented, one needs to consider also fundamental aspects like enabling OIM for indirect evaluation of signal quality or enabling direct OPM (Kilper et al. 2004b). In 2004, ITU-T defined a list of OPM parameters that might be used for impairment-aware RWA (ITUG697): (a) residual GVD, (b) total EDFA input and output powers, (c) a channel's optical power budget, (d) OSNR and (e) Q-factor – as an estimator of the overall optical performance. An effective OIM/OPM strategy shall also support the control plane in performing lightpath establishment and rerouting. Proposals on how to integrate physical layer impairments in GMPLS can be found in (Cugini et al. 2008; Cugini et al. 2004). Monitoring of a path's end-to-end QoT can, in addition to providing the basis for improved routing, also contribute to the control of SLA fulfilment and may trigger autonomous rerouting whenever the headroom diminishes below a constraint specific threshold. If rerouting is done obeying the make-before-break strategy, seamless rerouting can support SLA levels else inaccessible.

Per se, the outlined approaches to IA-RWA are not applicable for packet switching due to the complexity that finding feasible paths introduces. Yet, some experts recently proposed to control the network dynamics by limiting routing freedom. For instance, MPLS exactly performs that per flow (per ingress/egress pair). The routing underlying a LSP typically does not change, and thus the effective effort introduced by finding feasible paths to transport a certain flow type inversely depends on the LSP lifetime. Complex set-up procedures for long living LSPs thus actually introduce less effort if thereby frequent rerouting can be evaded.

It should be highlighted that the control loop performance targets for the deployment of physical impairments compensation mechanisms in meshed transmission networks with dynamic traffic assignment cannot be specified independent of traffic characteristic (holding time distribution), traffic matrix (likelihood of flows per node pair) and network topology (average and maximum path length). Even for network architectures comprising restriction to a certain topology and stable traffic distribution, the specs need to be individually derived. Only if QoT stabilisation is reached within a fraction of a single optical pulse or if adjustments have per se no effect on QoT ($BER_{transient} = 0$), traffic assignment dynamics independent QoT could be achieved. Assuming that this is not possible today, the demand on stabilisation speed is directly related to traffic assignment dynamics. To achieve the same BER as for non-dynamic optical connections (lines), the target QoT needs to provide sufficient extra headroom. The slower the stabilisation, the more hops, the more dynamic traffic changes actually cause adjustments, the more QoT headroom is effectively required and that needs to be granted by constraint-based RWA.

Finally, a dynamically switched Layer 0 (analogous physical layer) lacking perfect transparent signal regeneration within every hop can, in general, not grant constant QoT and thus shall not be made responsible for end-to-end BER liability.

2.5 Conclusions

Advanced signal processing, management and monitoring techniques are being developed for the next generation of optical networks, which are quickly evolving towards dynamically reconfigurable optical transport networks in which impairment-aware control planes are mandatory in order to grant the required end-to-end QoT and QoS.

Simultaneously, a convergence between wide area and access optical networks is taking place. This actual trend in access and metro networks demands, significantly enhance the requirements of signal management and monitoring while maintaining the targets for low cost and passiveness of the network as much as possible. This is especially relevant as signal impairments link losses, Rayleigh Backscattering are significantly increased demanding the development of strategic deployments, monitoring systems and Extender Boxes, as proposed in new recommendations, as G.984.6.

Effective preventive maintenance methods are needed to help network operators reduce the operation-and-maintenance expenses. Most of the monitoring solutions have been based on the well-known OTDR technique. In the framework of the SARDANA project, an elaborated monitoring approach was studied for a converged metro/access network which is a combination of WDM bidirectional rings and TDM access trees. Alternative monitoring approaches based on OFDR and OCDM techniques have also appeared in the scientific literature. However, among many approaches proposed so far, there are no standardised and mature infrastructure monitoring methods that fulfil the operators' needs.

The need for performance monitoring tools (other than a simple BER measurement at line extremity) is reinforced by the evolution of optical transport networks towards larger transmission reach and flexible wavelength paths.

Several optical monitoring techniques have been used for more than one decade. Spectral techniques are well suited to the monitoring of basic parameters (e.g. wavelength value, out-of-band OSNR) of an entire WDM multiplex. Improvements are still required to achieve in-band OSNR measurement of a set of channels with suitable trade-off between measurement accuracy and speed.

Single channel techniques provide a deeper insight into the various optical impairments. Among them, single-tap asynchronous sampling offers many advantages, including simplicity, rapidity, high sensitivity to low impairment levels and bit rate independence. Two-tap sampling has been recently proposed and has attractive performances in presence of multiple impairments and complex modulation schemes. These techniques will have to be cost-effective compared to optical channel estimation which becomes available with advanced digital signal processing in the receivers.

CD monitoring is an important issue to be addressed in reconfigurable WDM optical networks. All-optical CD compensation using RR-based integrated optical filters can be an effective solution for WDM transmission in comparison to other approaches, such as FBG or dispersion-compensating fibres. Recently proposed optical RR filters with internal reflections using Sagnac loops are a promising architecture to achieve periodic and tunable dispersion-compensating modules for DWDM applications in compact and lightweight devices.

The prime targets of impairment control and compensation shall be improving the end-to-end QoT and stabilising fluctuating QoT caused by dynamic resource assignments. With dynamic transmission networks, impairment-aware routing becomes responsible for the provision of optical paths that end-to-end assure a signal-format-dependent QoT sufficient for the QoS requirements posted by the connection requesting services.

References

10Gb/s Ethernet Passive Optical Network standard, IEEE 802.3av (http://www.ieee802.org/3/av/)

Agrawal, G.P.: Nonlinear Fiber Optics, 4th edn. Academic, Boston (2007)

Ahmed, J., Monti, P., Wosinska, L.: Concurrent processing of multiple LSP request bundles on a PCE in a WDM network. In: Proceedings of OSA/IEEE Optical Fiber Communication/National Fiber Optic Engineers Conference OFC/NFOEC 2010, San Diego, Mar 2010

Ali, M., Ramamurthy, B., Deogun, J.S.: Routing algorithms for all-optical networks with power consideration: the unicast case. In: Proceedings of the 8th IEEE ICCCN 1999, pp. 335–340, (1999)

Anderson, T.B., Dods, S.D., Wong, E., Farrell, P.M.: Asynchronous measurement of chromatic dispersion from waveform distortion. Paper OWN4, OFC 2006

Arbab, V.R., Wu, X., Willner, A.E., Weber, C.L.: Optical performance monitoring of data degradation by evaluating the deformation of an asynchronously generated I/Q data constellation. Paper P3.23, ECOC 2009

ITU.: Architecture for the Automatically Switched Optical Network (ASON), ITU-TRec.G8090/Y.1304, Nov 2001

IETF.: A Path Computation Element (PCE)-Based Architecture, IETF RFC 4655, Aug 2006

Banerjee, A., Drake, J., Lang, J., et al. Generalized multiprotocol label switching: An overview of signaling enhancements and recovery techniques. IEEE Commun. Mag. 144–151. (2001) (citeseer.ist.psu.edu/banerjee01generalized.html)

Baveja, P.P., Maywar, D.N., Agrawal, G.P.: Optimization of all-optical 2R regenerators operating at 40 Gb/s: role of dispersion. J. Lightwave Technol. **27**, 3831–3836 (2009)

Bellman, R.E.: Dynamic Programming. Princeton University Press. Republished 2003, Dover, ISBN 978-0486428093 (1957)

Bendelli, G., Cavazzoni, C., Girardi, R., Lano, R.: Optical performance monitoring techniques. In: Proceedings of 26th European Conference on Optical Communication (ECOC), vol. 4, 3–7 September, Munich, Germany, pp. 113–116 (2000)
Bertsekas, D.P.: Dynamic Programming and Optimal Control, 3rd edn. Athena Scientific, Belmont (2005). ISBN 978-1886529083
Bocci, M., Bryant, S., Frost, D., Levrau, L., Berger, L.: A framework for MPLS in transport networks, draft-ietf-mpls-tp-framework-12. IETF MPLS Working Group, Nov 2010
Bock, C., Lazaro, J.A., Prat, J.: Extension of TDM-PON standards to a single-fiber ring access network featuring resilience and service overlay. IEEE/OSA J. Lightwave Technol. **2007**, 1416–1421 (2007)
Brandt-Pearce, J.H., Pointurier, M., Subramaniam, Y.: QoT-aware routing in impairment-constrained optical networks. Proc. GLOBECOM **2007**, 26–30 (2007)
Cahill, M., Bartolini, G., Lourie, M., Domash, L.: Tunable thin film filters for intelligent WDM networks. In: Ellison, M.J. (ed.) Proceedings of SPIE, vol. 6286, Advances in Thin Film Coatings for Optical Applications III, Aug 2006
Castoldi, P., Cugini, F., Valcarenghi, L., Sambo, N., Le Rouzic, E., Poirrier, M.J., Adriolli, N., Paolucci, F., Giorgetti, A.: Centralized vs. distributed approaches for encompassing physical impairments in transparent optical networks, Lecture Notes in Computer Science Optical Network Design and Modeling 4534/2007 (2007) 68–77
Caucheteur, C., Mussot, A., Bette, S., Kudlinski, A., Douay, M., Louvergneaux, E., Mégret, P., Taki, M., Gonzalez-Herraez, M.: All-fiber tunable optical delay line. Opt. Express **18**, 3093–3100 (2010)
Chen, L.K., Cheung, M.H., Chan, C.K.: From optical performance monitoring to optical network management: research progress and challenges. Proceedings of ICOCN 2004, (2004)
Chen, W., Tucker, R.S., Yi, X., Shieh, W., Evans, J.S.: Optical signal-to-noise ratio monitoring using uncorrelated beat noise. IEEE Photon. Technol. Lett. **17**, 2484–2486 (2005)
Chlamtac, I., Ganz, A., Karmi, G.: Lightpath communications: an approach to high-bandwidth optical WAN's. IEEE Trans. Commun. **40**, 1171–1182 (1992)
Choi, E., Na, J., Ryu, S., Mudhana, G., Lee, B.: All-fiber variable optical delay line for applications in optical coherence tomography: feasibility study for a novel delay line. Opt. Express **13**, 1334–1345 (2005)
Chua, D.B., Kolaczyk, E.D., Crovella, M.: Network kriging. IEEE J. Sel. Areas Commun. **24**(12), 2263–2272 (2006)
Chung, Y.C.: Optical performance monitoring techniques; current status and future challenges, ECOC-2008, invited paper. Bruxselles, 21–25 Sept 2008
Cressie, N.A.C.: Statistics for Spatial Data Wiley Series in Probability and Mathematical Statistics: Applied Probability and Statistics. Wiley, New York (1993)
Cugini, F., Andriolli, N., Valcarenghi, L., Castoldi, P.: A novel signaling approach to encompass physical impairments in GMPLS networks. In: Proceedings of IEEE GLOBECOM 2004 Workshops, Dallas, November 29 – December 3 Dallas, Texas, USA, pp. 369–373 (2004)
Cugini, F., Andriolli, N., Valcarenghi, L., Castoldi, P.: Physical impairment aware signaling for dynamic lightpath set up. In: Proceedings of ECOC2005, Glasgow, vol. 4, pp. 979–980. Sept 2005
Cugini, F., Paolucci, F., Valcarenghi, L., Castoldi, P.: Implementing a Path Computation Element (PCE) to encompass physical impairments in transparent networks. In: Proceedings of OFC/NFOEC 2007, 25–29 March, Anaheim, California, USA, pp. 1–3 (2007)
Cugini, F., Sambo, N., Andriolli, N., Giorgetti, A., Valcarenghi, L., Castoldi, P., Le Rouzic, E., Poirrier, J.: Enhancing GMPLS signaling protocol for encompassing quality of transmission (QoT) in all-optical networks. J. Lightwave Technol. **26**(19), 3318–3328 (2008)
Deng, T., Subramaniam, S.: Adaptive QoS routing in dynamic wavelength-routed optical networks. Proceedings of BROADNETS 2005, vol. 1, pp. 184–193, (2005)
Dilwali, S., Soundra Pandian, G.: Pulse response of a fiber dispersion equalizing scheme based on an optical resonator. IEEE Photon. Technol. Lett. **4**(8), 942–944 (1992)

Dods, S.D., Anderson, T.B.: Optical performance monitoring technique using delay tap asynchronous waveform sampling. Optical Fiber Communication Conference, Paper OthP5, OFC 2006

Effenberger, F., Meng, S.: In-band optical frequency domain reflectometry in pons. In: OFC/NFOEC, pp. 1–3. (2008)

Ezzahdi, M., Zahr, S., Koubaa, M., Puech, N., Gagnaire, M.: LERP: a quality of transmission dependent heuristic for routing and wavelength assignment in hybrid WDM networks. Proc. ICCCN **2006**, 125–136 (2006)

Frigo, N.J., et al.: Centralized in-service OTDR testing in a CWDM business access network. IEEE J. Lightwave Technol. **22**(11), 2641–2652 (2004)

Fu, L.B., Rochette, M., Ta'eed, V.G., Moss, D.J., Eggleton, B.J.: Investigation of self-phase modulation based optical regeneration in single mode As2Se3 chalcogenide glass fiber. Opt. Express **13**, 7637–7644 (2005)

G.7713.1/Y.1704.1, DCM signalling mechanism using PNNI/Q.2931

G.7713.2/Y.1704.2, DCM signalling mechanism using GMPLS RSVP-TE

G.8080/Y.1304, Architecture for the automatically switched optical network

Generalized Multi-Protocol Label Switching (GMPLS) Architecture, IETF RFC 3945, Oct 2004

Gnauck, A.H., Cimini Jr., L.J., Stone, J., et al.: Optical equalization of fiber chromatic dispersion in a 5-Gb/s transmission system. IEEE Photon. Technol. Lett. **2**(8), 585–587 (1990)

Halabi, S.: Metro Ethernet. Cisco Press, Indianapolis (2003)

Huang, Y., Heritage, J.P., Mukherjee, B.: Connection provisioning with transmission impairment consideration in optical WDM networks with high-speed channels. J. Lightwave Technol. **23**(3), 982–983 (2005)

Hult, J.: A fourth-order Runge-kutta in the interaction picture method for simulating supercontinuum generation in optical fibers. J. Lightwave Technol. **25**, 3770–3775 (2007)

IEEE 802.1Qay: Provider Backbone Bridging Traffic Engineering, work in progress, http://www.ieee802.org/1/pages/802.1ay.html.

ITU-T G.697 Optical Monitoring for DWDM Systems, Nov 2009 – Prepublished

ITU-T Recommendation X.200: Information Technology – Opens Systems Interconnection – Basic Reference Model: The Basic Model, (1994)

ITU-T G984.6, GPON Optical Reach extension

Jargon, J.A., Wu, X., Willner, A.E.: Optical performance monitoring by use of artificial neural networks trained with parameters derived from delay-tap asynchronous sampling. Paper OThH1, OFC 2009

Jukan, A., Franzl, G.: Path selection methods with multiple constraints in service-guaranteed WDM networks. IEEE/ACM Trans. Netw. **12**(1), 59–72 (2004)

Kashyap, R., Chernikov, S.V., McKee, P.F., Williams, D.L., Taylor, J.R.: Demonstration of dispersion compensation in all-fibre photoinduced chirped gratings. Pure Appl. Opt. **4**, 425–429 (1995)

Kilper, D.C., Bach, R., Blumenthal, D., Einstein, D., Landolsi, T., Ostar, L., Preiss, M., Willner, A.E.: Optical performance monitoring. J. Lightwave Technol. **22**, 294–304 (2004a)

Kilper, D.C., Bach, R., Blumenthal, D.J., Einstein, D., Landolsi, T., Ostar, L., Preiss, M., Willner, A.E.: Optical performance monitoring. IEEE/OSA J. Lightwave Technol. **22**(1), 294–304 (2004b)

Kilper, D.C., Fergunson, A., O'Sullovan, B., Korotky, S.K.: Impact of topology and traffic on physical layer monitoring in transparent networks, OFC, invited paper, (2009)

Kirstaedter, A., Wrage, M., Goeger, G., Fishler, W., Splinner, B.: Current aspects of optical performance monitoring and failure root cause analysis in optical WDM networks. Proceeding SPIE, vol. 5625, 362–373, (2005)

Kompella, K., Rekhter, Y.: Label Switched Paths (LSP) Hierarchy with Generalized Multi-Protocol Label Switching (GMPLS) Traffic Engineering (TE) (proposed standard track). IETF RFC4206. (2005)

Laming, R.I., Robinson, N., Scrivener, P.L., Zervas, M.N., Barcelos, S., Reekie, L.: Dispersion tunable grating in a 10-Gb/s 100–220 km step-index fiber link. IEEE Photon. Technol. Lett. **8**, 428–430 (1996)

Lazaro, J.A., Bock, C., Polo, V., Martinez, R.I., Prat, J.: Remotely amplified combined ring-tree dense access network architecture using reflective RSOA-based ONU. OSA J. Opt. Netw. **6**(6), 801–807 (2007a)

Lazaro, J.A., Arellano, C., Polo, V., Prat, J.: Rayleigh scattering reduction by means of optical frequency dithering in passive optical networks with remotely seeded ONUs. IEEE Photon. Technol. Lett. **19**, 64–66 (2007b)

Lazaro, J.A., Prat, J., Chanclou, P., Tosi Beleffi, G.M., Teixeira, A., Tomkos, I., Soila, R., Koratzinos, V.: Scalable Extended Reach PON, OFC/NFOEC 2008, invited paper OThL2 (2008)

Lee, K.I., Shayman, M.: Optical network design with optical constraints in multi-hop WDM mesh networks. IEICE Transactions on Communications, (2005)

Lee, J.H., Choi, H.Y., Shin, S.K., Chung, Y.C.: A review of the polarization-nulling technique for monitoring optical-signal-to-noise ratio in dynamic WDM networks. IEEE/OSA J. Lightwave Technol. **24**(11), 4162–4171 (2006)

Lenz, G., Madsen, C.K.: General optical all-pass filter structures for dispersion control in WDM systems. J. Lightwave Technol. **17**(7), 1249–1254 (1999)

Lima, M.A.C., Cesar, A.C., Araujo, A.F.R.: Optical network optimization with transmission impairments based on genetic algorithm. In: Proceedings of the SBMO/IEEE IMOC, vol. 1, pp. 361–365, (2003)

Liu, X., Kao, Y.-H.: A simple OSNR monitoring technique independent of PMD and chromatic dispersion based on a 1-bit delay interferometer. In: Proceedings of European Conference on Optical Communication, PaperMo4.4.5, Cannes, (2006)

Lizé, Y.K., Yang, J.-Y., Christen, L.C., Wu, X.-X., Nuccio, S., Wu, T., Willner, A.E., Kashyap, R., Séguin, F.: Simultaneous and independent monitoring of OSNR, chromatic and polarization mode dispersion for NRZ-OOK, DPSK and Duobinary. In: Proceedings of Optical Fiber Communication Conference, Paper OThN2, Anaheim, Mar 2007

Lopez, E.T., Lazaro, J.A., Arellano, C., Polo, V., Prat, J.: Optimization of rayleigh-limited WDM-PONs with reflective ONU by MUX positioning and optimal ONU gain. IEEE Photon. Technol. Lett. **22**, 97–99 (2010)

Luis, R., Andre, P., Teixeira, A., Monteiro, P.: Performance monitoring in optical networks using asynchronously acquired samples with non ideal sampling systems and intersymbol interference. J. Lightwave Technol. **2**(11), 2452–2459 (2004)

Madsen, C.K., Lenz, G.: Optical all-pass filters for phase response design with applications for dispersion compensation. IEEE Photon. Technol. Lett. **10**, 994–996 (1998)

Madsen, C.K., Zhao, J.H.: A general planar waveguide autoregressive optical filter. J. Lightwave Technol. **14**(3), 437–447 (1996)

Madsen, C.K., Lenz, G., Bruce, A.J., et al.: Multistage dispersion compensator using ring resonators. Opt. Lett. **24**(22), 1555–1557 (1999)

Madsen, C.K., Walker, J.A., Ford, J.E., et al.: A tunable dispersion-compensating MEMS all-pass filter. IEEE Photon. Technol. Lett. **12**(6), 651–653 (2000)

Mamyshev, P.V.: All-optical data regeneration based on self-phase modulation effect. In: Proceedings of ECOC, 475–476 (1998)

Mannie, E., Papadimitriou, D.: Generalized Multi-Protocol Label Switching (GMPLS) Extensions for Synchronous Optical Network (SONET) and Synchronous Digital Hierarchy (SDH) Control (standards track), IETF RFC3946. (2004)

Marin, E., Sánchez, S., Masip, X., Solé, J., Maier, G., Erangoli, W., Santoni, S., Quagliotti, M.: Applying prediction concepts to routing on semi transparent optical transport networks. In: Proceedings of ICTON 2007, pp. 32–36, (2007)

Markidis, G., Sygletos, S., Tzanakaki, A., Tomkos, I.: Impairment aware based routing and wavelength assignment in transparent long Haul networks. Optical network design and monitoring. In: Optical Network Design and Modeling. Lecture Notes in Computer Science, pp. 48–57. Springer, Berlin/Heidelberg (2007)

Martinez, R., Pinart, C., Cugini, F., Andriolli, N., Valcarenghi, L., Castoldi, P., Wosinska, L., Comellas, J., Junyent, G.: Challenges and requirements for introducing impairment-awareness into the management and control planes of ASON-GMPLS WDM networks. IEEE Commun. Mag. **44**(12), 76–85 (2006)

Militello, M., et al.: Optical dynamic monitoring in next generation networks. IEEE CONTEL Conference, vol. ISBN 978-953-184-130-6, pp. 289–291, Zagreb, June 2009

Montalvo, J.: Applications of ring resonators and fiber delay lines for sensors and WDM networks. Ph.D. thesis, Univ. Carlos III Madrid, 2008

Mueller, K., Hanik, N., Gladish, A., Foisel, H-M., Caspar, C.: Application of amplitude histograms for quality of service measurements of optical channels and fault identification, ECOC 98, pp. 707–708

Mukherjee, B., Ramamurthy, S., Banerjee, D., Mukherjee, A.: Some principles for designing a wide-area WDM optical network. IEEE/ACM Trans. Netw. **4**(5), 684–696 (1996)

Nguyen, T.N., Gay, M., Bramerie, L., Chartier, T., Simon, J.-C.: Noise reduction in 2R-regeneration technique utilizing self-phase modulation and filtering. Opt. Express **5**, 1737–1747 (2006)

Omella, M., Lazaro, J.A., Polo, V., Prat, J.: Driving requirements for wavelength shifting in colorless ONU with dual-arm modulator. J. Lightwave Technol. **27**(17), 3912–3918 (2009a). ISSN: 0733-8724

Omella, M., Papagiannakis, I., Schrenk, B., Klonidis, D., Lazaro, J.A., Birbas, A.N., Kikidis, J., Prat, J., Tomkos, I.: 10 Gb/s full-duplex bidirectional transmission with RSOA-based ONU using detuned optical filtering and decision feedback equalization. Opt. Express **17**(7), 5008–5013 (2009b)

Orta, R., Savi, P., Tascone, R., Trinchero, D.: Synthesis of multiple-ring-resonator filters for optical systems. IEEE Photon. Technol. Lett. **7**(12), 1447–1449 (1995)

Othonos, A., Kalli, K.: Fiber Bragg Gratings: Fundamentals and Applications in Telecommunications and Sensing. Artech House, Norwood (1999)

Ouelette, F.: Dispersion cancellation using linearly chirped Bragg grating filters in optical waveguides. Opt. Lett. **12**, 847–849 (1987)

Pachnicke, S., Luck, N., Krummrich, P.M.: Online Physical-Layer Impairment-Aware Routing with Quality of Transmission Constraints in Translucent Optical Networks, ICTON 2009 Tu.A3.5

Pan, Z., Yu, C., Willner, A.E.: Optical performance monitoring for the next optical communication network. Opt. Fiber Technol. **16**, 20–45 (2010)

Park, K.J., Youn, C.J., Lee, J.H., et al.: Performance comparisons of chromatic dispersion-monitoring techniques using pilot tones. IEEE Photon. Technol. Lett. **15**, 873–875 (2003)

Pavani, G.S., Zuliani, L.G., Waldman, H., Magalhães, M.F.: Distributed approaches for impairment-aware routing and wavelength assignment algorithms in GMPLS networks. Comput. Netw. **52**(10), 1905–1915 (2008)

Petersen, M.N., Pan, Z., Lee, S., et al.: Online chromatic dispersion monitoring and compensation using a single inband subcarrier tone. IEEE Photon. Technol. Lett. **14**, 570–572 (2002)

Pinart, C., Amrani, A., Junyent, G.: Design and experimental implementation of a hybrid optical performance monitoring system for in-service SLA guarantee. 9th IFIP/IEEE International Symposium on Integrated Network Management (IM 2005), Nice, 16–19 May 2005

Prat, J., Lazaro, J.A., Chanclou, P., Cascelli, S.: Passive OADM Network Element for Hybrid Ring-Tree WDM/TDM-PON, ECOC'09, Proceedings, paper: P6.23, Vienna, Sept 20–24, 2009

Provost, L., Finot, C., Mukasa, K., Petropoulos, P., Richardson, D.J.: Design scaling rules for 2R-optical self-phase modulation-based regenerators. Opt. Express **15**, 5100–5112 (2007a)

Provost, L., Finot, C., Mukasa, K., Petropoulos, P., Richardson, D.J.: Generalisation and experimental validation of design rules for self-phase modulation-based 2R-regenerators. In: OFC/NFOEC (2007b)

Provost, L., Parmigiani, F., Petropoulos, P., Richardson, D.J.: Investigation of timing jitter reduction in a bidirectional 2R all-optical Mamyshev regenerator. In: OFC/NFOEC (2008)

RFC 4202 – Routing Extensions for GMPLS. Oct 2005 (Standards Track)
RFC 4204 – Link Management Protocol (Standards Track)
RFC 4974 – Generalized MPLS (GMPLS) RSVP-TE Signaling Extensions in Support of Calls
Rossi, G., Dimmick, T.E., Blumenthal, D.J.: Optical performance monitoring in reconfigurable WDM optical networks using subcarrier multiplexing. IEEE/OSA J. Lightwave Technol. **18**, 1639–1648 (2000)
Rotwitt, R., Guy, M.J., Boscovik, A., Noske, D.U., Taylor, J.R., Kashyap, R.: Interaction of uniform phase picoseconds pulses with chirped and unchirped photosensitive fibre Bragg gratings. Electron. Lett. **30**, 995–996 (1991)
Saleh, M., Simmons, J.M.: Evolution toward the next-generation core optical network. IEEE/OSA J. Lightwave Technol. **24**(9), 3303–3321 (2006)
Salvadori, E., Ye, Y., Zanardi, A., Woesner, H., Carcagni, M., Galimberti, G., Martinelli, G., Tanzi, A., LaFauci, D.: A study of connection management approaches for an impairment-aware optical control plane. In: Proceedings of IFIP ONDM2007. Lecture Notes in Computer Science, vol. 4534, pp. 229–238. Springer, Athens (2007a)
Salvadori, E., Ye, Y., Zanardi, A., Woesner, H., Carcagni, M., Galimberti, G., Martinelli, G., Tanzi, A., LaFauci, D.: Signalling-based architectures for impairment-aware lightpath set-up in GMPLS networks. In: Proceedings of IEEE GLOBECOM 2007, Washington, pp. 2263–2268. Nov 2007b
Sambo, N., Giorgetti, A., Andriolli, N., Cugini, F., Valcarenghi, L., Castoldi, P.: GMPLS signaling feedback for encompassing physical impairments in transparent optical networks. In: Proceedings of IEEE GLOBECOM2006, Sanfrancisco, pp. 1–5. Nov 2006
Sambo, N., Pointurier, Y., Cugini, F., Castoldi, P., Tomkos, I.: Lightpath establishment in PCE-based dynamic transparent optical networks assisted by end-to-end quality of transmission estimation. In: Proceedings of ICTON 2009, Mo.D3.1, (2009a)
Sambo, N., Pointurier, Y., Cugini, F., Valcarenghi, L., Castoldi, P., Tomkos, I.: Lightpath establishment in distributed transparent dynamic optical networks using network kriging. ECOC 2009 proceedings, paper 1.5.3, (2009b)
Shake, I., Takara, H.: Averaged Q-factor method using amplitude histogram evaluation for transparent monitoring of optical signal-to-noise ratio degradation in optical transmission system. J. Lightwave Technol. **20**, 1367–1373 (2002)
Shake, I., Takara, H.: Chromatic dispersion dependence of asynchronous amplitude histogram evaluation of NRZ signal. J. Lightwave Technol. **21**(10), 2154–2161 (2003)
Shake, I., Takara, W., Kawanishi, S., Yamabayashi, Y.: Optical signal quality monitoring method based on optical sampling. Electron. Lett. **34**(22), 2152–2154 (1998)
Shake, I., Takara, H., Kawanishi, S.: Simple measurement of eye diagram and BER using high-speed asynchronous sampling. J. Lightwave Technol. **22**, 1296–1302 (2004)
Shieh, W., Tucker, R.S., Chen, W., Yi, X., Pendock, G.: Optical performance monitoring in coherent optical OFDM systems. Opt. Express **15**(2), 350–356 (2007)
Strand, J., Chiu, A.L., Tkach, R.: Issues for routing in the optical layer. IEEE Commun. Mag. **39**(2), 81–87 (2001)
Striegler, A.G., Schmauss, B.: Analysis and optimisation of SPM-Based 2R signal regeneration at 40 Gb/s. J. Lightwave Technol. **24**, 2835–2842 (2006)
Takiguchi, K., Okamoto, K., Moriwaki, K.: Dispersion compensation using a planar lightwave circuit optical equalizer. IEEE Photon. Technol. Lett. **6**(4), 561–564 (1994)
Teixeira, A., Costa, L., Frantzl, G., Azodolmolky, S., Tomkos, I., Vlachos, K., Zsigmond, S., Cinkler, T., Tosi Beleffi, G., Gravey, P., Loukina, T., Lázaro, J.A., Vazquez, C., Montalvo, J., Le Rouzic, E.: An integrated view on monitoring and compensation for dynamic optical networks from management to physical layer. Photon. Netw. Commun. (2009). doi:DOI: 10.1007/s11107-008-0183-5
Tomkos, I.: Transport performance of WDM metropolitan area transparent optical networks. In: Proceedings of OFC, Mar 2002, pp. 350–352

Tomkos, I., Vogiatzis, D., Mas, C., Zacharopoulos, I., Tzanakaki, A., Varvarigos, E.: Performance engineering of Metropolitan area optical networks through impairment constraint routing. IEEE Optical Communications Magazine, pp. 40–47, (2004a)

Tomkos, I., Vogiatzis, D., Mas, C., Zacharopoulos, I., Tzanakaki, A., Varvarigos, E.: Performance engineering of metropolitan area optical networks through impairment constraint routing. IEEE Commun. Mag. **42**(8), S40–S47 (2004b)

Tsuritani, T., Miyazawa, M., Kashihara, S., Otani, T.: Optical path computation element interworking with network management system for transparent mesh networks. In: Proceedings of OFC/NFOEC 2008. pp. 1–10. San Diego, Mar (2006)

Valenti, A., Bolletta, P., Pompei, S., Matera, F.: Experimental investigations on restoration techniques in a wide area Gigabit ethernet optical test-bed based on Virtual Private LAN Service. In: Proceedings of ICTON 09, Ponta Delgada, (2009)

Vargas, S., Vázquez, C.: Synthesis of optical filters using Sagnac interferometer in ring resonator. IEEE Photon. Technol. Lett. **19**, 1877–1879 (2007)

Vargas, S., Vázquez, C., Pena, J.M.S.: Novel tunable optical filter employing a fiber loop mirror for synthesis applications in WDM. 14th Annual meeting LEOS, 2:899–900, 2001

Vargas, S., Vázquez, C.: Synthesis of optical filters using microring resonators with ultra-large FSR. Opt. *Express* **18**, 25936–25949 (2010)

Wu, X., Jargon, J.A., Jia, Z., Paraschis, L., Skoog, R.A., Willner, A.E.: Optical performance monitoring of PSK data channels using artificial neural networks trained with parameters derived from delay-tap asynchronous diagrams via balanced detection. Paper P3.04, ECOC 2009

Wuilmart, L., et al.: A PC-based Method for the localisation and quantisation of faults in passive tree-structured optical networks using OTDR technique. In: Proceedings of IEEE LEOS'96, Boston, pp. 122–123, Nov 1996

Yan, S., Ali, M., Deogun, J.: Route optimization of multicast sessions in sparse light-splitting optical networks. IEEE GLOBECOM 2001, vol. 4, pp. 2134–2138, (2001)

Yang, X., Shen, L., Ramamurthy, B.: Survivable lightpath provisioning in WDM mesh networks under shared path protection and signal quality constraints. IEEE/OSA J. Lightwave Technol. **23**(4) (2005). pp. 1556

Yüksel, K., Moeyaert, V., Wuilpart, M., Mégret, P.: Optical layer monitoring in Passive Optical Networks (PONs): a review, International Conference on Transparent Optical Networks (ICTON), paper Tu.B1.1, Athens, 22/06-26/06, 2008

Yuksel, K., Wuilpart, M., Moeyaert, V., Mégret, P.: A novel monitoring technique for passive optical networks based on optical frequency domain reflectometry and fiber Bragg gratings, ICTON 2010, paper ThA2.2, 27 June – 1 July, Munich

Zang, H., Jue, J.P., Mukherjee, B.: A review of routing and wavelength assignment approaches for wavelength-routed optical WDM networks. SPIE/Baltzer Opt. Netw. Mag. **1**(1), 47–60 (2000)

Zou, N., Namihira, Y., Ndiaye, C., Ito, H.: Fault location for branched optical fiber networks based on ofdr technique using fsf laser as light source OFC/NFOEC, p. NWC2, (2007)

Chapter 3
Simulations of High-Capacity Long-Haul Optical Transmission Systems

Francesco Matera, Rebecca Chandy, Valeria Carrozzo, Karin Ennser, Guido Maier, Achille Pattavina, Marina Settembre, Domenico Siracusa, and Marcelo Zannin

3.1 Introduction

The wider and wider bandwidth required for users stimulates the increase in the bit rate transmission for optical fiber systems (Bergano 2005), and one of the current aims for long-haul links is channel transmission at 100 Gbit/s (Cai et al. 2010). To reach such a capacity, several technological approaches are required, both in terms of advanced modulation formats (Winzer and Essiambre 2006) and also methods to mitigate impairments due to in-line degradation effects, like the ASE noise of optical amplifiers (Agrawal 2007), and dispersive (Elrefaie et al. 1988) and nonlinear fiber effects (Chraplyvy 1990). Concerning advanced transmission and detection formats (Gnauk 2005), the combination of amplitude and phase modulation allows

F. Matera (✉) • M. Settembre
Fondazione Ugo Bordoni, Viale America, 201, 00144 Roma, Italy
e-mail: mat@fub.it; marina.settembre@elsagdatamat.com

R. Chandy
Ericsson, UK

Unit 4, Midleton Gate, Guildford, GU2 8SG, UK
e-mail: rebeccachandy@gmail.com

V. Carrozzo
ISCOM, MSE – Communication Department, Viale America, 201, 00144 Roma, Italy
e-mail: valeria.carrozzo@gmail.com

K. Ennser • M. Zannin
College of Engineering, Swansea University, Singleton Park, SA2 8PP Swansea, West Glamorgan, UK
e-mail: k.ennser@swansea.ac.uk; marcelo.z.rosa@gmail.com

G. Maier • A. Pattavina • D. Siracusa
Electronic and Information Department, Politecnico di Milano, Piazza L. da Vinci, 32, 20133 Milano, Italy
e-mail: maier@elet.polimi.it; pattavina@elet.polimi.it; siracusa@elet.polimi.it

A. Teixeira and G.M.T. Beleffi (eds.), *Optical Transmission: The FP7 BONE Project Experience*, Signals and Communication Technology,
DOI 10.1007/978-94-007-1767-1_3,

us to use a higher bit rate reducing the signal bandwidth, and therefore, these modulation formats have shown enormous advantages with respect to conventional IM-DD systems. In particular, DQPSK is assumed as one of the most interesting formats since in a bit time, four symbols can be transmitted with a consequent enormous advantage for reduction of dispersive impairments (Fuerst et al. 2006, 2008). Concerning the mitigation of fiber impairments at the end of 1980, the soliton propagation seemed the best solution for long transmission systems (Mollenauer et al. 1991); conversely, the introduction of dispersion compensating devices such as DCF and grating showed that the regime of the periodic chromatic dispersion compensation, also known as dispersion management, was the best method since it is also able to limit the accumulation of nonlinear effects (Lichtman and Evangelides 1994).

A fundamental thrust for progress in the field of optical communications was given by numerical simulations (Matera and Settembre 1996, 2000; Eramo et al. 2003) and, therefore, we dedicate a chapter to this subject, both describing the main routines and some results regarding the performance of high-capacity, long-haul optical transmission systems (Marcuse 1991a, b).

In particular, we first describe the numerical code to simulate optical transmission systems, mainly based on the split-step method (Iannone et al. 1998); subsequently, we focus on some specific cases regarding soliton transmission and dispersion management links, including the transmission formats of IM-DD, RZ-DPSK, and RZ-DQPSK systems.

The last part of the chapter is dedicated to the simulation of WDM burst systems.

3.2 The Role of the Simulation Tools in the Evolution of the Optical Transmission Systems

In the last years, numerical simulations of optical transmission systems have been a fundamental tool for investigating optical communication networks (Eramo et al. 2003), since many components are involved, manifesting several effects on signal propagation, and as a result, analytical approaches are useful only for a limited number of cases. Reliability and effectiveness in terms of time consumption have been important requirements for simulation tools, and a deep analysis of comparison with experimental tests has been carried out by research institutes, especially in the framework of international projects.

Several methods can be used to simulate the behavior of an optical transmission system. It is not within the scope of this book to compare all of them, but we will focus our attention on one of the most known technique, called ‘split step’ method, whose results are very interesting in terms of time effectiveness thanks to the use of a fast Fourier transform algorithm (Agrawal 2007). This method is very friendly in describing the evolution of a signal under the action of deterministic effects as chromatic dispersion and Kerr effects, also in the presence of stochastic processes as ASE noise and random mode coupling in optical fibers (Curti et al. 1990). A numerical code can be developed at very different levels of investigation to obtain a qualitative prediction of signal evolution as the shape evolution of the signal; furthermore, it is also important to have information about the correct possibility to transmit a signal, and therefore, we need to correctly estimate system performance. Nowadays, optical systems are required to operate at very low BER. To correctly evaluate the BER, a too long computational time is required if Monte Carlo methods are used. For example, if the BER is required to be lower than 10^{-9}, to correctly estimate its value by simulations more than 10^{10} bits have to be considered, even though, it has to be pointed out that by using FEC methods higher error probability could be accepted (Poggiolini et al. 2006). However, in this contribution, we always look for propagation conditions that guarantee very good performance (10^{-9} or less) thinking to the FEC how a further method to permit the transmission in worse real environment. Nevertheless, Monte methods operating with more than 10^9 bits make impractical any direct measurement of BER, and an important role is then played by indirect measurements. One of the most important indirect technique is the evaluation of *Q factor* (Bergano et al. 1993; Anderson and Lyle 1994) defined as:

$$Q = \frac{m_1 - m_0}{\sigma_1 + \sigma_0}, \tag{3.1}$$

where m_1 and m_0 are the mean values of the decision variable I_k obtained by sampling the current $I(t)$ at the electrical filter output of the optical receiver at the instant t_k, for the bit "1" and for the bit "0," and σ_1 and σ_0 are the corresponding standard deviations.

When the decision variable can be assumed Gaussian, the error probability, *PE*, depends on the mean values, on the standard deviations and on the threshold current, *D*, according to the equation (Agrawal 2007):

$$PE = \frac{1}{4} erfc\left[\frac{|I_1 - D|}{\sqrt{2}\sigma_1}\right] + \frac{1}{4} erfc\left[\frac{|I_0 - D|}{\sqrt{2}\sigma_0}\right], \tag{3.2}$$

In Eq. 3.2, we suppose equal probability for the bit "1" and the bit "0." By varying the threshold current, we can minimize the error probability and the minimum value is achieved when the threshold current is:

$$D = \frac{\sigma_0 I_1 + \sigma_1 I_0}{\sigma_1 + \sigma_0}, \tag{3.3}$$

and in such a case,

$$PE = \frac{1}{2} erfc\left[\frac{Q}{\sqrt{2}}\right], \tag{3.4}$$

where Q is defined in Eq. 3.1.

Therefore, if we assume the decision variable as Gaussian, the BER coincides with the error probability, but it can be evaluated with a relatively short bit sequence, avoiding the huge sequences requested by error probability of the order of 10^{-9} or less. Unfortunately, the decision variable can be rigorously considered as Gaussian in a very limited number of cases. However, when the Q factor cannot be directly related to the BER, it can be considered as a very reliable parameter to show the system performance since, as demonstrated in several experiments (Favre et al. 1999), it permits to achieve the error probability with a good approximation.

For example, concerning the reliability of a numerical code and the comparison with experimental measurements, in the framework of two European ACTS projects (ESTHER and UPGRADE), systems based on soliton transmission were investigated. The ESTHER project was based on the demonstration of 10 and 40 Gbit/s soliton transmission at 1.5 μm with erbium-doped fiber amplifiers both on G.652 (step-index) and G.653 (dispersion shifted, DS) fiber links (Favre et al. 1999). A polarization multiplexing format based on the transmission of adjacent pulses with orthogonal states of polarization was adopted to reduce the effect of soliton interaction (Franco et al. 1999).

The UPGRADE project was also based on the demonstration of 10 and 40 Gbit/s nonlinear transmission, but exploiting the 1.3 μm optical window of the fiber where G.652 fibers exhibit lower chromatic dispersion and by using SOAs. This project was mainly intended to face the nonlinear behavior of SOA, which is responsible both for pulse gain variations and pulse time jitter depending on the bit pattern (Settembre et al. 1997).

These Projects were a great opportunity since they allowed us to compare both experimentally and numerically evaluated system performances, in very different conditions: in presence of weak or strong pattern effects, in presence of deterministic and random processes, and stationary and nonstationary effects.

It was found that only when the Q factor was evaluated taking into account the patterning effect (Anderson and Lyle 1994) and optimizing the parameters for a good evaluation of statistical properties, a good agreement between experimental and numerical results was obtained.

3.3 Modeling of Optical Systems

Schematically, an optical communication system is composed by a transmitter, a transmission line, and a receiver (Agrawal 2007; Iannone et al. 1998). The transmission line is supposed to be composed by optical fibers and optical amplifiers. Optical amplifiers are cheaper and more flexible than electrical regenerators, but they introduce ASE noise and signal distortion limiting the system capacity. Many efforts have been developed towards the optical regeneration of the signal along the line, and the current technology is almost mature to include such devices in the networks. Even if sophisticated routines have been developed for transmitter (Lebref et al. 1999) and receiver (Betti et al. 1994) for taking into account their not ideal behavior, in this chapter, they will be considered as ideal since the propagation aspects along the line have a much more relevant role. As far as the optical amplifier is concerned, an accurate model has been developed, for instance, in the framework of the UPGRADE (Settembre et al. 1997; Goder et al. 1999) project for simulating the behavior of SOA (Agrawal and Olsson 1989). Conversely, for most of the systems, the erbium-doped fiber amplifier (EDFA) is simply simulated by multiplying the signal for a constant gain and introducing the ASE noise by adding in the frequency domain, a Gaussian noise term (Gordon and Mollenauer 1990), even though more realistic models are available in the literature (Lebref et al. 1999).

It is a common knowledge that high-capacity networks operate with digital signals. In this domain, several optical systems have been proposed and can be distinguished according to the modulation and detection schemes. A good compromise in terms of simplicity, cost, and robustness with respect to the degrading effects of a link is given by IM-DD systems (Agrawal 2007). Both NRZ and RZ formats have awakened a lot of interest (Matera and Settembre 1995); however, this chapter essentially deals with the RZ transmission since it allows us to achieve better transmission performance (Matera 1996). In the last years, other modulation formats (Gnauk 2006) have received much interest in terms of bandwidth saving to achieve very high bit rate ($\geq$40 Gbit/s) in long fiber links. In particular, several investigations, confirmed by experimental tests, have shown the importance of the phase/amplitude modulations. For this reasons, this chapter also shows results on DPSK and DQPSK transmission (Fuerst et al. 2006, 2008) that, among the several modulation formats, are the most interesting ones.

The signal degradation along the link can be described in terms of three different contributions: deterministic effects, random effects, and nonstationary processes (Matera and Settembre 2000). The impact of each contribution in a simulative code is very different, and it is not easy to find a code that can efficiently manage all of them at the same time. As it will be shown in the following paragraphs, it might be useful to consider more than one performance evaluation parameter to describe the impact of these different contributions. Here, let us briefly summarize some typical features of these effects. Deterministic processes are responsible of a distortion of signal shape that is always the same for each realization of the process. Typical examples of deterministic effects are those due to fiber chromatic dispersion and Kerr nonlinearity. When they act separately on a pulse, they are responsible of temporal and spectral broadening, respectively (Agrawal 2007). When they act together on the pulse, the evolution of the pulse shape can be very difficult to describe, and in some conditions, it is not easy to distinguish respective contributions; this is the case, for example, of four-wave mixing and cross-phase modulation. In particular conditions, a stable solution can be found, that is when the pulse shape propagates unaltered (soliton case). Beyond dispersion and nonlinearity, there are many other effects that lead to deterministic signal distortions, i.e. filtering or saturation of optical amplifiers. The signal shape distortion depends on the number of interacting bits or, in other words, from the memory of the effect under consideration (Settembre et al. 1997; Goder et al. 1999). In fact, generally, signal distortion depends not only on the characteristics of the single pulse and of the fibre parameters but, in principle, on the entire transmitted message. A typical example is given by the highly dispersive systems (i.e. 40 Gbit/s on G.652 links with dispersion management) where each pulse can interact with several pulses due to the strong broadening suffered (Zitelli et al. 1999).

We can say that for each propagation condition, we can define a time interval $(t_0, t_0 + T)$ in which the bit depends on the entire signal in time $(t_0 - nT, t_0 + T + nT)$, T being the bit time and n an integer defining the memory of the channel. Therefore, it is fundamental and necessary to correctly estimate the value of n to evaluate the system performance without losing too much performance in terms of computational time.

A typical example of random effect is given by ASE noise due to optical amplifiers. It perturbs the signal on a time scale faster than the bit time.

We refer to a nonstationary process if the temporal scale of the random process is much longer than the bit time. A typical example is PMD (Poole and Wagner 1986; Curti et al. 1990). In principle, an optical fiber with a constant and time-independent PMD behaves as a couple of unbalanced delay lines, so a deterministic pulse broadening can be observed due to the time delay between the two orthogonally polarized components of the signal. In a real fiber, the random fluctuations in the fiber structure induce a local birefringence and a random mode coupling that lead to a random power exchange between the polarized modes, and a pulse broadening, depending on the input state of polarization of the signal (Galtarossa et al. 2001). Moreover, local birefringence and random coupling are very sensitive to environmental conditions (De Angelis et al. 1992). Generally, PMD is measured through the relative group delay between two particular polarization states, *the principal states of polarization* (Curti et al. 1990), that substantially behave as the two polarization modes of an ideal birefringent fiber. The principal states and hence their relative group delay depends also on signal wavelength. So, in a real fiber, PMD is a statistical process that fluctuates in time and strongly depends on signal wavelength. The temporal scale of its fluctuation is typically longer than 1 min (De Angelis et al. 1992), and as a consequence, a signal operating with a bit rate higher than 622 Mbit/s sees the same PMD during the time interval necessary for the measurement of the BER at 10^{-9}. Conversely, different BER measurements can give different values because of PMD fluctuations.

3.3.1 System Performance Evaluation

The typical performance evaluation parameter for digital receiver is the BER, but it is not reasonable to accumulate all the statistical data for evaluating the extremely low required BER (10^{-9} or less) directly by simulations.

As mentioned in (Sect. 3.2), there is a correspondence between error probability and Q factor defined in Eq. 3.1, in the hypothesis of optimum decision threshold, only when the statistics of the decision variable is Gaussian (Agrawal 2007).

Rigorously speaking, the Gaussian approximation is not easy to be satisfied even for simple propagation conditions. As an example, let us consider the ideal propagation condition of a signal in an amplified link, in which all the degrading effects are neglected, except for the ASE noise. The ASE noise has a Gaussian statistics, but the photodiode performs a square operation, and thus, the PDF of the decision variable is not Gaussian anymore. Fortunately, along with the filtering process at the receiver, the resulting PDF turns slightly towards Gaussian, as can be understood by applying the central limit theorem (Humblet and Azizoglu 1991; Marcuse 1991a, b bis). This occurs when the optical filter in front of the receiver has a bandwidth, B, much larger than the bit rate, R, or in other words, if the number of degrees of freedom of the system, $m = B/2R$ is larger than 10, so that the decision variable can be assumed as the sum of m variables. If the Gaussian approximation is not verified, the BER obtained through the Q factor can be used only as a qualitative value useful in the optimization of the system design, but not for a quantitative estimation of BER.

To correctly estimate the BER, we need the calculation of the Q factor from a very long sequence of bits that takes into account all the pattern effects, and this could induce a long sequence and an infinite time for simulation. Therefore, especially some years ago, when the computing capacity was much slower with respect the current one, the main problem in simulations was to find the smallest, however, representative bit sequence, and the minimization was based on grouping the bits according to pattern effect.

The impact of pattern effects on the evaluation of the error probability can be taken into account by modifying the basic formulation of Q factor given in Eq. 3.1. In fact, that approach gives over-estimated standard deviations due to the intersymbol interference and hence also a pessimistic evaluation of the BER. In principle in any specific bit sequence, the total probability error should be calculated as a sum of the probability of error of each elementary bit instead of calculating only two error probability distributions for the "zero" and the "one" bit, respectively. The two procedures would lead to the same result in absence of pattern effect, but they will differ more and more as the number of interacting bits involved in the pattern effect increases. In order to save computational time and at the same time without losing too much in accuracy, it is important to estimate the number of interacting bits. If the interacting bits are $2n + 1$, there are p possible different configurations of $2n + 1$ bits, where $p = 2^{2n+1}$.For each of the p pattern, a correspondent Q_i factor can be evaluated on the central bit of the pattern according to the following formula (Iannone et al. 1998):

$$Q_i = \frac{|< I_i > - I_{\text{th}}|}{\sigma_i}. \tag{3.5}$$

Then, the overall Q factor is obtained as (Anderson and Lyle 1994):

$$Q = \sqrt{2} erfc^{-1} \left(2 \left[\min \{P_{\text{E}}(I_{\text{th}})\}\right]\right), \tag{3.6}$$

where min $\{\mathrm{BER}(I_{\mathrm{th}})\}$ means the minimum value of the function $\mathrm{BER}(I_{\mathrm{th}})$ that is given by:

$$P_{\mathrm{E}}(I_{\mathrm{th}}) = \frac{1}{2p}\sum_{i=1}^{p} erfc\left(\frac{Q_i}{\sqrt{2}}\right). \tag{3.7}$$

In the most of practical cases, it is sufficient to assume that pattern effect is mainly due to the interaction with the next neighbors of each bit, or it can be said that the channel has a three bit memory ($n = 1, p = 8$), but in some particular cases, i.e. the saturation of SOA amplifiers, the number of interacting bits should be increased at least up to five bits ($n = 2, p = 32$) to obtain an accurate result (Goder 1999).

Another important parameter both for the reliability and time effectiveness of a simulation code is the length of the bit sequence. For the estimation of σ_{i} in each of p in Eq. 3.5, l-independent realizations of the noise processes are simulated by changing the seed of the random generators. In this way, the evaluation of the Q_i can be obtained by propagating $2n + 1$ bits for each p pattern for l different realizations of the noise process, that means by propagating l times a sequence of $p(2n + 1)$ bits. However, due to the independence of the patterns, the same statistical results can be obtained by considering sequences with a number of bits N greater than $p(2n + 1)$, reducing the number of different realizations from l to l'.

It can be demonstrated that the Q factor can be mainly affected by some critical bit patterns, and as a consequence, in order to evaluate the system performance, it is not necessary to take into consideration an enormous quantity of bits, but it is sufficient a bit sequence that contains all the most critical patterns reducing the computational time. The number of required runs and the length of the bit sequences depend on the considered propagation case, and preliminary simulations must be done to find stable values of the system performance parameters.

It has to be pointed out that the computation time of the split-step method grows as the square of bit sequence, and therefore, especially with slow time PC, the number of bits should be lower than 128–256 to have reasonable simulation time. Today, due to the much faster computation time, sequences of 2,048–4,096 bits can be used, and as a consequence, also the procedure regarding the patterning effect (Eq. 3.7) can be avoided, and the Q factor can be calculated according to Eq. 3.1.

If nonstationary processes are present, a further statistical investigation is necessary. In fact, the BER evaluation can be considered valid only for a particular realization of the nonstationary process, and the physical fluctuation of the BER in time will correspond to different realizations of nonstationary process in the simulations. An example is given by an optical system strongly affected by PMD, where the nonstationary of the process is simulated by changing the seeds of random generator for the parameters representing the local birefringence and axes of the fiber slices (Way and Menyuk 1996). To this aim, it should be defined the maximum time of system out-of-service, i.e. the time interval in which a system can operate with a BER higher than 10^{-9}. Such a condition depends on the quality of service; this is not the subject of this chapter, but details can be found in (Galtarossa et al. 2001).

3.3.2 Fiber Simulation Routines

The signal propagation in optical fiber can be evaluated by means of the split-step method that permits to numerically solve the NLSE that describes the fiber effects on spatial and temporal optical field propagation (Yariv 1989). The main difficulties in the simulations are given by the fiber nonlinear effects, even though for most of optical transmission systems, the main fiber nonlinearity is given by the Kerr effect. Taking into account the Kerr nonlinearity, the equation that describes the signal evolution in a single mode fiber is given by (Agrawal 2007):

$$\frac{\partial A}{\partial z} + \beta_1 \frac{\partial A}{\partial t} + \frac{i}{2}\beta_2 \frac{\partial^2 A}{\partial t^2} - \frac{1}{6}\beta_3 \frac{\partial^3 A}{\partial t^3} + \frac{\alpha}{2}A = i\gamma |A|^2 A, \quad (3.8)$$

where $A(t)$ is the base bandwidth envelop signal, β_1 is the inverse of the group delay (its contribution is only a delay $\beta_1 \Delta z$), β_2 is the chromatic dispersion, β_3 is the third order chromatic dispersion, α is the fiber loss, and γ is the nonlinear coefficient (in $(Wkm)^{-1}$). To numerically solve such an equation, we suppose to subdivide the fiber in several segments of δz length. If we suppose that the nonlinear contribution in each δz is negligible (and we will explain later the conditions to adopt such an approximation), we can evaluate the dispersive and the nonlinear evolution in a separate way, as if the two effects act separately. In particular, to have a better approximation, we can assume to have the following segments:

1. A dispersive segment $\delta z/2$
2. A nonlinear segment δz
3. A dispersive segment $\delta z/2$

In segment (1), the equation to solve is the Eq. 3.8 without the right term (nonlinear term). In this case, the relationship between the output signal and the input signal is given by:

$$A(z,t) = \frac{1}{2\pi} \int_{-\infty}^{\infty} \tilde{A}(0,\omega) \exp\left(i\beta_1 \frac{\Delta z}{2}\omega + \frac{i}{2}\beta_2 \omega^2 \frac{\Delta z}{2} + \frac{i}{6}\beta_3 \omega^3 \frac{\Delta z}{2} - i\omega t \right) d\omega, \quad (3.9)$$

where $\tilde{A}(0,\omega)$ is the Fourier transform of the signal $A(t)$ at the step input.

In segment (2), the relationship between the input and output signal is given by:

$$A(\Delta z, t) = A(0,t) \exp\left(i\Delta z \gamma |A(t)|^2 \right). \quad (3.10)$$

The third step is equal to the first. At this point, we can consider the next δz. It is clear that the shorter is δz, and the most accurate is the simulation, but calculation time is longer, as well. A good rule for the maximum δz choice is that its length must be much shorter than the nonlinear length $1/\gamma P$, where P is the total power launched into the fiber.

More details on the impact of other nonlinear fiber effects (as Brillouin and Raman) in optical transmission systems can be found in this book in Sect. 1.2.

Equation 3.8 describes the signal evolution in absence of polarization mode dispersion, which can be assumed as a good approach when the signal does not manifest broadening due to PMD. We remember that the pulse broadening due to PMD grows as the square root of the length.

When this condition is not verified, we have to solve two coupled equations similar to Eq. 3.8 that take into account the values of the propagation constants for the two polarization components, and the random coupling between the two modes that produces power exchange on very short distances (also less than meter). If we have a δz segment so short that we can suppose the fiber parameters independent of z, therefore, we can define two local orthogonal birefringence axis, x and y, and assume the propagation of the two fiber modes according these two coupled equations (Shen 1984; Agrawal 2007):

$$\frac{\partial A_x}{\partial z} + \beta_{1x}\frac{\partial A_x}{\partial t} + \frac{i}{2}\beta_{2x}\frac{\partial^2 A_x}{\partial t^2} + \frac{\alpha}{2}A_x$$

$$= i\gamma\left(|A_x|^2 + \frac{2}{3}|A_y|^2\right)A_x + \frac{i}{3}\gamma A_x^* A_y^2 \exp\left(-2i\Delta\beta z\right)$$

$$\frac{\partial A_y}{\partial z} + \beta_{1y}\frac{\partial A_y}{\partial t} + \frac{i}{2}\beta_{2y}\frac{\partial^2 A_y}{\partial t^2} + \frac{\alpha}{2}A_y$$

$$= i\gamma\left(|A_y|^2 + \frac{2}{3}|A_x|^2\right)A_y + \frac{i}{3}\gamma A_y^* A_x^2 \exp\left(+2i\Delta\beta z\right). \tag{3.11}$$

The effect of the random mode coupling can be taken into account with a random rotation of the birefringence axis between two following δz segments with a further random phase variation between the modes.

Also these two coupled equations can be solved with the split-step method supposing dispersive-nonlinear-dispersive segments both for the two x-y components.

3.3.3 Simulation of Other Devices

3.3.3.1 Transmitter

The generation of pulses in our experiments can be obtained by means of a soliton source or of a CW laser with external modulation. In the former case, the analytical expression of the signal is simple since given by a sec-hyperbolic function; conversely, in the latter, a more complicated expression is required.

In the case of a transmitter composed by a CW laser with a negligible linewidth and a MZM, the signal can be described in terms of power P and phase ϕ as (Lebref et al. 1999):

$$P(t) = P_0 \cos^2\left(\frac{\pi}{2}\frac{V(t)}{V_\pi}\right), \tag{3.12}$$

$$\phi(t) = K\frac{\pi}{2}\frac{V(t)}{V_\pi}, \tag{3.13}$$

where $V(t)$ is the drive voltage, V_π is the switch voltage, $K = \frac{V_A - V_B}{V_A + V_B}$ is the chirp parameter, and V_A and V_B are the peak-to-peak voltages applied to the electrodes A and B of the modulator. In the case of small modulation depth, the field at the transmitter output, in the case of Gaussian shape, can be written as:

$$E(t) = E_0 \exp\left(-\frac{1 + iK}{2}\frac{t^2}{T_0}\right). \tag{3.14}$$

For $K = 0$, we have the ideal behavior of a transmitter with Gaussian pulses.

Further descriptions for phase modulated signals will be reported in Sect. 3.4.

3.3.3.2 Optical Receiver

The model that we adopted in this book is quite simple. First of all, we only used a detection scheme only based on the direct detection. Therefore, we assume a block for the receiver composed by a PIN photodiode (see Chap. 1 for better description of optical receiver), a preamplifier, an amplifier, and an electronic filter (Agrawal 2007; Iannone et al. 1998). The current at the photodiode output is:

$$I_P = R \cdot P = \frac{\eta \cdot q}{h \cdot \nu} P, \tag{3.15}$$

where η is the quantum efficiency of the photodiode, q is the electron charge, and P is the optical power. A thermal noise contribution is also added. As electronic filter, we used a Bessel-Thomson one, as described in ITU-T G.957 recommendation.

After the electrical filtering, and before the sampling, we need to synchronize the signal with the original one in order to perform the comparison between the transmitted and the received value. In this tool, the synchronization is achieved by making the cross-correlation between the electrical received signal and the transmitted signal after its elaboration with the same receiver. After this process, we are able to distinguish the pulses corresponding to the transmitted bits. By means of sampling process, we can evaluate mean values and standard deviations to calculate the Q factor. The sampling time can be varied to optimize the Q factor.

This is the scheme for an IM-DD or OOK systems and can be used also for other transmission formats based on the direct detection as DQPSK and DPSK, as we will show in the following.

Optical Amplifier

Optical amplification is mainly based on EDFA (Agrawal 2007). However, interesting devices for amplification, especially for reduced distance, are the SOAs whose cost is lower with respect to EDFA, even though they have a higher level of noise and induce a weak pulse distortion. This section mainly deals with EDFA; however, some characteristics of the simulation of SOA are reported in Sect. 3.6.1, where the all-optical wavelength conversion in SOA is numerically investigated, and details on the simulation routines for SOA can be found elsewhere (Settembre 1997). Furthermore, investigation on the transmission in links encompassing G.652 fibers with SOA amplifications was carried out in the framework of the IST UPGRADE project.

In a first approximation (Agrawal 2007), the EDFA behavior is quite simple to describe, and the output signal can be obtained as function of the input one, by means of the following expression:

$$A_{\text{out}} = A_{\text{in}}\sqrt{G} + n_{\text{ASE}}. \tag{3.16}$$

The ASE noise, n_{ASE}, can be assumed as a Gaussian, zero mean, broadband noise with an average power given by $Fh\nu(G\text{-}1)B_{\text{A}}$, where F accounts for incomplete population inversion, h is Planck's constant, ν is the carrier frequency, G is the amplifier gain, and B_{A} is the ASE bandwidth. The value of B_{A} depends on the characteristics of the system and in particular on the optical filtering along the line and at the receiver; for instance, in absence of in-line optical filters, and if the bandwidth of the optical filter before the photodiode is narrower than the amplifier bandwidth, B_{A} coincides with the receiver optical bandwidth.

The ASE noise has several detrimental effects. First, it is responsible for energy fluctuations, which can make difficult to distinguish a "one" from a "zero" level. Second, it accumulates along the line and eventually can compete with the signal in saturating the amplifiers located further down the link, thus reducing the signal to noise ratio at the receiver. Third, it can nonlinearly interact with the signal. In the case of soliton signal, the nonlinear interaction between ASE and signal is responsible of a frequency shifting that, by means of chromatic dispersion, is converted into fluctuations of the arrival time of the pulses, referred as Gordon-Haus effect (Gordon and Hauss 1986). Moreover, in absence of ASE noise, solitons suffer a deterministic attractive or repulsive force depending on their separation and their relative phase. Due to ASE, noise phase and separation can fluctuate, resulting in a further contribution to time jitter.

It has to be pointed out that another way to achieve optical amplification is the exploitation of the Raman effect. In particular optical amplification can be achieved by launching a pump signal at the output of the fiber with a counter propagating direction (Agrawal 2007). Details on simulation of Raman amplification can be found for an example in (Eramo et al. 2003).

3.4 Simulation Models of Long-Haul Optical Transmission Systems

In this book, we report the results concerning three main kinds of optical transmission systems based on direct detection: IM-DD (or OOK), DPSK and DQPSK. The simulation of other transmission formats, and in particular for coherent systems, can be achieved with simple variations of the transmission and detection routines. We take into considerations the most general link that uses all the main optical in-line components, i.e. the long-haul link.

3.4.1 IM-DD Systems

In Fig. 3.1, we illustrate a complete IM-DD system. We suppose the link composed by n-spans. Each span is composed by a fiber segment, with an amplifier (EDFA) to compensate the fiber loss and DCU to compensate the chromatic dispersion of the fiber segment. As we will show in the following, the transmission performance can be improved by the introduction of a partial precompensation, by using a prechirp device, as reported in Fig. 3.1.

3.4.2 RZ-DPSK

The DPSK system is an interesting modulation formats since it does not need the coherent detection, even if it is based on a phase modulation. DPSK needs a predecoder to achieve the differential code, and therefore, the initial bit sequence is transformed in another bit sequence. For long-haul transmission systems, RZ pulses better behave with respect to NRZ signals, and therefore, we investigate on a DPSK with RZ pulses (RZ-DPSK). The RZ-DPSK transmitter can be thought as a system where a sequence of RZ pulses are phase modulated according to the expression:

$$s(t) = \sum_{k} A s_0(t - kT_B) \exp\left(i\pi\phi_k(t)\right), \tag{3.17}$$

where A is the amplitude, $s_0(t - kT_B)$ is the shape of the RZ pulse, $\exp\left(i\pi\phi_k(t)\right)$ is the phase modulation, and in the DPSK can be ± 1.

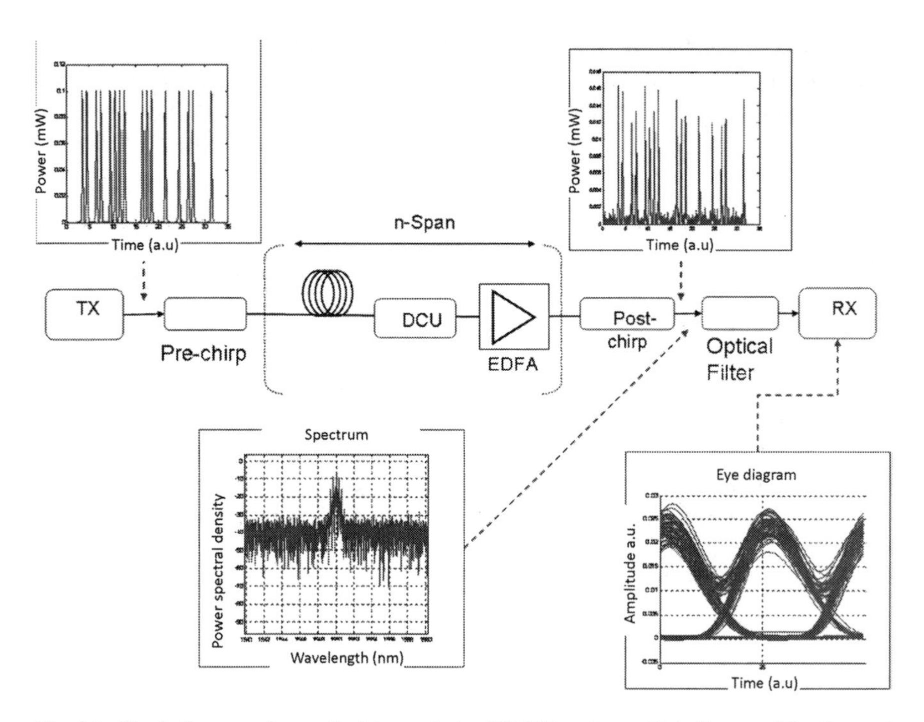

Fig. 3.1 Block diagram of an optical transmission IM-DD system with in-line amplification and chromatic dispersion compensation. Pre- and postcompensation of the chromatic dispersion is also present

In Fig. 3.2, we illustrate the behavior of $\phi_k(t)$, based on raised cosine shape.

In Fig. 3.3, we illustrate the effects of the phase modulation of Fig. 3.2 on a train of Gaussian pulses.

The complete scheme of the DPSK system is reported in Fig. 3.4; in particular, the receiver is based on the combination of the signal with its replica delayed by a bit time T and detected by a balanced receiver. The shape of the input signal, the output spectrum and the eye diagram are illustrated in Fig. 3.4, as well.

For such a system the Q factor can be given by:

$$Q = \frac{\bar{I}_0 - \bar{I}_\pi}{\sigma_0 + \sigma_\pi}, \tag{3.18}$$

where $\bar{I}_0$ and $\bar{I}_\pi$ are the average of the current samples associated to the symbol "1" (phase equal to zero) and to the symbol "0" (phase π), respectively, and σ_0 and σ_π are the standard deviations.

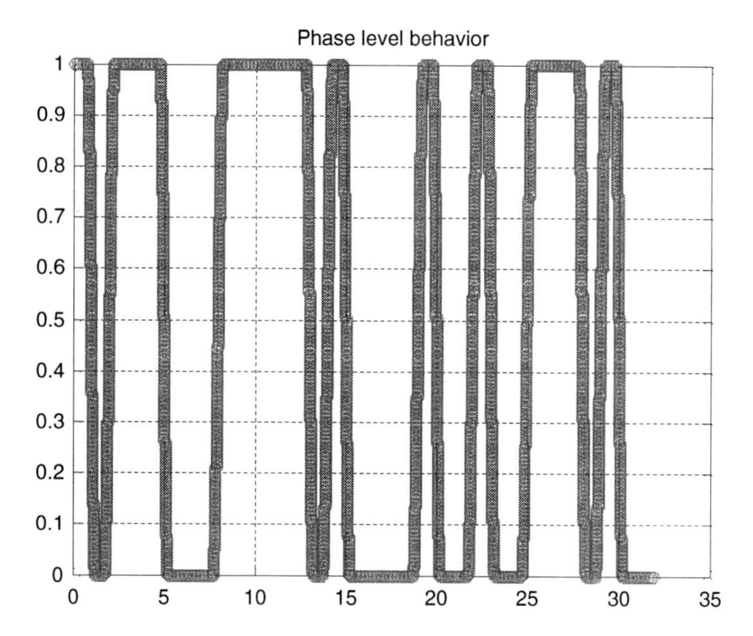

Fig. 3.2 Typical behavior of the phase $\phi_k(t)$

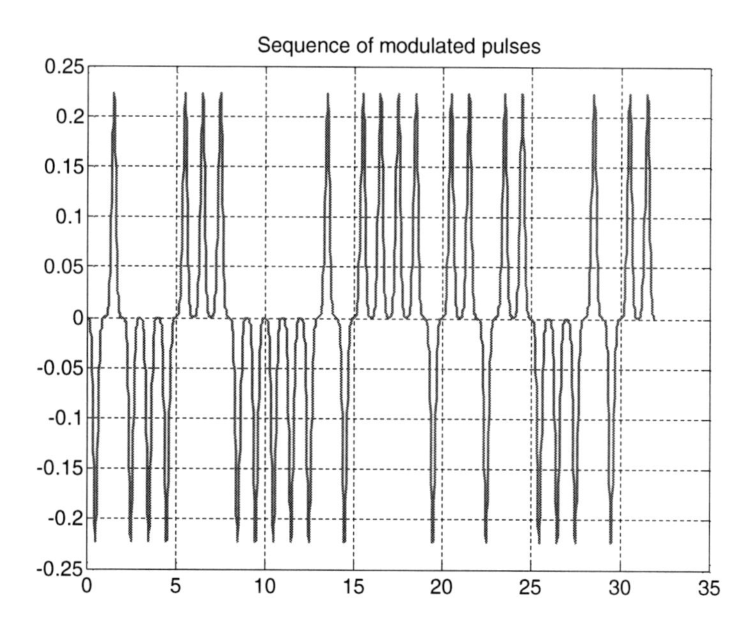

Fig. 3.3 Sequence of RZ Gaussian pulse with phase modulation

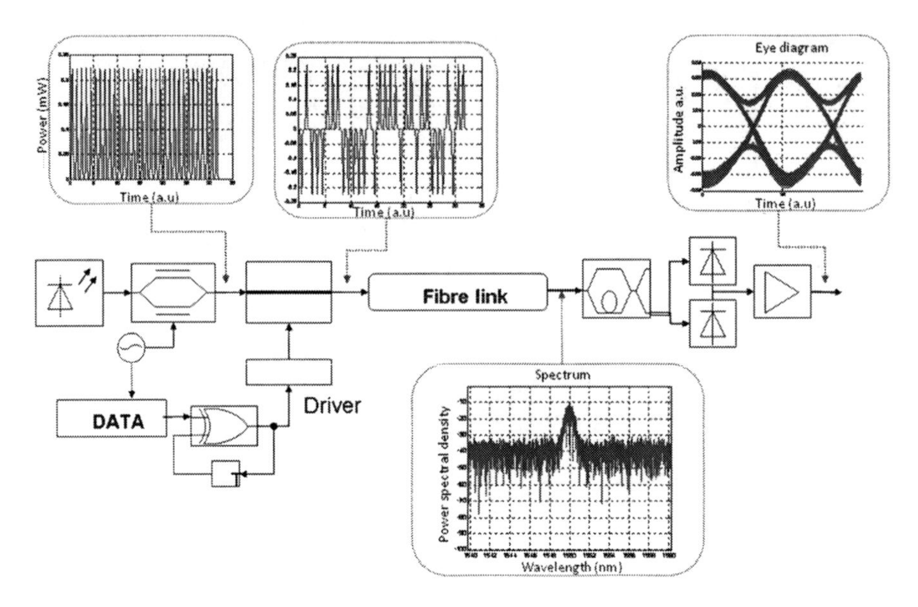

Fig. 3.4 Scheme of a RZ-DPSK system

3.4.3 RZ-DQPSK Systems

DQPSK is assumed as one of the most interesting modulation-detection technique since in a bit time four symbols can be transmitted with a consequent enormous advantage of dispersive impairments reduction. Furthermore, the direct detection scheme allows us to have a quite simple receiver with respect to a coherent one. As in the DPSK case scheme, RZ pulses have the advantage to have a better control of the interplay between dispersive and nonlinear effects with respect to NRZ formats.

The RZ-DQPSK signal can be written as:

$$s(t) = \sum_k A s_0(t - kT_B) \exp\left(i\pi\left(\frac{2\phi_k(t) + 1}{4}\right)\right), \tag{3.19}$$

where $\exp\left(i\pi\left(\frac{2\phi_k(t)+1}{4}\right)\right)$ is the phase of the pulse sequence, and it assumes the following values: $\pm\frac{1}{\sqrt{2}} \pm j\frac{1}{\sqrt{2}}$. In Fig. 3.5, we report the typical behavior of the term $\phi_k(t)$, based on raised cosine function.

The real and imaginary part of the signal $s(t)$ is shown in Fig. 3.6.

In Fig. 3.7, we report a complete scheme for RZ direct detection DQPSK system (RZ-DD-DQPSK). The detection scheme is based on a double balanced detection. The signal at the output of the fiber channel is subdivided in two branches, and in each one, the signal is combined with its replica delayed of bit time with a phase difference equal to $\pi/4$. In such a way in each arm, we have two DPSK signals in phase and quadrature.

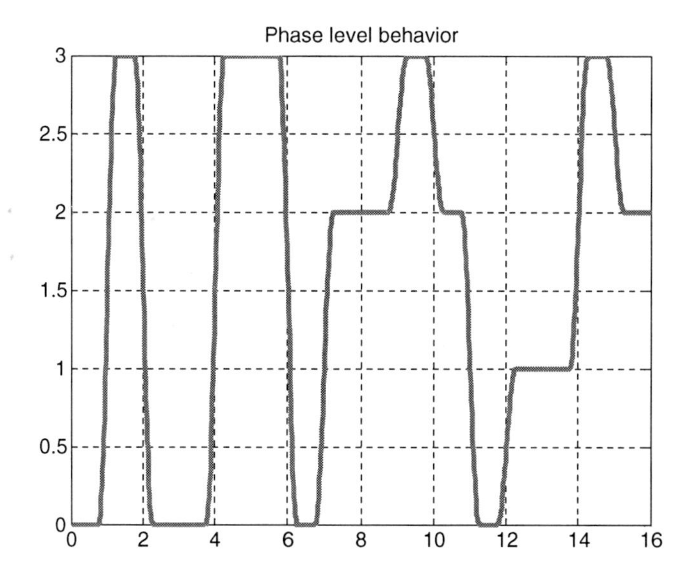

Fig. 3.5 Behavior of $\phi_k(t)$

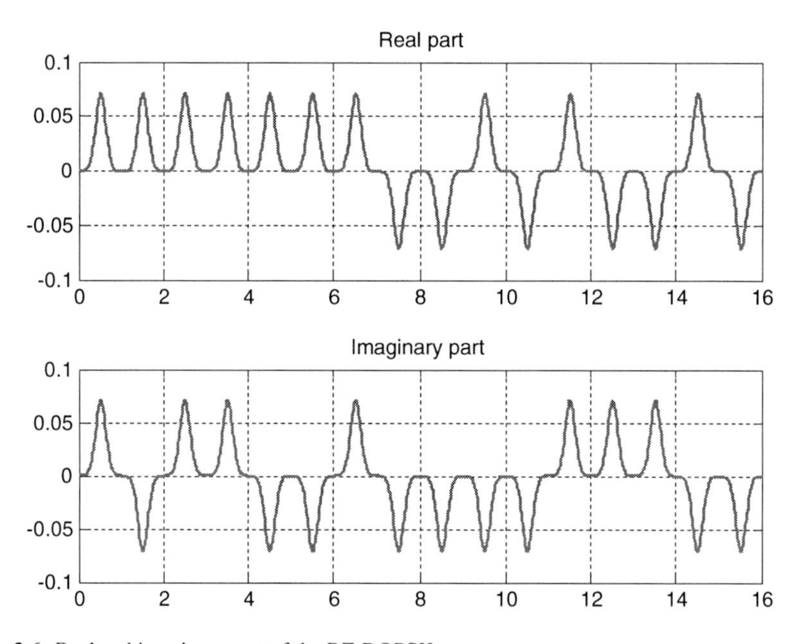

Fig. 3.6 Real and imaginary part of the RZ-DQPSK

To better understand the performance of phase modulated systems, it is interesting to report the sampled field of each bit in a *Q-I* plane, as reported in Fig. 3.8 (Tabacchiera et al. 2010) where we show the case of 2,048 bits (1,024 symbols)

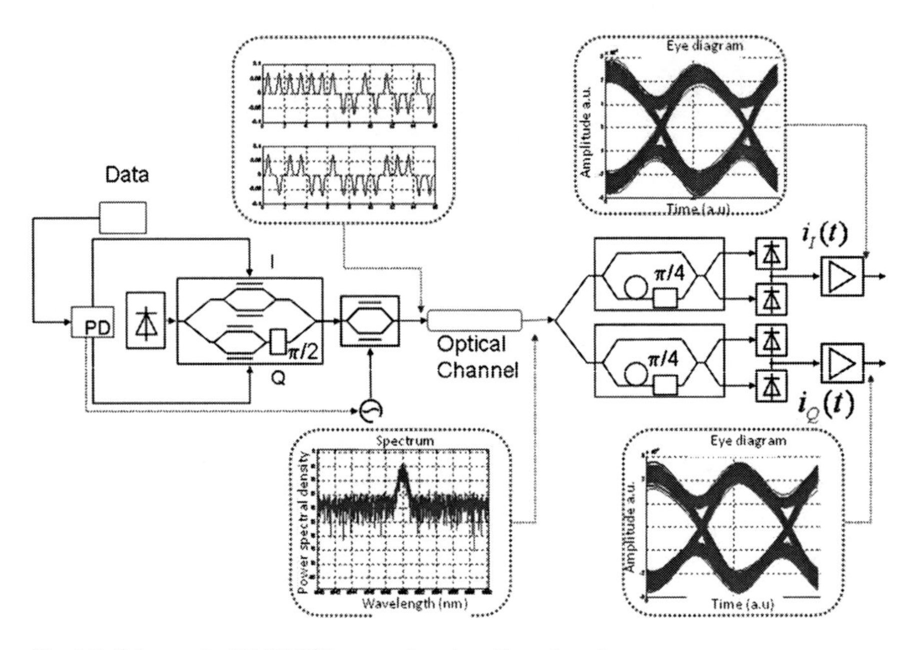

Fig. 3.7 Scheme of a RZ-DQPSK system based on direct detection

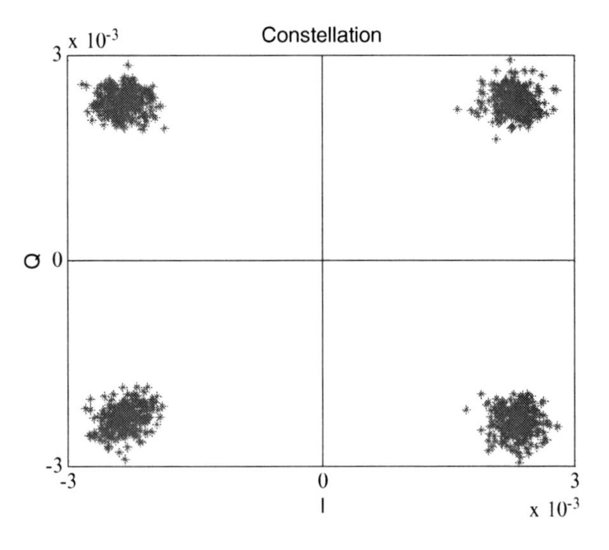

Fig. 3.8 Constellation of the bits in a *Q-I* plane for RZ-DQPSK system with direct detection

in the case of ideal propagation. The four bits for each pulse are well distinct, permitting an easy detection of the transmitted pulse.

For RZ-DQPSK system with direct detection, for each arm, we can evaluate a Q factor as defined in Eq. 3.1. Such operation corresponds to the evaluation of the distance $D1$ and $D2$ in Fig. 3.9. It has to be pointed out that such a Q factor

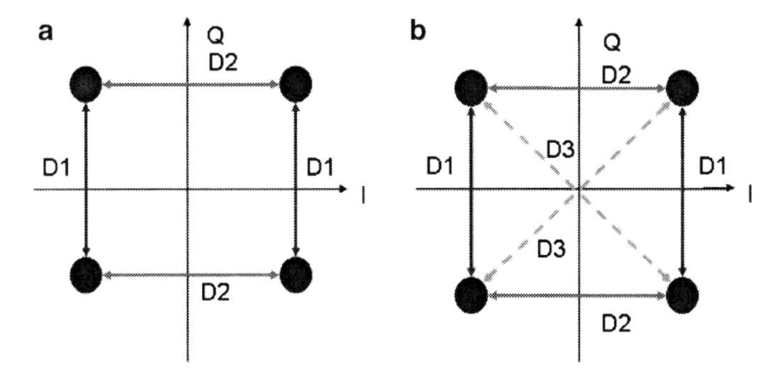

Fig. 3.9 Distances among the bits in RZ-DQPSK with direct detection

evaluation method is based on the detection of the points in the *Q-I* plane located at the lateral part of the square (with length $D1$ and $D2$). Better performance could be achieved considering detection schemes that look for points located at the end of the diagonal $D3$ (see Fig. 3.9), and in principle, it is possible with the coherent detection.

3.5 Simulation Results for Long-Haul Transmission Systems

In this section, we report a comparison among different transmission systems (IM-DD, DPSK, and DQPSK) in different propagation conditions, looking for the best solution, according to the parameters of the link under consideration.

3.5.1 The Limits of the Chromatic Dispersion

In order to clarify the propagation limits in fiber links, we start with an investigation about links with G.652 fibers. The maximum propagation distance, only due to the dispersive effects, in case of a source with high spectral quality (i.e. DFB laser), can be evaluated in a simple way (Elrefaie et al. 1988) for IM-DD system according to the equation:

$$L_{\max} \approx \frac{1}{10\,|\beta_2|\,R^2}, \tag{3.20}$$

where $R$is the signal bit rate.

In the following, we report the transmission behavior of IM-DD systems at different bit rates, in a links with amplifier spacing equal to 100 km, and the fiber parameters are reported in Table 3.1.

Table 3.1 Link parameters

Fiber	EDFA
$\alpha = 0.25$ dB/km	$G = 25$ dB
$D = 16$ ps/(nm·km)	$F = 6$ dB
$\gamma = 1.3\ (W \cdot km)^{-1}$	

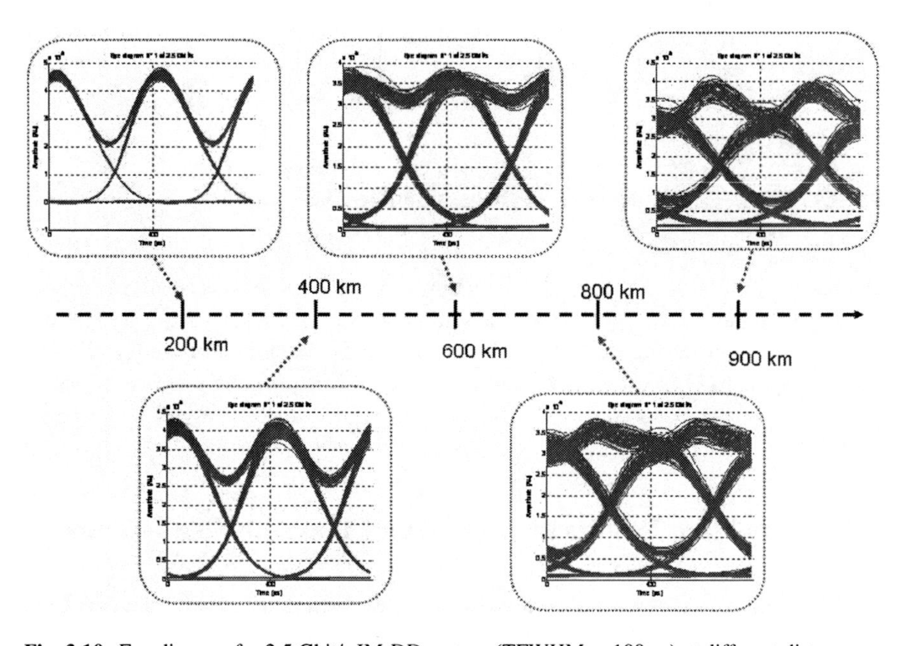

Fig. 3.10 Eye diagram for 2.5 Gbit/s IM-DD system (TFWHM = 100 ps) at different distances

In Fig. 3.10, we report the eye diagram of a 2.5 Gbit/s IM-DD system operating in a link with the parameters reported in Table 3.1. At 2.5 Gbit/s, as shown by Eq. 3.20, the transmission can be achieved with good performance up to 800 km, while high degradation is evidenced after 900 km, where the eye gets closer.

Another investigation on the same link is reported in Fig. 3.11 for a 10 Gbit/s IM-DD system (TFWHM = 25 ps). In such a case, the maximum propagation distance is limited to 60 km (less than one amplifier spacing).

If we consider a transmission at 40 Gbit/s, in the same propagation conditions, the maximum transmission distance is 2.5 km.

For DPSK systems, the limitation in terms of maximum propagation distance is almost the same of IM-DD systems. Conversely, a wide better improvement in terms of maximum propagation distance can be achieved in case of DQPSK systems, since the signal bandwidth can be one half lower, and as a consequence, according to Eq. 3.20, the maximum propagation distance can be four times longer. This is confirmed by the Fig. 3.12, where we report the eye diagram evolution for a 10 Gbit/s RZ-DQPSK system.

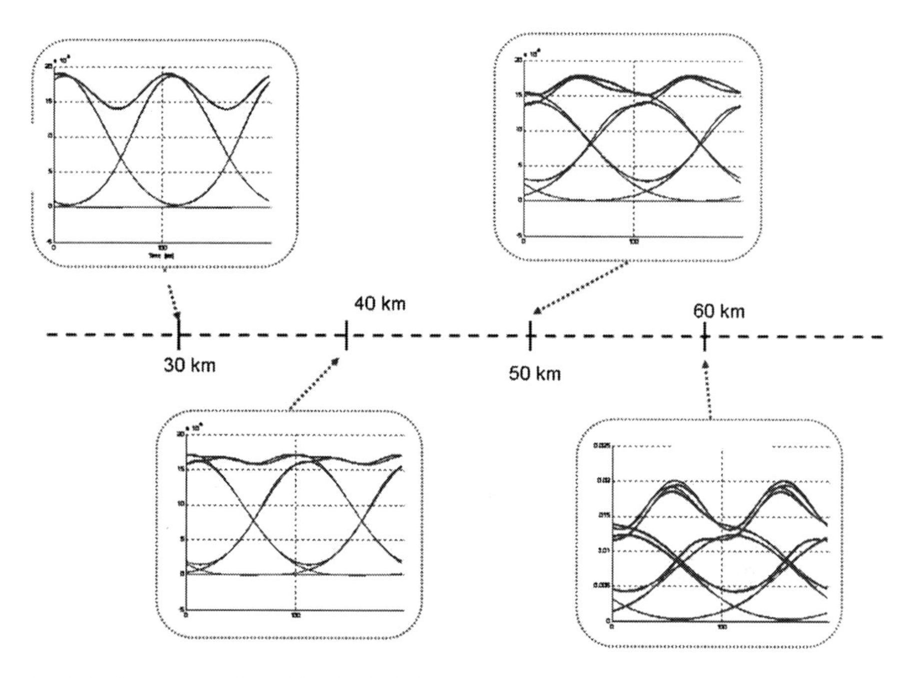

Fig. 3.11 Eye diagram for a 10 Gbit/s IM-DD system

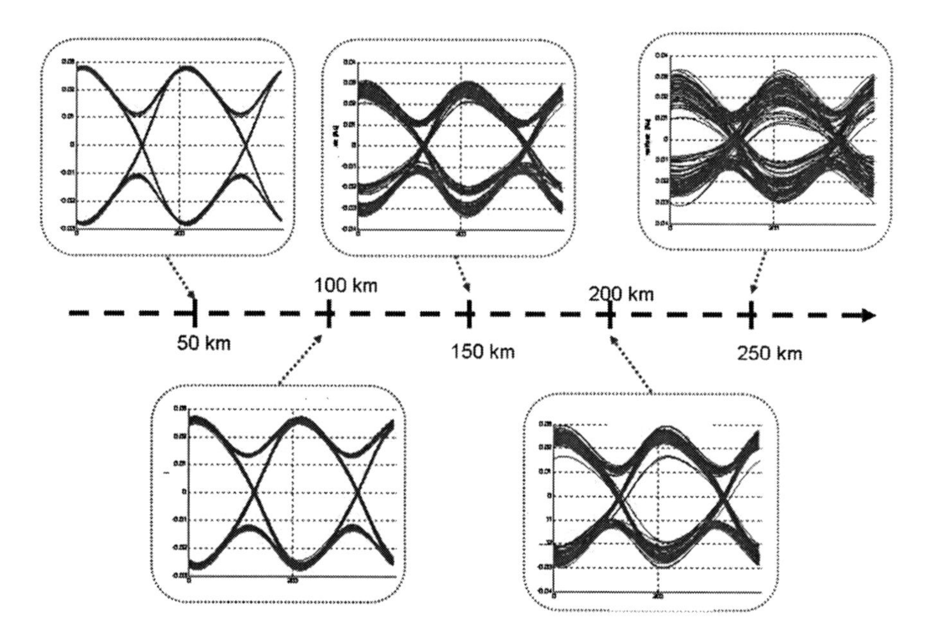

Fig. 3.12 Eye diagram of a DQPSK system at 10 Gbit/s in a link encompassing a G.652 fiber

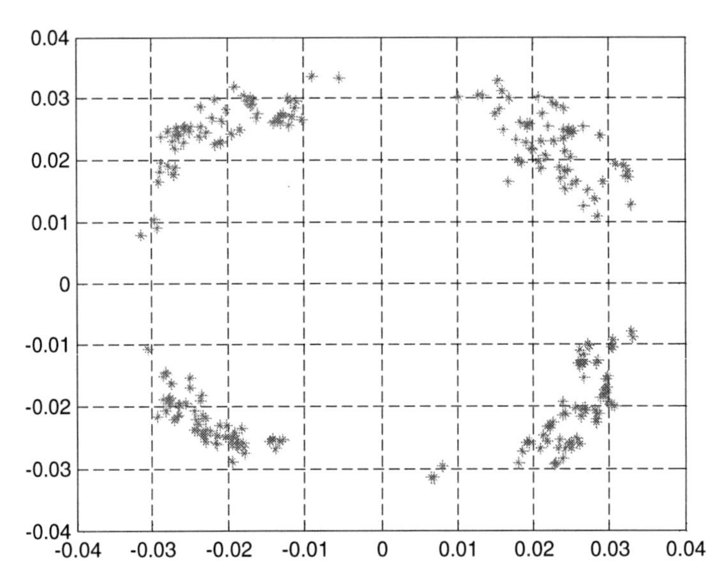

Fig. 3.13 Constellation of the points representing bits of a RZ-DQPSK at the output of a link 250 km long

To better understand the DQPSK signal evolution in Fig. 3.13, we report the constellation of points representing the symbols.

Results show the difficulties to achieve very long distances in links encompassing G.652 fibers without using chromatic compensation techniques. The most simple and efficient chromatic dispersion method is the dispersion management that is based on chromatic dispersion compensation with linear devices. In principle, another method to compensate the chromatic dispersion in the anomalous regime of the fiber is based on the use of the Kerr nonlinearity. As reported in Sect. 1.2.2, when some conditions on the pulse duration and peak power are satisfied, pulses can propagate for enormous distances without varying the shape; this is the well-known case of soliton propagation. Soliton systems were much appealing at the end of the last century since they seemed the correct solutions for transoceanic systems. Surely soliton systems have contributed to the development of the optical transmissions; however, some problems, mainly due to the severe control of link parameters that were required, heavily limited the diffusion of the solitons at advantage of other propagation regimes as the dispersion management. Furthermore, it has to be pointed out, that in some case also the soliton propagation can take advantage from the dispersion management. Soliton is still a fundamental concept of optical communication, and therefore, we dedicate the following section.

3.5.2 *Chromatic Dispersion Compensation by Means of Kerr Nonlinearity: The Soliton Case*

In the last decades, among the most interesting investigations about the fiber propagation, the ones on the soliton effects (Hasegawa and Tapper 1973; Mollenauer et al. 1980) surely received very much interest, and the literature is full of fundamental outputs. Soliton propagation is also one of the methods to check the reliability of the simulation tool since it allows us to check if all the fiber effects are correctly taken into account.

We remember that soliton propagation occurs if:

1. The signal has a sec-hyperbolic shape

$$A(t) = \frac{2}{e^{t/T_s} + e^{-t/T_s}}. \tag{3.21}$$

   However, it has to be pointed out that the soliton is a stable solution, and therefore, soliton propagation occurs also for an input similar shape, but different from the sec-hyperbolic one, as for an instance for a Gaussian pulse.
2. The peak power has to satisfy the following relationship

$$P_k = \frac{\beta_2}{\gamma T_s^2}. \tag{3.22}$$

3. If fiber loss is relevant, in-line optical amplification is necessary; in such a case, in order to have a very weak shape variation of the soliton along the link, it is necessary to have the amplifier spacing much shorter than the soliton period ($\frac{\pi}{2}\frac{T_s^2}{\beta_2}$).

In the following, we report some cases of a sequence of Gaussian pulses (the shape of a Gaussian pulse is quite similar to a soliton one) in a link without loss with the following characteristics: $D = 16$ ps/nm/km, $\gamma = 1.3\ (\text{Wkm})^{-1}$, and $L = 20$ km. Concerning the IM-DD system, parameters are $T_{\text{FWHM}} = 5$ ps (that corresponds to a $T_s = T_{\text{FWHM}}/1.763 = 2.83$ ps) and bit time of 25 ps (corresponding to a 40 Gbit/s IM-DD system). The step is 10 m, and the bit sequence was chosen in order to analyze the nonlinear pulse interaction, and therefore, we introduce long "0" and "1" sequences.

In Fig. 3.14, we report a sequence of pulses at the link output for different input peak powers. For these IM-DD parameters, the corresponding soliton peak power is 1.97 W; however, the input pulse requires a bit higher pulse to modify its Gaussian shape to reach the soliton one, and in fact simulations show that the pulses maintain the same shape and duration (soliton propagation) for an input power between 2.2 and 2.4 W (case b). For lower power, the pulses show wide temporal broadening (case a). For $P = 3.5$ W, the power is too high that the pulses show narrowing;

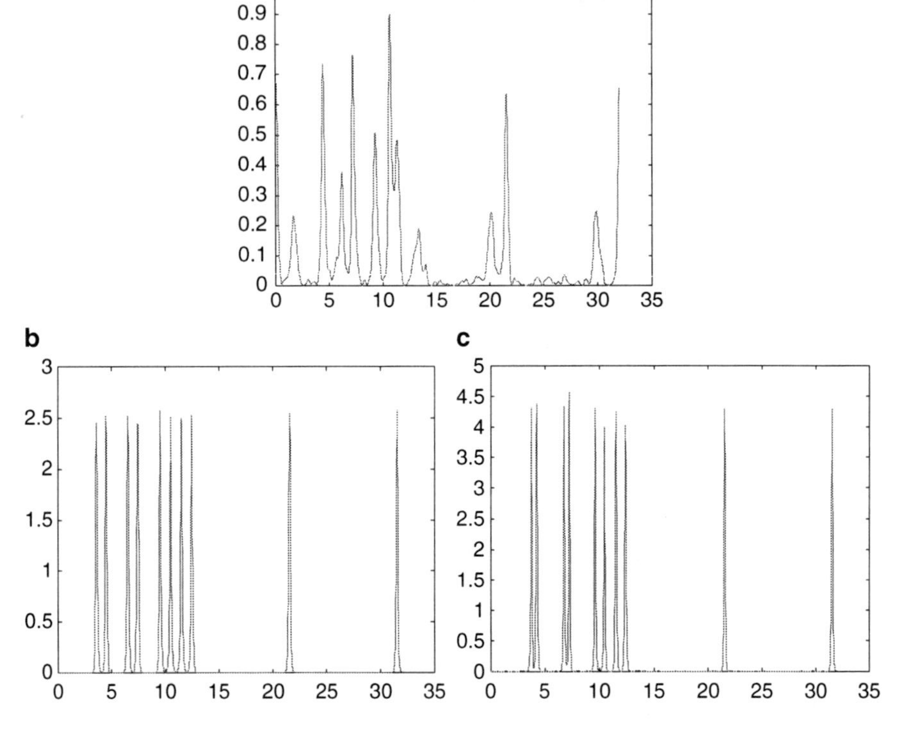

Fig. 3.14 Gaussian pulse propagation in a G.652 fiber link 20 km long in absence of loss for three different input peak powers: case (**a**) $P = 1.4$ W, case (**b**) representing the soliton case, and case (**c**) $P = 3.5$ W

in fact, the pulse time duration tends to reduce to satisfy the input power according to Eq. 3.22. The figure also shows that when the pulses are close, they tend to attract each other.

Figure 3.14 clearly illustrates the importance of soliton propagations in telecommunications. However, fiber loss and other link imperfections introduce limits in soliton diffusion, and this is the cause of the limited number of installations of soliton systems.

Let us see a short overview about the soliton characteristics in different propagation conditions.

3.5.2.1 Fiber Loss

Fiber loss reduces peak power, and therefore the pulse, trying to satisfy Eq. 3.22, would tend to increase its duration with a consequent pulse broadening. It has to be pointed out that making a comparison between linear and nonlinear regime, the Kerr effect tends to limit the pulse broadening with respect to an only dispersive regime, and therefore, in anomalous region, Kerr effect always tends to improve the system performance. This behavior was also shown in case of NRZ systems (Matera 2000). Obviously, optical amplifiers permit to compensate the fiber loss; however, the pulse would tend to increase its duration in each fiber span. It was demonstrated that introducing an extra gain to the amplifiers, with the aim to induce a pulse narrowing in first part of the propagation in each span (higher peak power in Eq. 3.20), it is possible to have an adiabatic propagation of the pulses for transoceanic distances. This adiabatic propagation is obtained when the amplifier spacing is much shorter than the soliton period, and this occurs in the presence of pulses that do not have a time duration too short, and therefore, it regards system with a limited bit rate and mainly for transoceanic links where the chromatic dispersion can be carefully controlled and maintained with low value (lower than 1 ps/nm/km). However, it has to be pointed out that such low chromatic dispersion is not suitable for very long WDM systems since FWM effects can be much detrimental (Chraplyvy 1990).

3.5.2.2 ASE Noise

Optical amplifiers produce noise (ASE) that reduces SNR; in case of solitons, ASE noise induces further degradation due to its nonlinear Kerr interaction with the signal. If we analyze the signal in the frequency domain, the ASE noise fluctuation can cause the variation of the baricenter of the signal; in fact, since in soliton regime, the signal always tends to maintain its Sech shape (also in the frequency domain), and therefore, a power fluctuation due to ASE noise is compensated by the nonlinear effect along the line with a variation of the signal baricenter, in such a way, the power variation due to the ASE is included in the signal. Such variation of the signal baricenter, due to the presence of chromatic dispersion, induces a time variation of pulse arrival with a consequent jitter of the pulses (Gordon-Haus effect) (Gordon and Haus 1986). Such jitter can be reduced with in-line filtering, and the improvement can be much higher if a frequency shift of the filter is made at each amplifier position (sliding filters) (Hasegawa and Kodama 1991; Mollenauer et al. 1992). A soliton sliding filter is the best method to achieve the highest performance; however, it results a quite complex technique, and this is one of the reasons that limited the soliton transmission installation, inducing a preference for systems based on dispersion management. To better illustrate the role of the soliton propagation in a regime of constant low chromatic dispersion, in Fig. 3.15, we report a comparison in terms of maximum capacity among different NRZ and soliton systems. The link is composed by fiber span 100 km long having a chromatic dispersion equal to 1 ps/nm/km. For each distance and system, we looked for the best Q performance at different bit rates, varying the input peak power. By increasing the bit rate, the Q factor will tend to decrease, due to increased penalty. From these type of results the maximum bit rate can be derived and is the one corresponding to $Q > 6$.

Such a figure clearly shows the advantages of the soliton transmission, especially in the regime with in-line sliding filters.

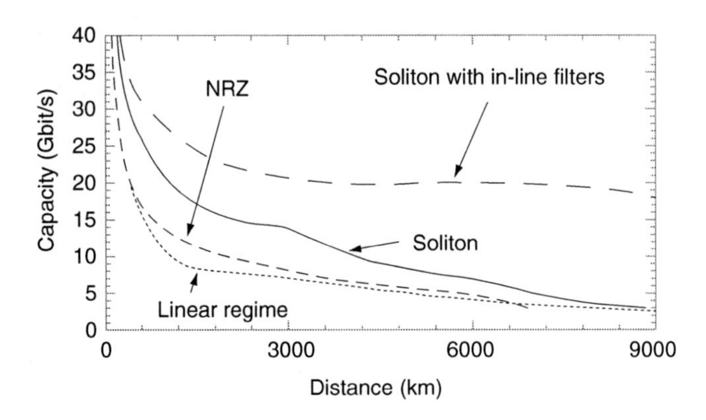

Fig. 3.15 Comparison of long-haul optical transmission systems

3.5.2.3 Polarization Mode Dispersion

Pulse broadening due to PMD could destroy soliton propagation. However, the pulse splitting due to PMD can be limited by the attraction that the soliton pulse manifests. As demonstrated in (Menyuk 1988; Mollenauer et al. 1989), if some power conditions are satisfied no pulse broadening is observed, and therefore soliton propagation can be also seen as a method to compensate PMD. In (Matera 1995), it was also shown that the nonlinear Kerr effect in the anomalous dispersion can limit PMD broadening effect also for NRZ signals.

3.5.2.4 Chromatic Dispersion Fluctuation and Dispersion Management

Stable soliton propagation can be achieved even in presence of small chromatic dispersion fluctuation, if it occurs on distances much shorter than the soliton period. In presence of strong chromatic dispersion fluctuation, the pulse can have local variations, even if they can periodically maintain the same shape and duration when the average chromatic dispersion is the same in each amplifier spacing. In particular, an interesting topic is the soliton propagation in the dispersion management regime, and several investigations are present in literature. However, especially in of strong chromatic dispersion fluctuations, it is difficult to assume the existence of sokiton pulses in their normal conditions since the temporal broadening, even if temporarily, can be quite high. This topic of nonlinear dispersion management regime has been broadly studied and as a rule of thumb the best performance can be achieved in conditions corresponding to the zero average chromatic dispersion with precompensation.

3.5.2.5 WDM Systems

Several studies and experiments have shown the importance of WDM soliton systems, even though the control of the in-line parameters is quite complex. Also for WDM systems, a general approach based on the dispersion management is much better.

3.5.3 Simulations for Links with Dispersion Management

Conventional single mode fibers, G.652, are characterized by a large amount of chromatic dispersion. Consequently, optical pulses with a time width of few picoseconds undergo a strong broadening that takes place in some hundreds of meters (Zitelli et al. 1999). This propagation regime has already been analyzed in literature and is usually divided in two branches: strong DM (Srivastava et al. 1996) that is based on periodic dispersion compensation, and propagation of highly dispersive pulses (HDP), which considers the dispersion compensation all at the end of the system (Zirngibl 1991).

In (Zitelli et al. 1999), it was shown that the best configuration to achieve the highest capacity with G.652 fibers was the technique of the postcompensation with prechirp (or precompensation). It is based on the periodic compensation of GVD, by means of a fiber grating at each span out (before the optical amplifier) with a precompensation (and an opposite postchirp) that can be achieved with the introduction of a suitable fiber grating having a small value of dispersion at the link input. In such a work, it was found that the optimum value of the prechirp is given by:

$$\frac{\exp\left(\alpha\left|\frac{T_0^2 C_{\mathrm{pr}}}{\beta_2}\right|\right)-1}{\alpha\left|\frac{T_0^2 C_{\mathrm{pr}}}{\beta_2}\right|}\frac{C_{\mathrm{pr}}^2}{\left(1+\frac{C_{\mathrm{pr}}^2}{4}\right)^{3/2}}=0.77, \tag{3.23}$$

where α is the loss, T_0 is the time duration of the pulse and C_{pr} is the value of the prechirp.

In Fig. 3.16, we report the considered dispersion compensation schemes. Inset (a) refers to the periodic post dispersion compensation (by locating the DCU unit at the output of the amplifier, the precompensation scheme is obtained), which can be eventually modified by adding an initial prechirp and a final postchirp, as shown in inset (b). Inset (c) corresponds to the all at the end dispersion compensation scheme.

In Fig. 3.17, we report the value of the Q factor versus input average power for different configurations, assuming the time duration of the pulses equal to 5 ps. In particular, we report the following configurations: PC, PR, and PC plus prechirp; ideal means a link in which the only fiber effect that is present is the ASE noise and END the configuration in which the dispersion is all compensated at the end of the link.

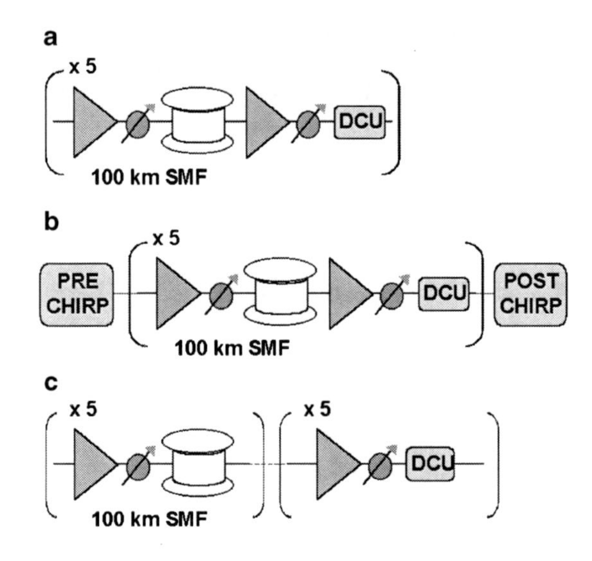

Fig. 3.16 Scheme of the link. DCU stands for Dispersion Compensation Unit. Variable optical attenuators are placed after the amplifiers to change input power. Inset (**a**) periodic postcompensation scheme. Inset (**b**) periodic postcompensation with prechirp and postchirp. Inset (**c**) all at the end compensation

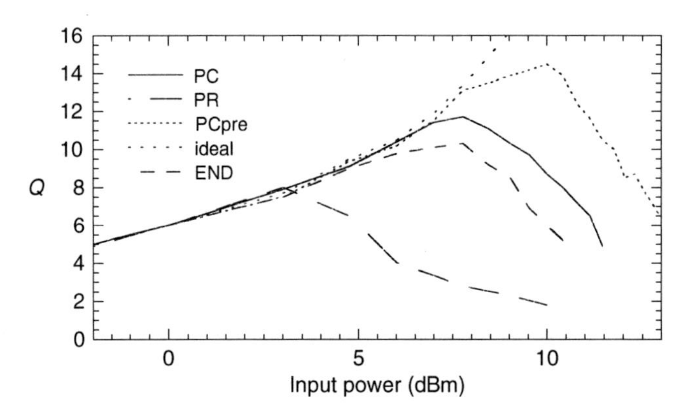

Fig. 3.17 Q factor versus input power for different DM configurations

According to Eq. 3.23, the value of the prechirp is 100 ps^2, and it was also verified by simulations that it was the best value of the prechirp. Looking at the results, we can see that the best result, as foreseen in (Zitelli et al. 1999), is the one due to the scheme of PCpre, while the worst is the one due to the precompensation. How already explained in (Zitelli et al. 1999), we can see the detrimental role of the Kerr nonlinearity for the PR curve that mainly manifests in terms of nonlinear pulse interaction, and in curve PC as SPM, while in the case PCpre, the impact of Kerr nonlinearity is strongly limited as shown by the fact that for power lower than 7 dBm, the behavior is quite ideal.

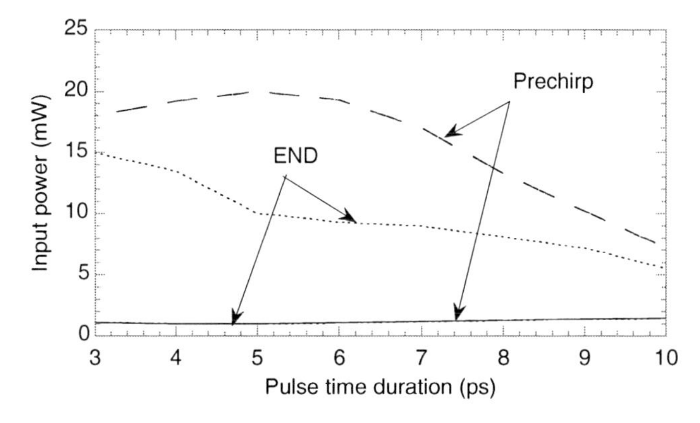

Fig. 3.18 Maximum (*upper*) and minimum (*lower*) input power for dispersion scheme based on the prechirp plus postcompensation (PCpre) and all at the end (END)

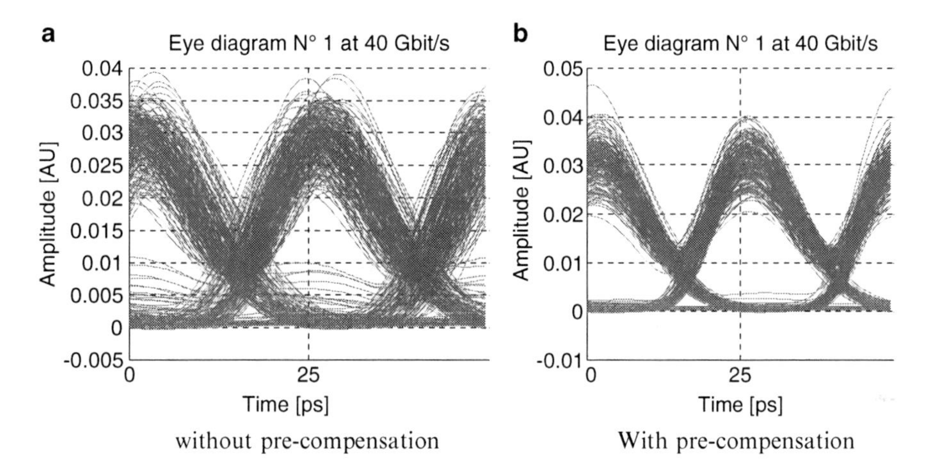

Fig. 3.19 Eye diagram for PC and PC-PRE (**a**) without precompensation and (**b**) with precompensation

Also the behavior of the curve END is very interesting, in fact the performances are lower with respect to the case of PCpre and PC, but the scheme is much simpler since the compensation is performed all at the end of the link.

Making a comparison between PCpre and END, we can see that for PCpre, the interval of acceptance ($Q > 6$) for the input power is between 0.4 and 13 dBm, while for END, between 0.4 and 10 dBm. In Zitelli et al. (1999), the behavior of such acceptance power interval, assuming different time duration of the pulses, was reported for the scheme PCpre and END (Figs 3.18 and 3.19).

The results show that the best configuration can be obtained when $T_0 = 5$ ps. It is important to underline that the difference in terms of performance for the two schemes (PCpre and END) gets less relevant if, from Eq. 3.23, Cpr tends to zero (short pulses) or to infinity (long pulses).

Some considerations have to be done comparing the results of the curves PR and END; in fact in both the cases, the pulses propagate along the link in conditions of very large time broadening. In fact in PR, the pulses have already been totally broadened at each span input, while in END, the pulse is broadened after some kilometer of the first span, and the GVD is compensated only at the end of the link. However, some differences are present in two cases: In PR, the way of interacting among the pulses can be considered as a periodical process, and the detrimental effects tend to accumulate; conversely, in END, the interaction among the bits change in each span, and as a consequence, the detrimental effects tend to average.

3.5.3.1 Theory on Dispersion Management for Phase Modulated Signals

In this short description, we will try to simply illustrate the nonlinear behavior in the pulse propagation in links with dispersion management for different modulation formats.

If the transmitted pulses are so short that the spacing between dispersion compensating stations greatly exceeds the dispersion length, a large number of pulses belonging to the same channel overlap during propagation, and their nonlinear interaction leads to transmission impairments. When all transmitted pulses are in-phase, each overlapping pulse contributes to the total nonlinear distortion with an in-phase and an out-of-phase component (Mecozzi et al. 2000). The in-phase component affects the pulse shape more than the out-of-phase component because it adds linearly to the pulses intensity, whereas the out-of-phase component adds quadratically. If the dispersion profile in the span between dispersion compensating stations is carefully designed, the in-phase component of the nonlinear distortion added in the first half of the span is exactly compensated by an opposite amount of distortion in the second half, therefore reducing the total in-phase component to zero (Tabacchiera et al. 2010). In the ideal case in which the line loss is compensated by an equal amount of distributed gain, this symmetric condition is achieved when dispersion compensation is equally split between the beginning and the end of the span (Tabacchiera et al. 2010). This effect is clearly shown in Fig. 3.20, where we report the points corresponding to the electrical field (real and imaginary part) of each pulse of a DPSK signal (512 pulses) after 700 km of the link in the absence of loss for an input peak power of 0.02 W, without prechirp in the left, with prechirp on the right. The squeezing effect due to the prechirp allows us to well distinguish between "−1" and "1."

When lumped erbium amplifiers are used, then the amount of precompensation that reduces the in-phase component to zero is smaller because pulse attenuation leads the final part of the span to contribute less than the initial to the nonlinear interaction (Tabacchiera et al. 2010). These results have been used in recent years as guiding rules for a careful optimization of the dispersion profile in high bit rate links based on intensity modulation.

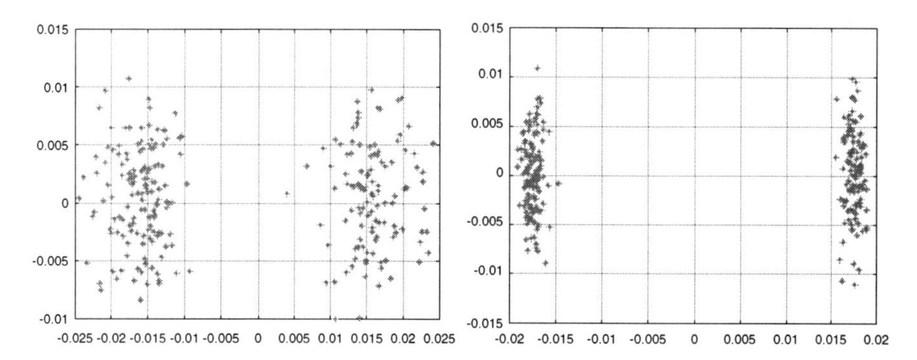

Fig. 3.20 Field of each pulse of a DPSK signal (512 pulses) after 700 km of the link in the absence of loss for an input peak power of 0.02 W, without prechirp in the *left*, with prechirp on the *right*

While the above results were originally derived for intensity modulation only, they apply to antipodal DPSK as well. With this modulation scheme, the pulses are either in-phase or out-of-phase. Being the nonlinear interaction phase independent, it is possible to show that the qualitative picture that we have given above does not change. Therefore, the dispersion profiles that reduce the nonlinear impairments to a minimum in system based on intensity modulation will reduce to a minimum the nonlinear impairments in systems based on DPSK as well.

The scenario is different with DQPSK systems; in fact, the nonlinear interaction takes place between pulses whose relative phases are multiple of 90°, and the squeezing effect of Fig. 3.20 cannot help to distinguish the four levels. Therefore, no benefit in adjusting the predispersion of the line to maximize the link capacity. The nonlinear impairments of DQPSK are therefore higher than the nonlinear impairments of DPSK, because there is no way to eliminate the in-phase component of the nonlinear distortion by acting on dispersion profile.

This fact, while explaining the generally recognized property that DQPSK is more prone to nonlinear impairments than DPSK, it does not imply that the capacity of DQPSK systems is lower than that of DPSK systems. Nonlinear impairments lead to an amplitude jitter on the detected eyes that, with a large number of interacting pulses, may be approximated with a Gaussian noise. This extra noise produces a logarithmic decrease of the capacity with the variance of the jitter, according to Shannon's celebrated capacity formula (Tabacchiera et al. 2010). At the same time, however, the number of signal degrees of freedom of DQPSK is double than DPSK. The proportional doubling of the linear channel capacity more than compensates for the logarithmic decrease caused by the higher nonlinear noise, thus making the capacity of DQPSK systems generally higher than the corresponding capacity of DPSK systems, also in the presence of nonlinear impairments.

These affirmations will be confirmed in the following by means of numerical simulations for different optical high bit rate transmission systems.

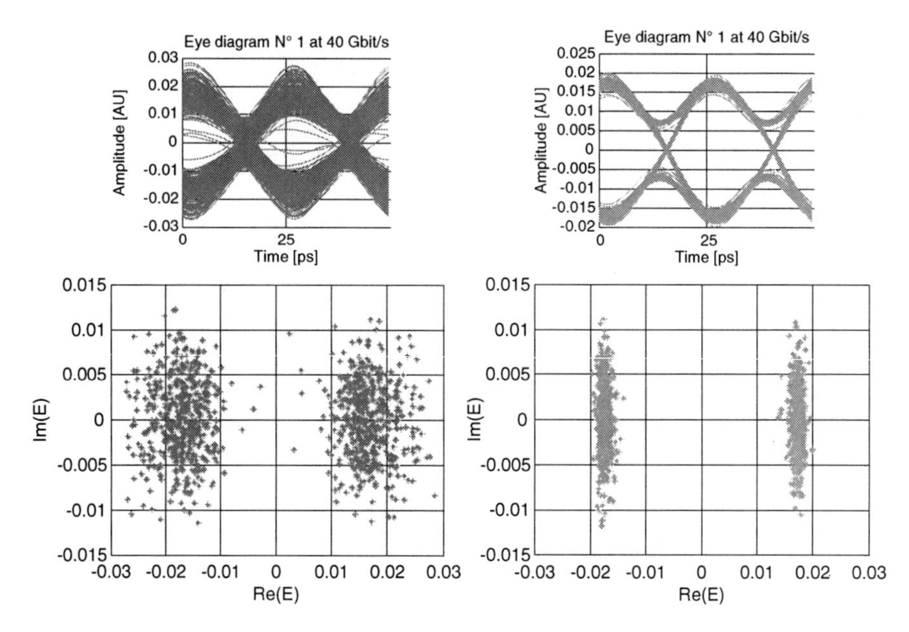

Fig. 3.21 Eye diagram and field distribution (real and imaginary part) of a 40 Gbit/s RZ DPSK system in a link encompassing G.652 fibers with dispersion management after 700 km. *Left* is without precompensation, and *right* is with precompensation (78 ps/nm/km)

3.5.3.2 Results of Precompensation Effect on DPSK and DQPSK

In this paragraph, we report some simulation results regarding DPSK and DQPSK in dispersion management links both with and without precompensation using links having the same parameters, as reported in (Zitelli et al. 1999; Tabacchiera et al. 2010) and in Table 3.1.

As described in Sect. 3.5.3.1, we foresee that the use of the precompensation can be very useful in terms of system performance for the DPSK case, and in fact in Fig. 3.21, we can see the benefit of the precompensation both in terms of pulse field distribution (real and imaginary part) and in terms of eye diagram. We used Gaussian pulse with a TFWHM = 5 ps.

In Fig. 3.22, we report the behavior of the Q factor versus input average power, assuming different distances for a 40 Gbit/s RZ DPSK with PC technique.

Conversely, in case of 80 Gbit/s RZ-DQPSK (Fig. 3.23), we can see how the effect of the precompensation is negligible, even though using the same pulse duration the bit rate of RZ-DQPSK is double with respect to DPSK.

Also in terms of Q factor, as shown by the comparison between the Figs. 3.24 and 3.25, the difference between the cases with and without precompensation is negligible (Tabacchiera et al. 2010).

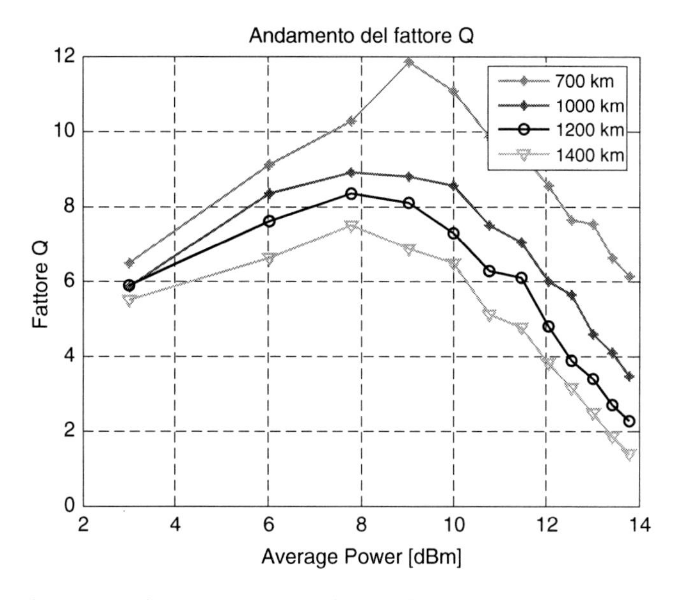

Fig. 3.22 Q factor versus input average power for a 40 Gbit/s RZ-DPSK with PC technique

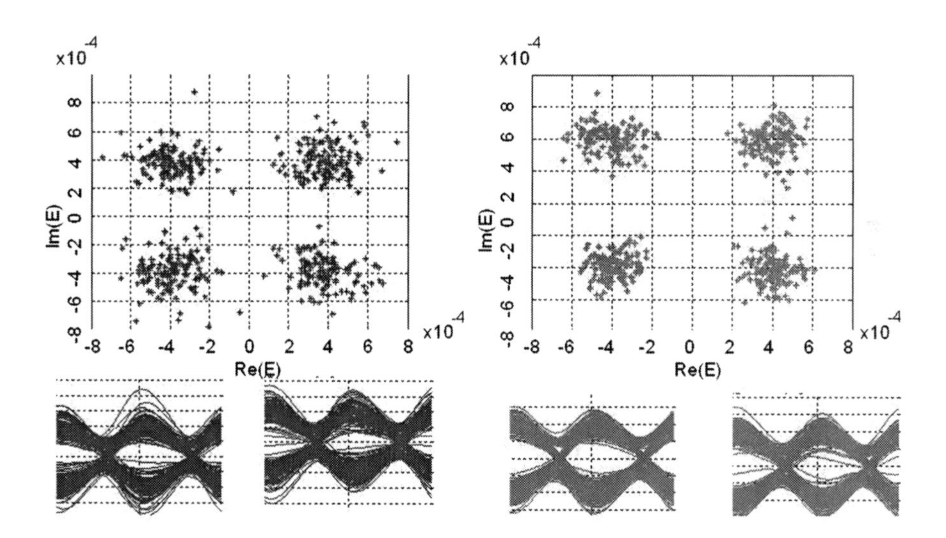

Fig. 3.23 Field distribution (real and imaginary part) and eye diagrams (for both the arms of the receiver) of a 40 Gbit/s RZ-DQPSK system in a link encompassing G.652 fibers with dispersion management after 700 km. *Left* is without precompensation, and *right* is with precompensation

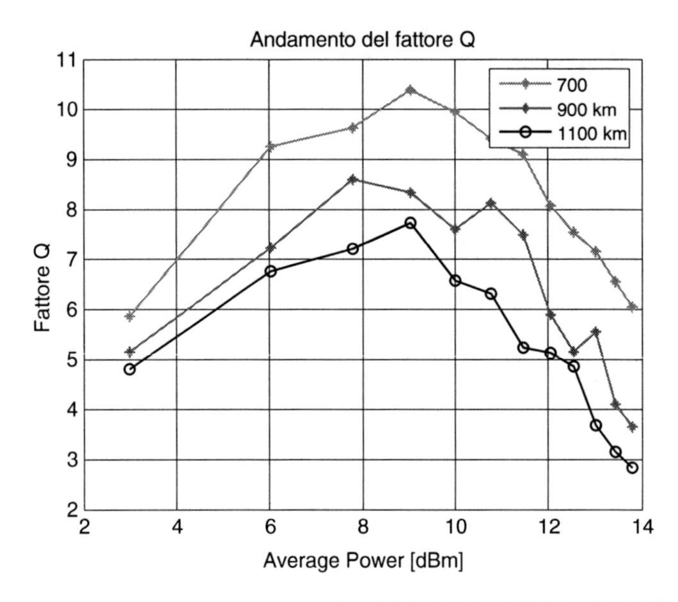

Fig. 3.24 *Q* factor versus input average power for a 80 Gbit/s RZ-DQPSK with POC technique

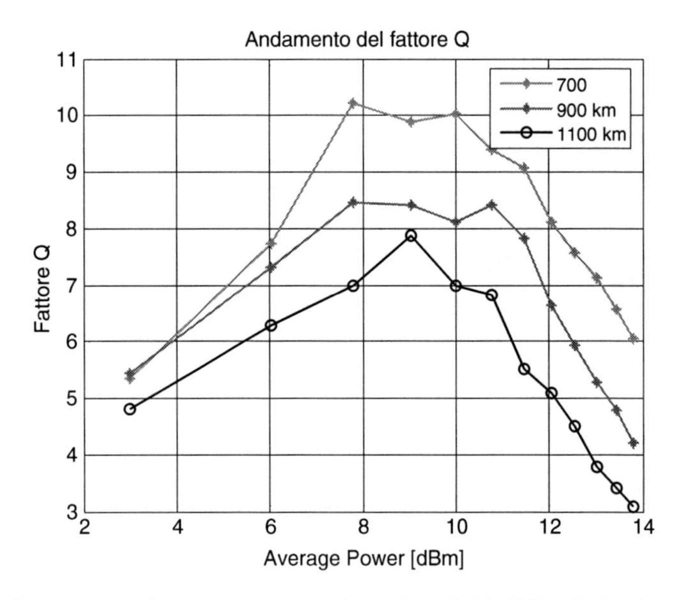

Fig. 3.25 *Q* factor versus input average power for a 80 Gbit/s DQPSK with POC technique and precompensation

3.5.3.3 Comparison at 100 Gbit/s

By means of Fig. 3.26, we illustrate the comparison in terms of capacity among a 40 Gbit/s RZ-DPSK, a 40 Gbit/s IM-DD, and a 80 Gbit/s RZ-DQPSK system for a link 700 km long with and without precompensation (78 ps/nm/km).

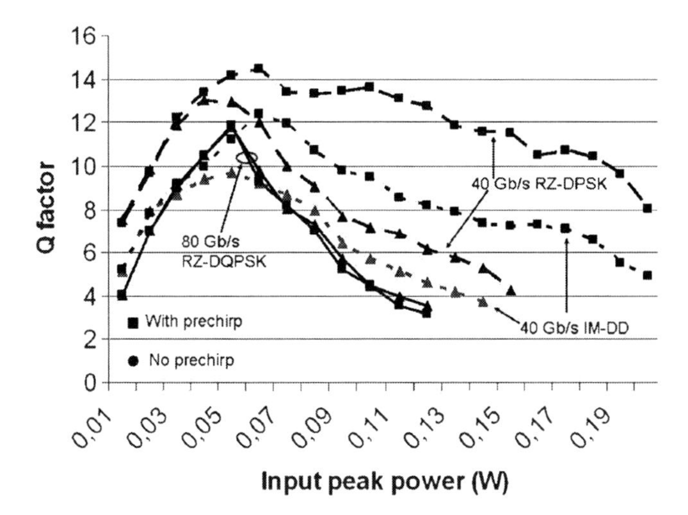

Fig. 3.26 Q factor versus input power for 40 Gbit/s IM-DD, 40 Gbit/s DPSK, and 80 Gbit/s DQPSK

All the three systems have the same symbol time equal to 25 ps and pulse duration equal to 5 ps. As reported in Fig. 3.26, the improving effect of prechirp is much evident for IM-DD and DPSK, while its use for DQPSK is null, as foreseen from the reported theory.

The figure also shows that DPSK manifests advantages with respect to IM-DD in terms of performance, which can be simply explained by the absence of jitter effect for the presence of a pulse in each time slot. Conversely, DQPSK allows us to double the capacity in the same link conditions.

Figure 3.26 already shows the importance of RZ-DQPSK for 100 Gbit/s long transmission systems. Furthermore, to operate at such a bit rate, the time duration of pulse for IM-DD and DPSK must be reduced. We investigate on 100 Gbit/s IM-DD, RZ-DPSK, and DQPSK both in absence and in the presence of prechirp, looking also for the best pulse duration. At 100 Gbit/s, the advantages of the DQPSK are evident both with respect to IM-DD and DPSK. For instance, in Fig. 3.27, we report the comparison between RZ-DQPSK and IM-DD at 100 Gbit/s. For each curve, we indicate a number that is the FWHM time duration of pulses, and a letter that indicates the presence of precompensation (s is without and p is with precompensation).

To conclude the investigation on the 100 Gbit/s DQPSK, we report some results on the numerical simulation of WDM nx100 Gbit/s RZ-DQPSK systems. In particular in Fig. 3.26, we report the maximum distance versus number of RZ-DQPSK channels at 100 Gbit/s (TFWHM = 5 ps). In the inset of the figure, we report the spectrum of the 16 channels system.

Concerning the role of the PMD in 100 Gbit/s systems, a deep investigation was carried out in the framework of the WP15 FP7 BONE project, and the results showed that the influence of PMD is negligible for value lower than 0.1 ps/$\sqrt{\text{km}}$;

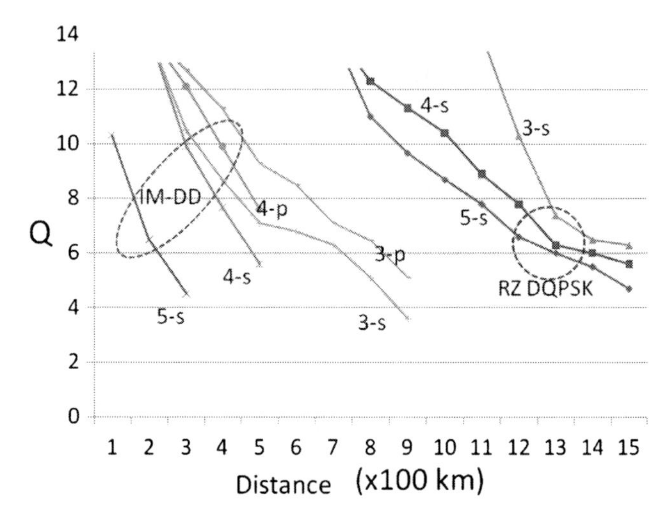

Fig. 3.27 Q factor for 100 Gbit/s IM-DD and DQPSK

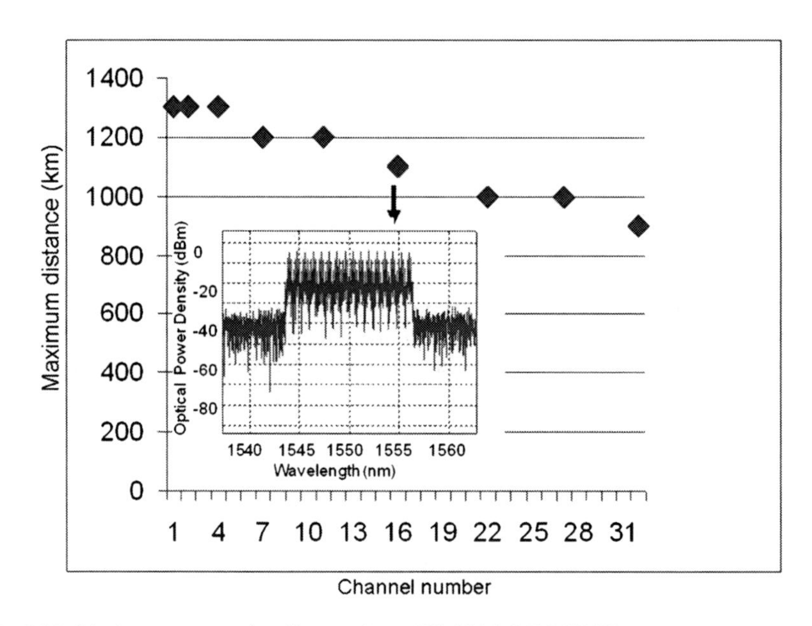

Fig. 3.28 Maximum propagation distance for nx100 Gbit/s RZ-DQPSK systems

furthermore, it has to be taken into account for distances longer than 500 km when it assumes values higher than 0.5 $ps/\sqrt{km}$, and for distances longer than 300 km in case of values higher than 1 $ps/\sqrt{km}$.

Simulations of Fig. 3.28 show how a Tbit/s capacity can be transmitted over 1,000 km in G.652 fibers by means of the dispersion management technique by using the RZ-DQPSK format with a channel bit rate equal to 100 Gbit/s.

In conclusion, we have shown that prechirp induces benefits when the information is carried on binary formats as IM-DD and DPSK, while no effect is manifested in DQPSK, even though the advantages of DQPSK are wide with respect to the other formats even without the contribution of precompensation.

3.6 Propagation in the Presence of In-line All Optical Device

Optical networks are composed by other devices as OXC, OADM, ROADM, and other novel devices will be introduced as the ones that will be based on optical processing as AOWC and All Optical 3 R regenerations (Agrawal 2007). As a consequence, the simulations have to take into account all the devices for optical processing.

3.6.1 Reconfigurable Optical Add-Drop ROADM

Concerning the links with ROADM, several simulations and investigations were carried out in the framework of the FP7 BONE project, and the simulation plan was based on different platforms (e.g., VPI photonics). In Fig. 3.29, we report a scheme of ROADM.

ROADM is mainly composed by optical filters that are fabricated to introduce very weak degradations. For an instance in the FP7 BONE project, a numerical investigation on the presence of ROADM in links for nx100 Gbit/s systems was carried out [Deliverable Year 1-WP15 FP7 BONE Project], and here we report their characteristics, in the scheme in Fig. 3.29:

- **ROADM.** MUX/DEMUX characteristics:
    - Filter type: pass band
    - Transfer function: Gaussian
    - Filter order: 3

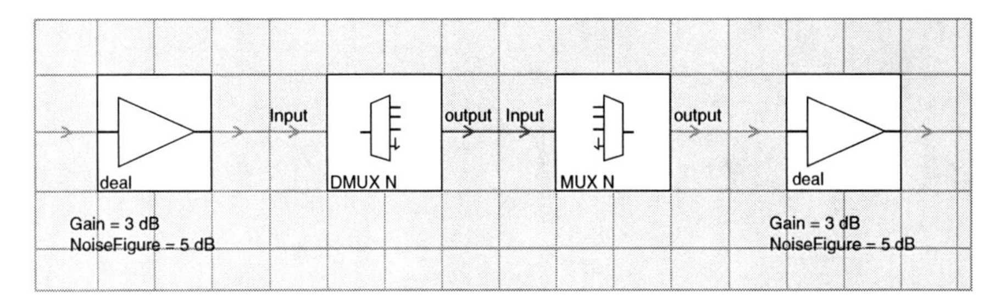

Fig. 3.29 Schemes for ROADM

- Insertion loss: 3 dB
- Bandwidth: 0.75*bit rate
- Channel spacing:
  - 40 Gbps: 50 GHz
  - 100 Gbps: 100 GHz

The results reported in the Deliverable Year 1-WP15 FP7 BONE Project show that the ROADM with the listed characteristics can induce a degradation that can reduce the maximum propagation distance up to 20%.

3.6.2 All-Optical Wavelength Conversion Based on SOA

Concerning the AOWC, several models have been introduced, and the difficulty of the model depends on the adopted device. For instance, by using the four-wave mixing process in optical fibers, the simulation can be obtained by means of the split-step method. Furthermore, due to the optimum behavior of an optical fiber as AOWC, especially operating at the zero dispersion wavelength (in this case, the AOWC can be assumed as ideal), the converted signal can be simply simulated by only inverting the sign of the imaginary part of the input signal.

Wavelength conversion can be also obtained by exploiting nonlinear processes in SOA, and one of the most interesting in terms of simplicity and efficiency is the FWM in SOA. Simulations on wavelength conversion based on FWM in SOA were obtained by using the numerical routine described in par. 2.5 of D211 of IST UPGRADE deliverable that is based on the theory reported in (Tabacchiera et al. 2010). Such a routine is based on the solution of a set of ordinary differential equations that allow us to evaluate the gain. In such a model, the input-output relation for field is given by:

$$E_{\text{out}}(t) = E_{\text{in}}(t) \exp\left\{ \frac{1}{2} g_m \left[P_{\text{in}}(t) + i\phi P_{\text{in}}(t) \right] \right\}. \tag{3.24}$$

With $g_m(t) = h_N(t) + h_{\text{SHB}}(t) + h_{\text{CH}}(t)$ and $\phi(t) = -\frac{1}{2}\alpha_N \left[h_N(t) - g_0 \right] - \frac{1}{2}\alpha_T h_{\text{CH}}$ and the h functions are obtained by solving the ordinary differential equations:

$$\begin{aligned}
\frac{dh_N}{dt} &= -\frac{h_N}{\tau_s} - \frac{1}{P_s \tau_s} \left[G(t) - 1 \right] P_{\text{in}}(t) + \frac{g_0}{\tau_s} \\
\frac{dh_{\text{SHB}}}{dt} &= -\frac{h_{\text{SHB}}}{\tau_1} - \frac{\varepsilon_{\text{SHB}}}{\tau_1} \left[G(t) - 1 \right] P_{\text{in}}(t) - \frac{dh_{\text{CH}}}{dt} - \frac{dh_{\text{N}}}{dt} \\
\frac{dh_{\text{CH}}}{dt} &= -\frac{h_{\text{CH}}}{\tau_h} - \frac{\varepsilon_{\text{CH}}}{\tau_h} \left[G(t) - 1 \right] P_{\text{in}}(t).
\end{aligned} \tag{3.25}$$

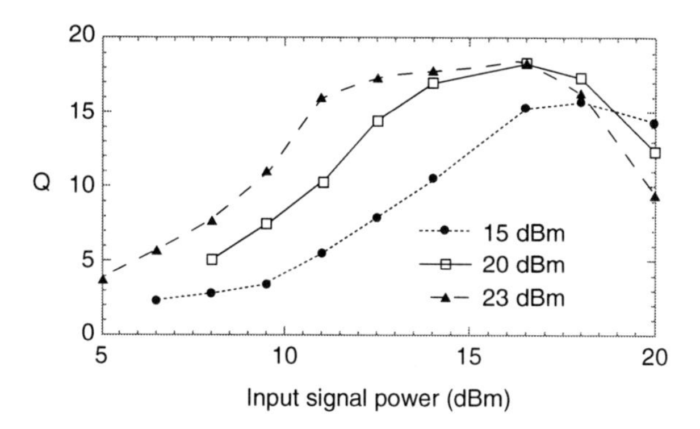

Fig. 3.30 Q factor at the output of the SOA AOWC vs input signal power for different pump powers. The frequency spacing between signal and pump is 200 GHz

Here $g_0$is the unsaturated gain, τ_s is the carrier lifetime, τ_1 is the spectral hole burning lifetime, τ_h is the carrier heating lifetime, P_{in} is the power at the input SOA, ε_{CH} is the nonlinear gain compression due to the carrier heating, and ε_{SHB} is the nonlinear gain compression due to the saturation power P_S.

The SOA ASE noise is taken into consideration by introducing an equivalent source of noise at the input of the ideal device.

Several tests of the SOA routine are reported in (Matera et al. 2005) where the numerical results were compared with results about a wavelength conversion obtained by means of a FWM in a SOA.

The SOA parameters used in that work were: saturation power equal to 13 dBm, unsaturated gain 27 dB, carrier lifetime 100 ps, carrier heating lifetime 700 fs, and noise factor $n_{sp} = 5$. We assume single polarization propagation, but the presence of the two fiber modes, with polarization evolution along the line, can be taken into account by considering a polarization independent scheme with two equal SOAs, as described in (Martelli et al. 1999).

As an example in Fig. 3.30, we report the Q factor versus input signal power for a 40 Gbit/s IM-DD system after the wavelength conversion for different values of the pump power. The values of the Q factor show a degradation for high power due to SPM and XPM in SOA; in fact, the growth of the Q factor for input power higher than 13 dBm is lower with respect to the one that we should expect by simply considering the growth of the SNR; furthermore, for power higher than 16 dBm, a Q decreasing was also manifested.

In Fig. 3.31, we report the Q factor versus average signal power at the input of a link 200 km long with the same characteristics of dispersion management, as reported in Sect. 3.3, with a SOA AOWC located in the middle of the link. We assume an EDFA in the AOWC with a gain to launch a signal power at the coupler (C) in front of the SOA equal to 14 dBm, while the pump power was considered equal to 15 and 20 dBm. We compare the system performance with and without the presence of the SOA AOWC. The limitations due to the AOWC in the transmission performance were quite evident; however, by optimizing the signal power and the pump power (20 dBm), good performance can be achieved in a wide signal power interval at link input. Further results shown in Matera (2005) illustrated the ways

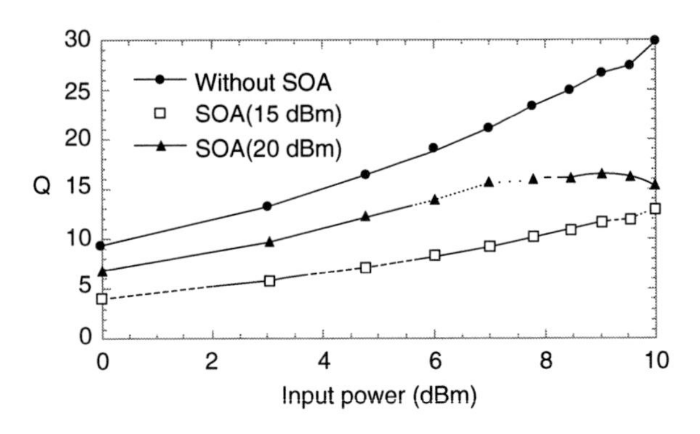

Fig. 3.31 *Q* factor vs. input power in a dispersion management link 500 km long, encompassing G.652 fibers, in the presence and in the absence of SOA AOWC

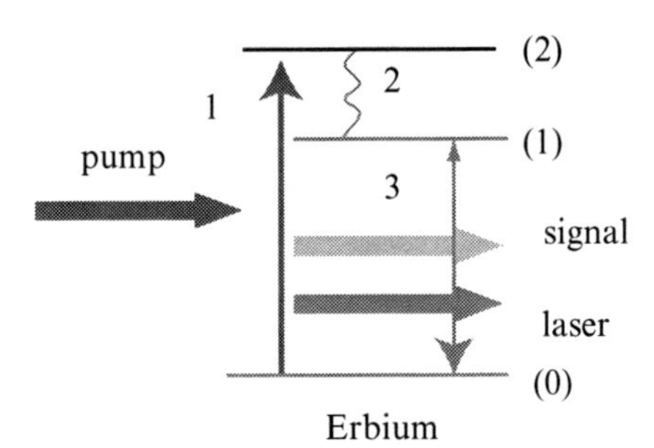

Fig. 3.32 Three-level erbium system

to use the AOWC based on FWM in SOA in 40 Gbit/s links 500 km long, and therefore, we can assume that AOWC based on FWM in SOA can be used for long links at 40 Gbit/s.

3.6.3 Optical Gain Clamped Erbium-Doped Fiber Amplifier Model

In the next subsection, we will investigate on the performance of burst signal amplification. In order to investigate the robustness of optical gain clamping technique to stabilize the erbium-doped fiber amplifier gain (Zirngibl 1991), we need the EDFA dynamic model for clamped and unclamped configuration.

The physical behavior of an optical gain clamped amplifier may be described through a laser model. The two-level energy system for erbium is a simplified representation of the optical gain clamped EDFA. According to the three-level system shown in Fig. 3.30, the pump – a high power light beam – excites the erbium ions included in the core of the fiber to a higher energy state: level 2 in Fig. 3.32. Due to instability, the ions immediately drop to level 1 with lifetime of τ_{21}. This lifetime is neglected, and the system is analyzed as a two-level system. When the incoming photons from the signal at a different wavelength meet the excited erbium

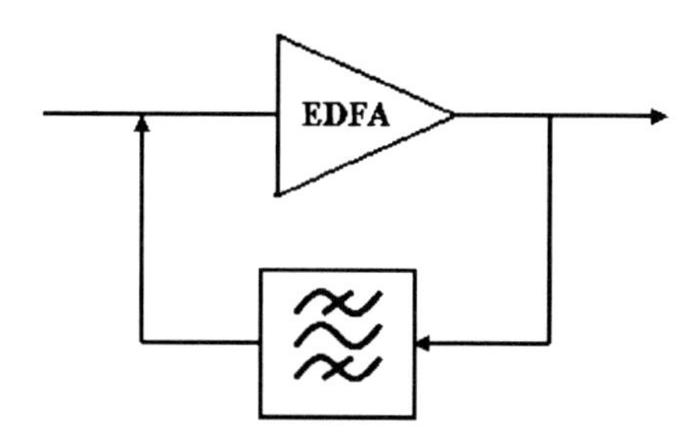

Fig. 3.33 EDFA in all-optical laser ring configuration

ions, the erbium atoms give up some of its energy in the form of additional photons and return to the initial state: level 0. The photons are exactly in the same phase and direction of the incoming photons from the signal.

The EDFA in an all-optical feedback stabilization configuration is shown in Fig. 3.33. This configuration is also called laser ring, and the lasing wavelength used to clamp the amplifier gain is defined by the narrowband filter in the optical feedback loop. Alternatively, one could use a linear cavity with narrowband FBGs to act as a mirror. This layout was first investigated to demonstrate the advantages of short cavity (Ennser et al. 2005a, b). Indeed, this is particularly advantageous when combined with short active medium as the case of EDWA (Ennser et al. 2005a, b). Other configurations were investigated using FBGs and variable optical attenuators to reduce the SHB offset (Ennser et al. 2005b) and later to avoid lasing wavelength contra-propagating and bidirectional operation of the OGC optical amplifier (Ennser et al. 2006; Della Valle et al. 2006). For the sake of modeling the OGC-EDFA we focus on an all-optical feedback ring configuration for stabilizing the gain of the erbium-doped fiber amplifier. A simple and more comprehensive way to model optical gain-clamped erbium-doped fiber amplifier (EDFA) dynamics is to describe it as a perturbed laser system (Ennser et al. 2005a, b). Here we neglect the Spectral Hole Burning (SHB) effects and the Amplified Stimulated Emission (ASE) noise due to its small contribution. For simplicity we focus on the erbium-doped system with negligible dissipative process. The 2-level ($I_{15/2}$ and $I_{13/2}$) rate-equation system is expressed on laser parameters:

$$\frac{d\Delta}{dt} = -\frac{N_{\mathrm{Er}} + \Delta}{\tau_{\mathrm{Er}}} - \frac{2}{V_a}\sum_{i=1}^{N} F_{\mathrm{i}}^{S,P}\left(G_{S,P}\right) - \frac{2}{V_a}F^{L}\left(G_L - 1\right), \tag{3.26}$$

$$\frac{dF_L}{dt} = F_L\frac{G_L\alpha - 1}{\tau_R}, \tag{3.27}$$

where Δ is the erbium population inversion, G_{x} and F^{x} are the signals, pump and laser gain and fluxes, respectively, τ_R is the roundtrip time, τ_{Er} is the lifetime of Er $\mathrm{I}_{13/2}$ upper laser level, V_{a} is the active volume, N_{Er} is the erbium concentration, and α is the laser loss. The pump flux term is expressed in terms of the laser threshold pump flux $F^P = x_{\mathrm{p}}\, F^{P,\mathrm{th}}$. Note that for a non-clamped case it is only necessary to set $x_{\mathrm{p}} = 1$ and the initial laser flux to zero to impede the lasing action.

Equation 3.26 shows that the signals add/drop perturbation is similar to the pump power perturbation with opposite sign. Note that 980-pumping, that excites erbium ions to $I_{11/2}$ level, can be described by the above equation set under assumption of the instantaneous decay from erbium $I_{11/2}$ level.

There are few differences in unclamped and clamped amplifier expressions. There is no laser expression in unclamped amplifier expression. On the other hand, in the clamped amplifier formula, there is a laser expression (Eq. 3.27). The laser gain and fluxes expression in the equations describes the feedback loop that enables to control the amplifier gain.

3.7 WDM Burst Traffic Amplification in Optically Gain Clamped Amplifier

3.7.1 Introduction

Improvements on transmission capacity of optical networks and on broadband services have increased the traffic to be processed on current point to point networks. The consequence is that a more effective use of the resources is necessary and therefore several alternative techniques have been studied. One of the proposed solutions is the migration of switching technologies from the electronic to the optical domain, with information gathered in packets or bursts. Optical burst switching (OBS) network (Qiao and Yoo 1999) is a proposed design whose goal is to carry data bursts transparently through the network as an optical signal. This allows different clients to use and share wavelength resources.

A characteristic of burst transmission is that the power of a single channel is constantly changing, depending whether it is transmitting or not. Considering a wavelength-division multiplexing (WDM) system with several burst channels, the total power oscillates according to the channels' traffic at each instant.

Power variations of the OBS network may cause problems when considering amplifier dynamics. When an EDFA operates in saturated regime, variations of the input power affect its gain. An increase of input power results in more stimulated emission, a phenomenon that reduces population inversion in the amplifier cavity, reducing, therefore, the gain of the amplifier. Analogously, a decrease of the input power of an amplifier in saturation reduces stimulated emission, increasing population inversion and the gain.

The influence of power variation on the EDFA gain depends, among other factors, on the burst inter-arrival time. While the amplifier is transparent to gigabit modulation rates, the random transmission of WDM burst channels impacts on the amplifier gain. Gain changes translate into dynamic power fluctuations and degradation of the signal-to-noise ratio in all channels, which accumulate with cascades of amplifiers (Qiao and Yoo 1999). The EDFA gain spectrum slope is also affected, meaning that different WDM channels will face different gains.

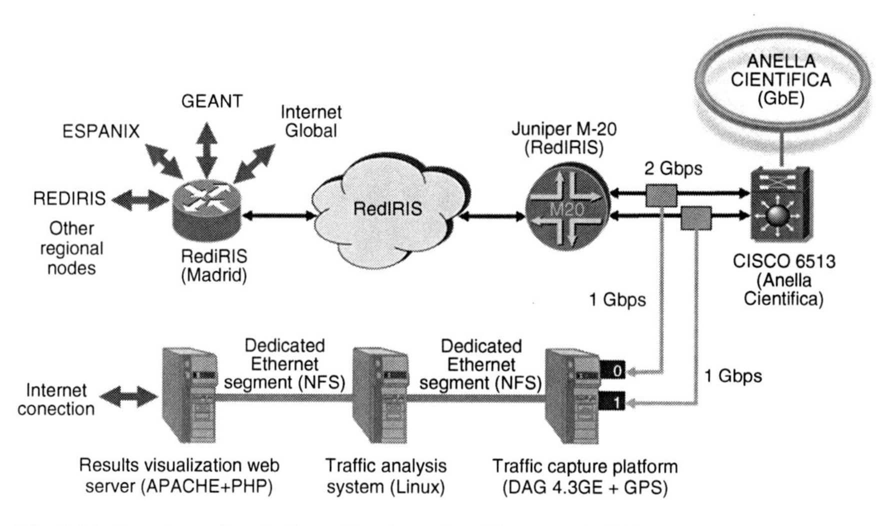

Fig. 3.34 Experimental testbed to collect burst data (Ennser et al. 2008)

Stabilization of the amplifier gain is a possible solution to minimize bit error rate (BER) penalties. There are several techniques proposed based on electronics (Srivastava et al. 1996) or optical control (Zirngibl 1991). Here we focus on an optical gain clamping (OGC) technique, which is a simple, passive, and fiber compatible solution that efficiently stabilizes the amplifier gain.

3.7.2 *Numerical Studies of WDM Burst Traffic Amplification*

In order to evaluate OGC-OA performance in OBS system, the burst data is collected from an experimental testbed illustrated in Fig. 3.32. We consider a scenario where several client networks are attached to a single OBS node, which performs packets aggregation and bursts generation. For the client traffic, we use real packet traces captured with a tailor-made measurement platform designed to operate at gigabit speed without packet losses and ns-precision in the packet time-stamp measurements. The point of measurement is a pair of full-duplex gigabit Ethernet links (two per each traffic direction) that connects the Catalan R&D network (about 50 universities and research centers) with the Spanish R&D RedIris network and to the global Internet (Barlet et al. 2006). A hybrid timer-length threshold burst assembler (Yu et al. 2004) is used to aggregate the packets; the timer threshold is set to 5 ms, and the maximum burst length to 250 kb. We assume that the OBS network domain is composed of 30 nodes; the destination of the bursts is therefore obtained by aggregating the IP addresses of the packet traces according to the geographical location (Fig. 3.34).

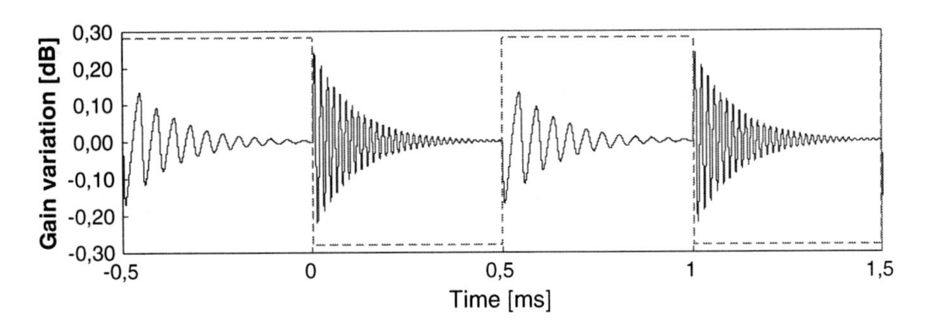

Fig. 3.35 Gain variation under on/off of all channels

The numerical investigation is based on the two-level equation set to simulate the dynamics of the OGC-EDFA. This model is simpler than the one used in Della Valle et al. (2007) but, after comparison, proved to be effective for this kind of simulation and was therefore chosen to reduce the computational time. The amplifier is assumed to have 20 dB gain, and the cavity length is 21 m. The maximum total input power is −1 dBm when all 16 WDM burst channels are simultaneously entering the amplifier. The pump power level is set to $x = 1.15$, i.e. at 15% higher level than for unclamped EDFAs (Ennser et al. 2005a, b). Figure 3.33 shows the OGC-OA dynamics when all channels are switched on/off (Fig. 3.35).

The gain excursion is limited to below 0.25 dB, and this is the maximum gain variation related to eventual sudden failure or network restoration. The typical laser ROF for full channel load (on) or no channel (off) is 22 kHz and 66 kHz, respectively. This sets the limits of ROFs characteristics of our amplifier. In a previous work, we have shown that if the burst frequency interplays with the characteristic OGC-OA relaxation oscillation frequencies, we may have enhanced oscillation, with interplay starting from $\nu_{\mathrm{on}}/3$ to about 2 times ν_{off}, corresponding in our case to 7 kHz to 100 kHz interval (Della Valle et al. 2007). We have also demonstrated that high gain variation was induced after a few burst sequences. The real data burst distribution fall well within this interval; however, different channels will combine in a random mode to generate a total power time variation, as stated below.

The aim of this numerical studies is to evaluate the effect on OGC-OA of real burst data traffic and verify if there are still instabilities or oscillations due to the interplay between burst and OGC-OA dynamics (Della Valle et al. 2007). To address this issue, we collect a 95 k burst trace (duration >14 s) in the testbed described in the prior section. We elaborate a WDM system with 16 channels by using uncorrelated data in different time slots for different channels to generate an about 1-s long trace. To enable an easy visualization and significant representation of the input data, here we show 3-ms trace.

Figure 3.36 shows the total power (i.e. sum of all burst channels). We see that we never have less than 2 channels and more than 13 channels at the same time. Along

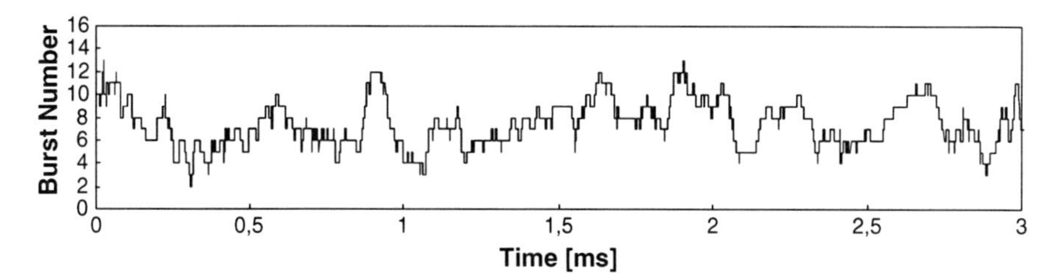

Fig. 3.36 Total burst number along time

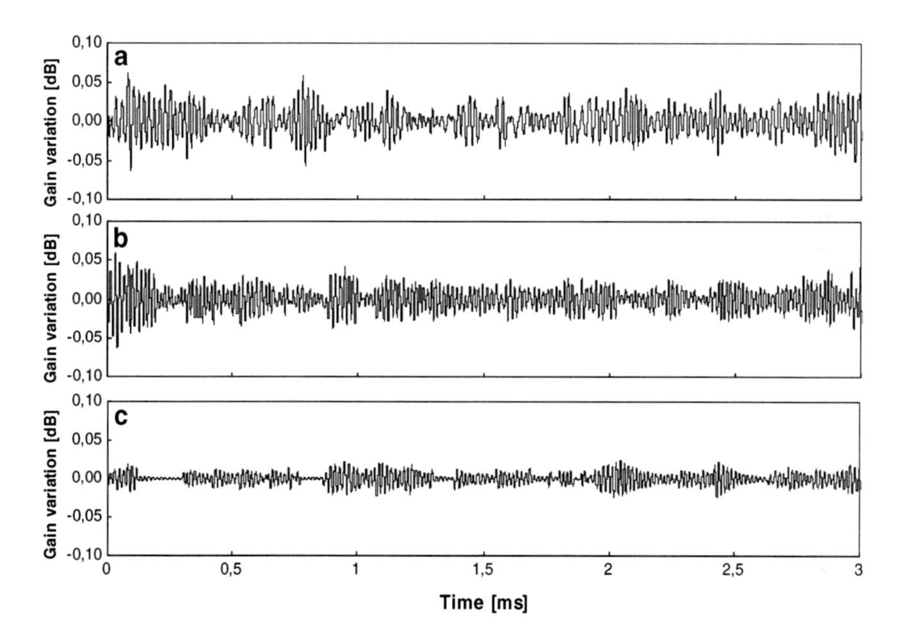

Fig. 3.37 Gain variation in case of (**a**) 16 channels and $x = 1.15$, (**b**) 16 channels and $x = 1.15$, and (**c**) 4 channels and $x = 1.15$

the 1-s trace, we had a minimum and maximum of 1 and 15 channels, respectively. We also note that the average power is about −4 dBm (8 channels out of 16), and we miss large variation of channel number in times comparable with laser dynamics (<50 μs). This suggests that the OGC-OA will work above threshold and is not subject to abrupt input power variation like in the on/off case.

Figure 3.37a shows the gain variation induced by the trace of Fig. 3.36 for an OGC-OA working at $x = 1.15$. Figure 3.37b shows the same results but in case of $x = 1.5$. We do not see any appreciable variation as the OGC-OA in burst mode works with average power of −4 dB and therefore is typically far more above threshold than $x = 1.15$. However, we note that at the same time interval in Fig. 3.37b, sometime we have stronger gain variations than in Fig. 3.37a despite a

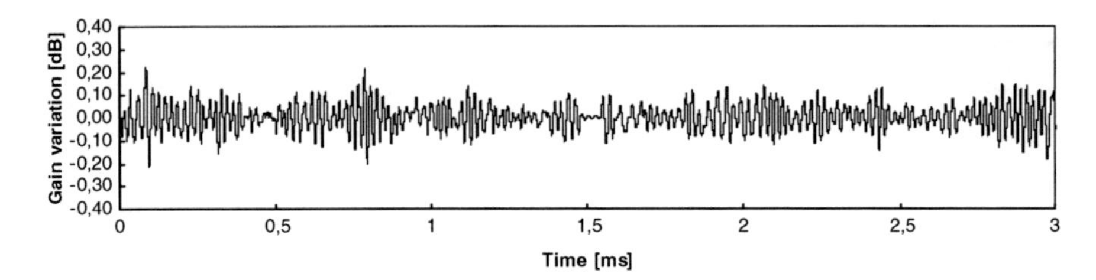

Fig. 3.38 Gain variation after four cascaded OGC-EDFAs for 16 channels and $x = 1.15$

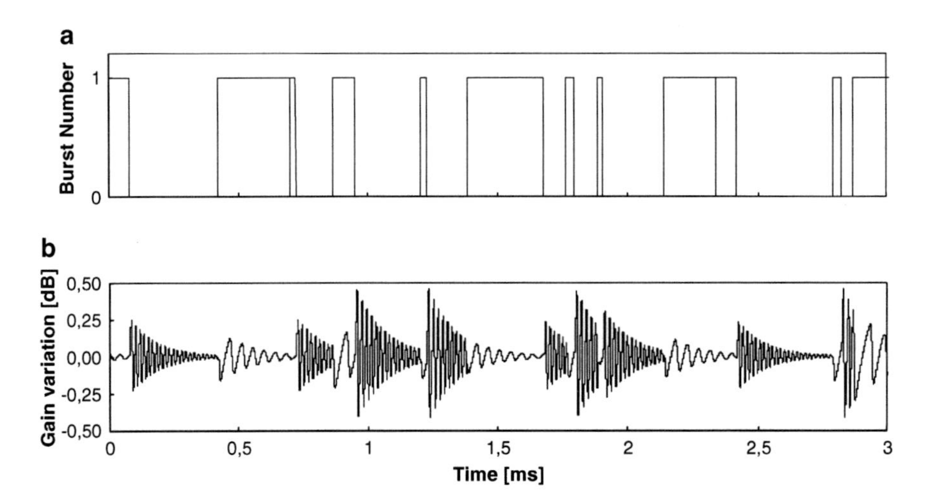

Fig. 3.39 (**a**) Burst trace, (**b**) gain variation for input power of −1 dBm and $x = 1.15$

larger x value. This is simply due to the fact that different x values corresponds different ROF, and therefore the burst sequence interplay can be different. This feature is therefore the proof that the burst sequence interplay with OGC-OA ROF may influence the gain dynamics, still with negligible impact for WDM systems. Figure 3.37c shows the results with only 4 channels (maximum power −7 dBm). We note that the high effective x value helps to suppress any gain dynamics.

To evaluate performance in a cascaded link, Fig. 3.38 shows the accumulated gain variation after four OGC-OAs with 16 burst channels and $x = 1.15$. By comparing with Fig. 3.37, we note that an almost linear accumulating effect as expected considering the small signal variation induced in each OA. We may conclude that the OGC-OA will effectively perform in the case of WDM burst traffic.

A different scenario is if the OGC-OA is used in few channel links with high channel power. As example, we simulate a link with one channel link with −1 dBm input. Figure 3.39a illustrates the trace, and Fig. 3.39b shows the dynamics of gain variation. As results, we note the interplay between burst sequence inter-arrival times and OGC-OA ROFs. In fact, we have stronger gain variation than in the on/off case (see Fig. 3.35) as expected according to experimental verification in Della Valle et al. (2007). The burst traffic data may therefore induce significant gain variation. To minimize the gain excursions, the OGC-EDFA should operate with a larger x value of about 1.5 or with shifted ROFs (Della Valle et al. 2007).

In conclusion, we investigated the optical burst amplification for a single and cascaded amplified WDM transmission. Overall optical gain clamped amplifier shows beneficial properties compared to unclamped configuration.

3.8 Propagation in the Presence of Arrayed Waveguide Gratings

The arrayed waveguide grating (AWG) is an imaging device since it creates an image of the field of an input waveguide onto an array of output waveguides in a dispersive way (Smit and Van Dam 1996). AWGs have been also commonly referred as phased-array gratings or waveguide grating routers.

This section will analyze the device principle of operations and its features; then, it will focus on its possible applications, and finally, an example of switching-system architecture based on AWGs will be provided.

3.8.1 Device Principle of Operations and Features

The typical structure of the arrayed waveguide grating implements a FIR filter with all stages in a serial connection. The imaging property described before is given by an array of waveguides of different lengths. Figure 3.40 shows the structure of such a device. The input consists of a several channels (N) carried on different frequencies.

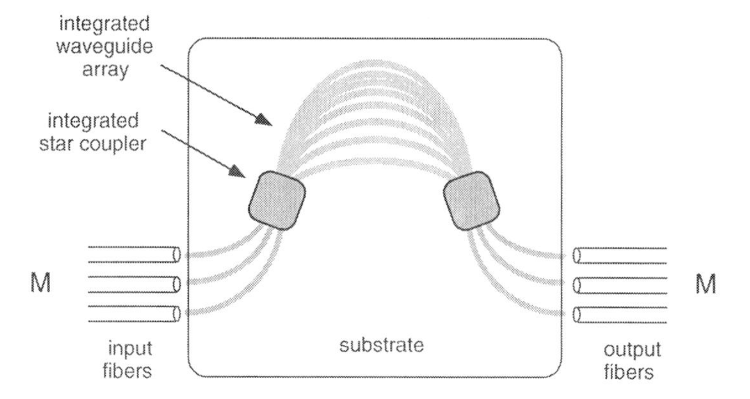

Fig. 3.40 Structure of an arrayed waveguide grating (AWG)

Thanks to the coupler, it is possible to implement the FIR filter with parallel delay lines. This concept is the generalization of the MZI one. In the AWG structure, two couplers are present, with size NM and MN, respectively, connected by an array of M guides of increasing length. The final coupler conveys the channels to the N output lines. The length of the waveguides array is chosen such that the optical path length difference between adjacent waveguides equals an integer multiple of the central wavelength of the device.

For this wavelength, the fields in the individual waveguides will arrive at the output with a constant phase profile. Signals with different wavelengths will exhibit different amounts of phase change, and thanks to the different lengths of the M guides, the phases will change on the output plane so that the beam will be focused on a specific output.

The transfer function of the device is strongly dependent on the number of waveguides of the array. This function has a periodic trend linked to the phase shift that occurs on the signal between the input and the output of the device. Each period includes a peak (primary lobe) and M-2 secondary lobes, separated by M-1 zeros. For values of the input-output phase shift that are integer multiples of 2π, the channel will be centered on the primary lobes of the transfer function. We call Q this integer number. In these cases, the signal will be entirely transferred from the input port to the output port without attenuation. This happens when:

$$\lambda pq = \lambda 0 - (p + q) \cdot \Delta\lambda, \tag{3.28}$$

where p and q are the input and output waveguides, respectively, whereas $\lambda 0$ is the central wavelength that depends on the effective index of the guide, on the differential length of the array guides ($\Delta\lambda$), and on Q. As stated before, the function has a periodic transfer function. This periodicity corresponds to the free spectral range (FSR) of the device, which is commonly referred as:

$$\text{FSR} = N \cdot \Delta\lambda, \tag{3.29}$$

where N is the number of input channels. The FSR is independent of the parameter M, but is dependent on the difference of optical path between two adjacent guides ($\Delta\lambda$). Figure 3.41 shows the periodical trend of the transfer function and its FSR when the number of managed channels is equal to N.

The behavior of an AWG can be described also by referring to a simple "rule of the thumb," which highlights its cyclic routing function:

$$j = (i + \lambda) \bmod N. \tag{3.30}$$

Equation 3.30 shows that, given the number of inputs of the AWG (N), the channel at wavelength λ entering the AWG on the port i will be redirected on the output j. We refer to the wavelength λ that is equivalent to indicate the corresponding frequency f. It is easy to understand that this rule makes sense when the number

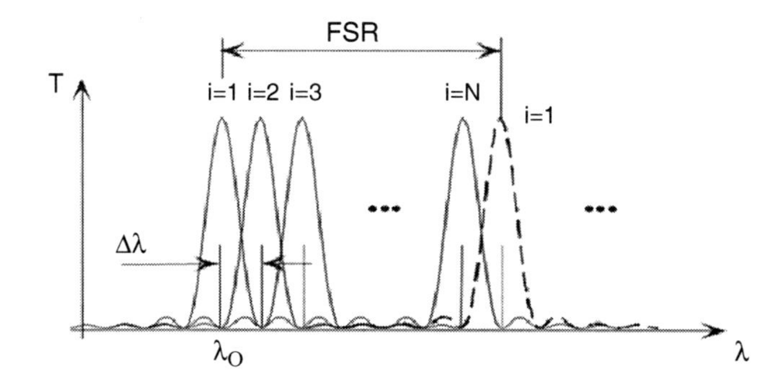

Fig. 3.41 Example of AWG transfer function

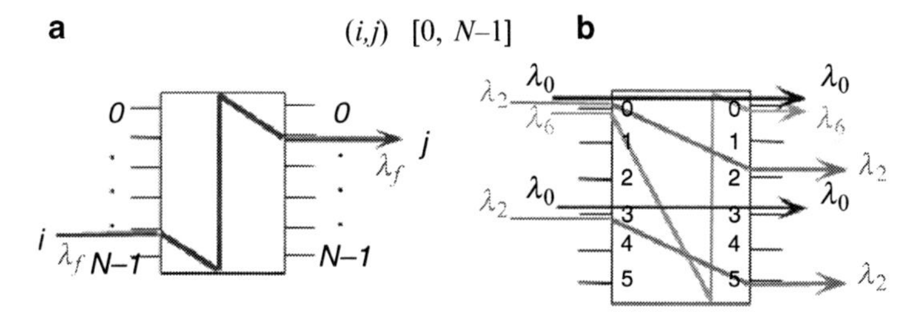

Fig. 3.42 AWG cyclic behavior in the case of (a) one channel or (b) more channels

of input channels is higher than the number of input lines in the AWG. This rule is an alternative form of Eq. 3.28, which requires the previous knowledge of the device parameters in order to be understood. The application of this easy rule can be appreciated also in Fig. 3.42, where a simple example with one channel (a) and another one with more channels (b) are depicted.

The main issue affecting the AWG performance is the crosstalk. As it can be appreciated also in Fig. 3.41, the transfer function has different side lobes, which means that, when a signal exits the AWG, part of the power of the other N-1 channels can be forwarded by the device on the same exit, thus creating a noise component for the signal that will be revealed at the receiver. This is called out-of-band, or incoherent, crosstalk. Moreover, in particular configurations (AWG used as wavelength router), and also signal components with the same wavelength can constitute noise for the signal taken into account. This kind of impairment is called in-band, or coherent, crosstalk. Different strategies have been studied to overcome crosstalk impairment, e.g., increasing the number of waveguides M: The produced effect is a larger number of lateral lobes with less amplitude. On the other side, this solution shrinks the available bandwidth for the transmission peak, so it is necessary for the designer to ensure that the peak of the signal will not be distorted.

Another important issue in AWGs is the insertion loss (IL). The primary cause for it is the inefficient coupling at the interface between the first coupler and the

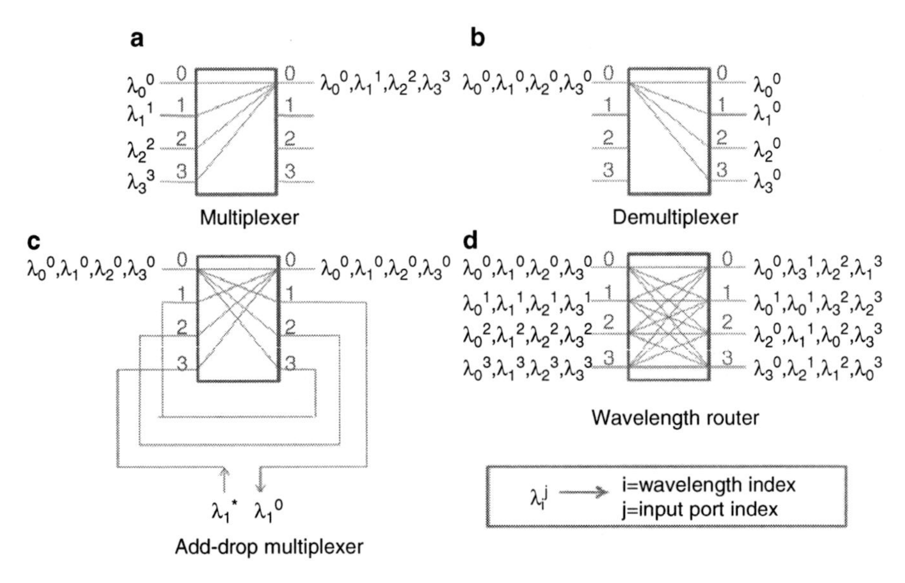

Fig. 3.43 Basic functions of an N·N AWG, with $N = 4$

waveguides of the central array. Due to reciprocity, the same happens at the second coupler. Other reasons that can generate losses are: material losses, scattering due to fabrication imperfections, and waveguide roughness.

The AWG is a complex structure; hence, different properties and issues of the device should be considered and analyzed. We now focus more on AWGs applications; for more information about AWGs properties, the reader can refer to (Agarawal 2002; Smit and Van Dam 1996; Munoz et al. 2002).

3.8.2 *Applications and Switching-System Architectures*

We will now describe the basic applications of the AWG device (Maier and Reisslein 2004). As shown in Fig. 3.43, an NN AWG provides four basic functions: (a) N1 multiplexer, (b) 1N demultiplexer, (c) add-drop multiplexer (OADM), and (d) NN full-interconnect wavelength permutation router.

In Fig. 3.43, for each application, signals are indicated as λ_i^j, where i is the wavelength index, and j is the input port index. As a wavelength router, the device accepts N wavelengths from each input port and routes each of them to a different output port. Each wavelength gives routing instructions that are independent of the input port, as previously specified in Fig. 3.30. For example, signal λ_0 goes from input port 0 to output port 0 and also from input port 3 to output port 3. Similarly, signal λ_2 entering on input port 0 is directed to output port 2, whereas if it enters on input port 1, it is routed to output port 3; analogously, if signal λ_2 enters on

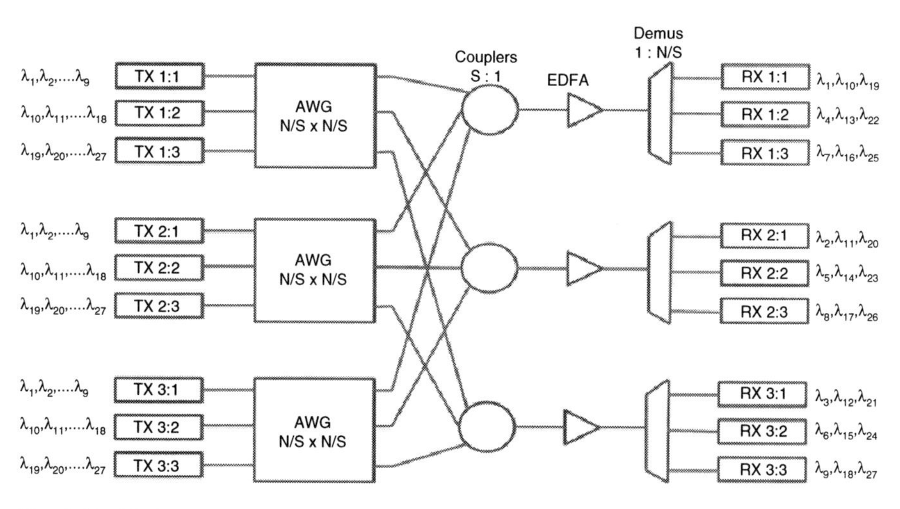

Fig. 3.44 AWG-based architecture

input port 3, it is routed to output port 1. Note that each output port receives N different wavelengths, one from each input port (Fig. 3.43). In this illustration, the wavelengths at a given output port are sorted from left to right according to the corresponding input port number (e.g., λ_2 exiting at output port 3 originates at input port 1). Moreover, this kind of AWG allows the spatial reuse of all wavelengths at all input ports, without resulting in channel collision at the output ports (Maier and Reisslein 2004).

3.8.3 An AWG-Based Architecture

AWGs can be used as the central stage of the architecture represented in Fig. 3.44, which has been proposed in (Cuda et al. 2009).

In this architecture, the N transmitters are divided into S switching planes, each one comprising N/S transmitters, so that the size of each AWG is $N/S{\cdot}N/S$. An example with $N=9$ and $S=3$ is provided in Fig. 3.44. After this stage, S:1 couplers collect signals coming from each single plane. They are needed to achieve connectivity between input/output ports belonging to different switching planes. The well-known routing properties of the AWG are exploited to steer each WDM channel from the source switching plane (the one containing the source transmitter) to the appropriate destination switching plane (the one containing the target transmitter). S EDFA amplifiers are then placed after the couplers, and finally, S demultiplexers 1:N/S send the signal to the filters placed in front of the receivers.

It is worth noting that if the cyclic property of the AWG is exploited, line cards do not reuse the same wavelength in different parts of the fabric. By doing so, the total transmitters tunability range is equal to *N*. The advantage is that no coherent crosstalk is introduced in the optical fabric, and only out-of-band crosstalk is present. The drawback is that the AWGs give an almost identical transfer function over the *N*/*S* FSRs; this means that the amplification bandwidth of the EDFA has to be flat and significantly large. On the other hand, if the cyclic property of AWGs is not exploited, and the system operation is limited to a single FSR, coherent crosstalk will be introduced. It can be shown that the contribution of the coherent crosstalk can limit the scalability of the architecture more than the incoherent crosstalk does (Gaudino et al. 2008).

3.8.4 Reliability of Arrayed Waveguide Gratings for Long-Haul Optical Transmission Systems

Reliability of optical components strongly depends on the detailed design of the structures, processes, and technologies used to fabricate them. Reliability testing and investigation must start at the beginning of the design phase to expose any potential weaknesses of the design. This approach eliminates many failure modes early in the design process.

Characterization tests and lot-to-lot controls are put in place to ensure that the device is a form, fit, and functionally compliant device. Screening is also performed on the components that include temperature cycling to assure satisfactory cumulative early life and long-term reliability.

The intention of qualification and reliability testing of photonic components is to assess the failure modes or degradation mechanisms in a design in the early part of the component's lifecycle. Components are exposed to various mechanical and thermal shocks during their lifetime. Components also age through applied electrical bias and other signals. Therefore, reliability tests expose devices for a shorter period to various stresses that simulate these conditions. Qualification testing is only part of a broader reliability program that identifies and eliminates problems in the design and manufacturing process. Without this program, qualification testing is time-consuming and expensive and still does not ensure reliability.

To qualify components and subsystems for use in telecommunications systems, Telcordia Generic Requirements are often used as a guideline. These include a detailed set of environmental accelerated life tests. Acceleration will normally be applied by increasing the temperature of the device or increasing the humidity it is exposed to. These tests are designed to guarantee that a particular component will remain within its specification for its design lifetime. The actual tests vary depending on the type of component. Lifetime estimation can be made from accelerated life tests or endurance tests and/or from field data. Accelerated life tests increase the stress levels above those used in normal operation to reduce the time needed to expose any defects in components. Endurance tests include long-term damp heat exposure, high-temperature aging, and low-temperature storage.

Mechanical robustness tests include mechanical shock, mechanical vibration, humidity, thermal cycling, and thermal shock. If the device has been in the field for a reasonable amount of time, lifetime can be estimated using the field failures based on information on the quantity shipped, in service device hours and number of failures.

Vendors of optical multiplexers and demultiplexers used in dense wavelength division multiplexing systems have aimed to achieve better reliability by improving three key characteristics: insertion loss, filter profiles, and thermal properties. Multiplexers based on AWG are temperature sensitive and require the use of heaters that keep the waveguide chip at a controlled temperature. (Chandy 2009). Heaters introduce several costs and reliability issues. A heater failure has a significant network impact from a reliability point of view since all wavelengths will be affected. A thermal AWGs are therefore preferred, though they have higher insertion losses than the AWGs with heaters. The demand for network reliability has favored the use of passive devices that do not require power or temperature control elements.

3.9 Micro-Ring Resonators

The MRR is operationally similar to the Fabry-Perot resonator: The resonant cavity is achieved with a ring-closed waveguide, while the two mirrors are replaced by two directional couplers. Due to internal optical feedback, MRRs can achieve high-frequency selectivity (Klein 2007). They also offer the possibility to be actively tuned providing them with an attractive functionality. Moreover, MRRs can be adopted to realize WC functionalities. This chapter will be divided into two parts: The first one will analyze the characteristics of the passive MRRs, and the second one will consider the active MRR devices and the realization of tunable ring-based wavelength converters.

3.9.1 Device Principle of Operations and Applications

According to the realization technology, ring resonators can have diameters of few tens of μm, giving them the potential to be integrated on a large scale. Ring resonators can be adopted for different applications; among them, the active or passive filtering function with high selectivity is the most implemented one.

Ring resonators can be realized as two- or four-port devices (Klein 2007). In the two-port configuration, the resonant cavity is coupled to a single-port waveguide and therefore has only one input and one output. This configuration is shown in Fig. 3.45.

This component can be used to compensate dispersion or to introduce delay. It can also be adopted to make lasers. Another configuration is the four-port ring,

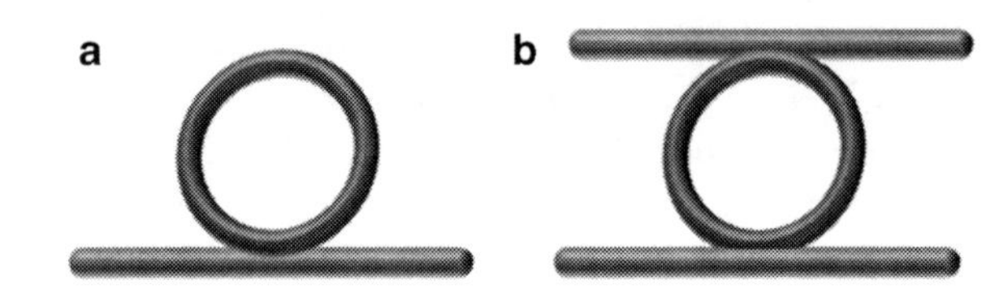

Fig. 3.45 Ring resonators structures with (**a**) two ports or (**b**) four ports

which is ideal to realize the wavelength filtering function. This configuration is shown in Fig. 3.45. It is created by adding a waveguide port to the two-port configuration. The device presents one input port and two different output ports, called through and drop. The filtering and routing property is related to the signal wavelength; as a matter of fact, if the input signal wavelength corresponds to the resonance wavelength of the ring, the signal will enter the ring and will be routed to the drop port otherwise it will proceed towards the through port. Eventually, the fourth port can be used as an additive input to add signals of the same wavelength of the resonance one at the through output. This solution can be adopted to realize add/drop functions.

There are many important parameters in ring design, but one of the most important is the coupling coefficient between the waveguide and the ring. It is conventionally referred as k, its value is set between zero and one, and it represents the fraction of light that is transferred from the waveguide to the ring or vice versa. The non-coupled fraction of light which goes on is usually called μ and is related to k. Another important parameter is the roundtrip loss coefficient; it is usually expressed in linear unit as χr. For the description of all parameters, the reader can refer to (Klein 2007) or (Cusmai et al. 2005). The mentioned parameters are the most important ones to derive the transfer function of the ring resonator between the input port and the drop output. The model will not be presented here, but it can be found on (Klein 2007).

Figure 3.46 shows the trend of the drop transfer function of the ring. In particular, it has been obtained using a ring radius $R = 50\ \mu\text{m}$, $k = 0.5$, and a modal wavelength group refraction index $ng = 1.5$, without considering losses.

By looking at the figure, two main design parameters can be highlighted:

FSR: it is the measure of the periodicity of the transfer function, and it is usually measured as the distance between two consecutive peaks. It can be derived as:

$$\text{FSR} = \frac{\lambda 0}{2\pi Rng}, \tag{3.31}$$

where $\lambda 0$ is the working wavelength of the ring.

The bandwidth of the filter $\Delta\lambda{-}3$ dB at the drop port: it is the interval of frequencies in which at least half of the total power is present at the drop port. It can be derived as the ratio between FSR and the finesse F of the ring, which is a

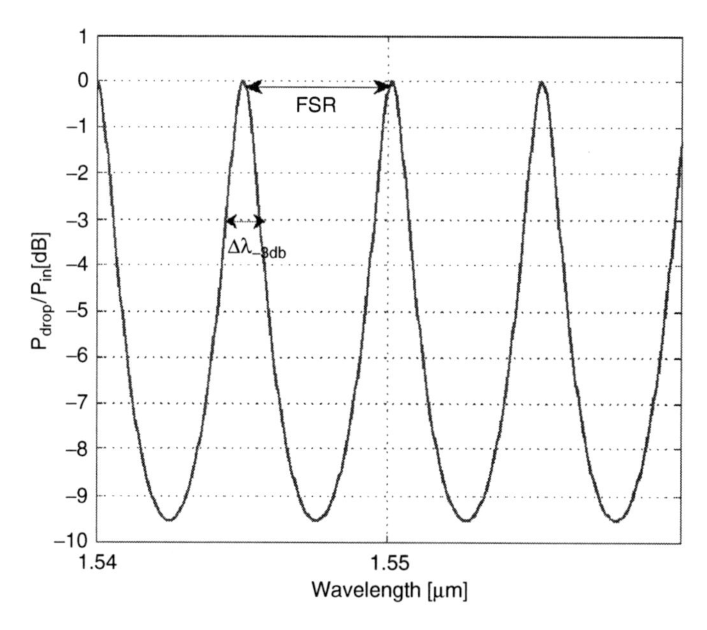

Fig. 3.46 Transfer function of a ring resonator at the drop port

measure of the quality of the resonator and can be related to the storage capacity (it is linked to the quality factor Q of the resonator):

$$\Delta\lambda - 3dB = \frac{\text{FSR}}{F}. \tag{3.32}$$

The transfer function of the ring at the through port is represented in Fig. 3.47; the parameters are the same as the ones taken into account for the drop port.

3.9.2 Issues Affecting MRR Performance

As it can be appreciated in Figs. 3.46 and 3.47, these devices, unlike the previously described AWGs, exhibit a transfer function without side lobes. This represents an undisputed advantage on the performance of the device. When the device is considered as a part of a system, different parameters and issues must be taken into account: (1) IL at the drop port, (2) IL at the through port, (3) filter bandwidth and crosstalk.

Insertion loss at the drop port is a measure of device transferring efficiency between input port and drop port, when a specific wavelength is considered. This parameter is particularly important since the drop output is the port in which filtering

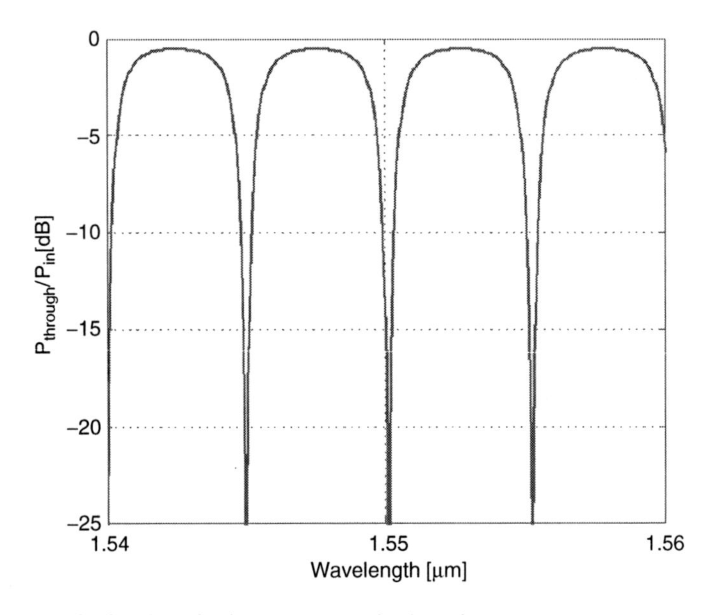

Fig. 3.47 Transfer function of a ring resonator at the through port

"happens." IL at the drop port is strongly influenced by the coupling coefficients (k) values, and it is usually the dominant factor on the total losses for these devices, especially when they are connected in cascade. IL at the through port is equally influenced by coupling coefficients. A more accurate analysis is needed for the bandwidth of the filter ($\Delta\lambda - 3$ dB) at the drop port. Taking into account, this port, $\Delta\lambda - 3$ dB, is also dependent on the rejection ratio (RR), which is the ratio between the maximum transmitted power in the resonance condition and the power that is anyway transmitted to the same port, when the ring is completely out of the resonance condition. It is shown that the rejection ratio is proportional to the finesse of the ring, and thus, it is inversely proportional to the filter bandwidth (Klein 2007). This means that it is important for the ring designer to search for an optimal tradeoff between the bandwidth of the filter, which ensures that the input channel will not be distorted, and the rejection ratio, which ensures that the filtering action of the device is effective. As a matter of fact, the more the rejection ratio, the less the impact of the crosstalk on device performance. This design issue has also relation with the coupling coefficients because low values of them are not sufficient to obtain high bandwidths, but, on the other side, high coupling coefficients lower the rejection ratio. An easy solution to the problem can be the reduction of the ring radius, but the possibility to accomplish this goal depends on the building technology. Moreover, reducing the ring radius strongly influences the FSR. An interesting alternative can be the design of multiple ring resonators connected in a cascade.

3.9.3 *Tunable Devices*

The previous paragraphs described ring resonators with their passive filtering functionality. It also possible to actively vary some of the parameters of such devices, thereby altering their response (Klein 2007). The tuning process can have a double aim: to compensate fabrication errors or (more commonly used) to add some form of active functionality. In the second case, two main types of tuning can be distinguished: (1) alter the resonator with the aim of shifting its resonant wavelength without changing the shape of its response or (2) do not shift the resonant wavelength, while changing the shape of the response.

There are several ways of tuning a ring resonator; the most straightforward between them consists of changing the optical index in one or all the materials of the resonator waveguide. This will affect the effective index of the waveguide which, in turn, will alter the roundtrip phase of the light in the resonator, and therefore the wavelength for which it achieves the resonance. This can be realized thanks to few effects that are able to change the optical index of a waveguide. The most important of them are the thermo-optic effect (Heimala et al. 1996) and the electro-optic effect (Soref and Bennett 1987) (for more information, the reader can refer to (Klein 2007)).

3.9.4 *Tunable Wavelength Converters*

Optical TWCs are devices that transfer information from one wavelength to another without entering the electrical domain. They improve network management and internetworking operations, enabling them to be scalable and modular (Wang et al. 1997).

To obtain all-optical wavelength conversion, it is possible to utilize several techniques; among them are passive fibers, semiconductor lasers and SOAs (Sabella et al. 1996). Historically, conversions based on SOAs are promising because they can be operated at high speed with high conversion ranges (Wang et al. 1997). In addition, wavelength converters using SOAs have modest input signal power requirements (Sabella et al. 1996).

As it can be expected, TWCs are complex devices that should be characterized by various features. Some of them are: bit-rate transparency, no extinction ratio degradation, high SNR at the output, moderate input power levels, large wavelength range, and insensitivity to input signal polarization.

We will briefly introduce the possibility of making WCs with MRRs. In fact, using the FWM effect, MRRs based on GaAs or InP can enable efficient wavelength conversion, combining the advantages of nonlinearity with high gains. The maximum conversion can be achieved when both the signal wavelength and the converted one fit the resonance wavelength of the ring. In Absil et al. (2000), the authors examine and demonstrate wavelength conversion by FWM in a semiconductor MRR

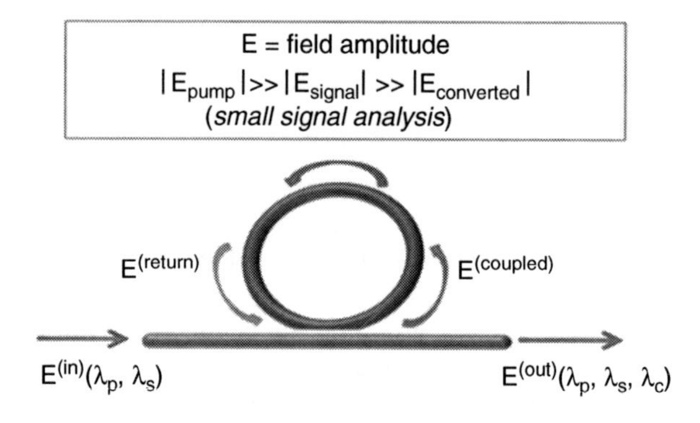

Fig. 3.48 Ring-based wavelength converter working scheme

with an improved conversion efficiency compared with that of an equivalent straight waveguide. The increase in efficiency is attributed to the increase of interaction length and the enhancement of optical power inside the cavity. The working scheme is shown in Fig. 3.48 and can be summarized in the following way: A pump wave at wavelength λp and a signal wave at wavelength λs are launched into the ring at two different resonant wavelengths. A new wave (the converted one) is generated by FWM at wavelength $\lambda c = \lambda p - \lambda s$. Figure 3.48 also shows the small signal analysis for the fields amplitude in the ring (Absil et al. 2000).

Considering that in the absence of second- or high-order dispersions, the resonant frequencies are evenly spaced, and λc will lie precisely on one of them. If second-order dispersion is present, it will lead to a mismatch between the converted and the resonant wavelengths of the ring. Further conversion improvements are also possible (Hamacher et al. 2002).

The coexistence and the internetworking of different technologies and devices are possible, as it has been shown in (Stampoulidis et al. 2010). In this work, the authors present a ring-resonator-assisted WC, in which a single SOA followed by a compact and integrated micro-ring-resonator-ROADM are employed.

References

Absil, P., Hryniewicz, J., Little, B., Cho, P., Wilson, A., Joneckis, L., et al.: Wavelength conversion in GaAs micro-ring resonators. Opt. Lett. **25**(8), 554–556 (2000)

Agrawal, G.P.: Fiber-Optic Communication Systems. Wiley, New York (2002)

Agrawal, G.P.: Nonlinear Fiber Optics, 4th edn. Academic, Boston (2007)

Agrawal, G.P., Olsson, N.A.: Self-Phase Modulation and Spectral Broadening of Optical Pulses in Semiconductor Laser Amplifiers. J. Quantum Electron. **25**, 2297 (1989)

Anderson, C.J., Lyle, J.A.: Technique for evaluating system performance using *Q* factor in numerical simulation exhibiting intersymbol interferences. Electron. Lett. **3000**, 71–72 (1994)

Barlet, P., Sole Pareta, J., Barrantes, J., Codina, E., Domingo, J.: SMARTxAC: a passive monitoring and analysis system for high-speed networks. In: Proceedings of the Terena Networking Conference, Catania (2006)

Bergano, N.S.: Wavelength division multiplexing in long-haul transoceanic transmission systems. IEEE/OSA J. Lightwave Technol. **23**, 4125–4139 (2005)

Bergano, N.S., Kerfoot, F.W., Davidson, C.R.: Margin measurements in optical amplifier systems. IEEE Photonics Technol. Lett. **5**, 304–306 (1993)

Betti, S., De Marchis, G., Iannone, E.: Coherent Optical Communication Systems. Wiley, New York (1994)

Cai, J.-X., Cai, Y., Davidson, C., Foursa, D., Lucero, A., Sinkin, O., Patterson, W., Pilipetskii, A., Mohs, G., Bergano, N.: Transmission of 96 × 100 G pre-filtered PDM-RZ-QPSK channels with 3000% spectral efficiency over 10,608 km and 400% spectral efficiency over 4,368 km. In: Proceedings of the Optical Fiber Communication Conference 2010:PDPB10, San Diego (2010)

Chandy, R.: Reliability analysis of optical modules for future optical networks. In: ICTON, Angers. Ponta Delgada, Azores (2009)

Chraplyvy, A.R.: Limitation on lightwave communications imposed by optical fiber nonlinearities. J. Lightwave Technol. **8**, 1548–1557 (1990)

Cuda, D., Gaudino, R., Gavilanes, G., Neri, F., Maier, G., Raffaelli, C., et al.: Capacity/cost tradeoffs in optical switching fabrics for terabit packet switches. ONDM **2009**, 1–6 (2009)

Curti, F., De Marchis, G., Daino, B., Matera, F.: Statistical treatment of the evolution of the PSP in single-mode fibers. J. Lightwave Technol. **6**, 704–709 (1990)

Cusmai, G., Morichetti, F., Rosotti, P., Costa, R., Melloni, A.: Circuit-oriented modelling of ring-resonators. Opt.Quantum Electron. **37**(1–3), 343–358 (2005)

De Angelis, C., Galtarossa, A., Gianello, G., Matera, F., Schiano, M.: Time evolution of polarization mode dispersion in long terrestrial links. J. Lightwave Technol. **10**, 552–555 (1992)

Della Valle, G., Sorbello, G., Festa, A., Taccheo, S., Ennser, K.: High-gain rbium-doped waveguide amplifier for bidirectional operation. In: Proceedings of the European Conference Optical Communication 2006:We2.6.5, Cannes, Tokyo (2006)

Della Valle, G., Festa, A., Taccheo, S., Ennser, K., Aracil, J.: Nonlinear dynamics induced by burst amplification in optically gain-stabilized erbium-doped amplifiers. Opt. Lett. **32**, 903–905 (2007)

Elrefaie, A.F., Wagner, R.E., Atlas, D.A., Daut, D.G.: Chromatic dispersion limitation in coherent lightwave transmission systems. J. Lightwave Technol. **6**, 704–709 (1988)

Ennser, K., Della Valle, G., Ibsen, M., Shmulovich, J., Taccheo, S.: Erbium-doped waveguide amplifier for reconfigurable WDM metro networks, IEEE Photon. Technol. Lett. **17**, 1468–1470 (2005a)

Ennser, K., Della Valle, G., Mariani, D., Ibsen, M., Shmulovich, J., Taccheo, Laporta, P.: Erbium-doped waveguide amplifier insensitive to channel transient and to spectral-hole-burning offset. In: Proceedings of the Advanced Solid-State Photonics 2005:MB32, Vienna (2005b)

Ennser, K., Taccheo, S., Careglio, D., Sole-Pareta, J., Aracil, J.: Real burst traffic amplification in optically gain clamped amplifier. In: Proceedings of the Optical Fiber Communication Conference, 2008: JThA16, San Diego (2008)

Ennser, K., Taccheo, S., Rogowski, T., Shmulovich, J.: Efficient erbium-doped waveguide amplifier insensitive to power fluctuations. Opt. Express **14**, 10307–10312 (2006)

Eramo, V., Matera, F., Schiffini, A., Guglielmucci Settembre, M.: Numerical investigation on design of wide geographical optical transport network based on nx40 Gb/s transmission. J. Lightwave Technol. **21**, 456 (2003)

Favre, F., Matera, F., Settembre, M., Tamburrini, M., Le Guen, D., George, T., Henry, M., Franco, P., Schiffini, A., Romagnoli, M., Guglielmucci, M., Cascelli, S.: Field demonstration of 40 Gbit/s soliton transmission with alternate polarizations. J. Lightwave Technol. **17**, 2225–2234 (1999)

Franco, P., Matera, F., Settembre, M., Tamburrini, M., Favre, F., Le Guen, D., George, T., Henry, M., Schiffini, A., Romagnoli, M., Guglielmucci, M., Cascelli, S.: Limits of polarization mode dispersion on a field demonstration of 40 Gbit/s soliton transmission over 500 km with alternate polarizations. Electron. Lett. **35**, 407–408 (1999)

Fuerst, C., Elbers, J.P., Camera, M.: 43Gbit/s RZ-DQPSK DWDM field trial over 1047 km with mixed 43 Gb/s and 10.7 Gb/s channels at 50 and 100 GHz channel spacing. In: Proceedings of the European Conference Optical Communication 2006:Th4.1.4, Cannes (2006)

Fuerst, C., Wernz, H., Camera, M., Nibbs, P., Pribil, J., Iskra, R., Parsons, G.: 43 Gb/s RZ-DQPSK field upgrade trial in a 10 Gb/s DWDM ultra-long-hual live traffic system in Australia. In: Proceedings of the Optical Fiber Communication Conference 2008:NTuB2, San Diego (2008)

Galtarossa, A., Pizzinat, A., Matera, F.: Statistical description of optical systems performance due to random coupling on the principal states of polarization. Photonics Technol. Lett. **3**, 1307–1309 (2001)

Gaudino, R., Gavilanes Castillo, G., Neri, F., Finocchietto, J.: Simple optical fabrics for scalable terabit packet switches. IEEE ICC **2008**, 5331–5337 (2008)

Goder, N., Settembre, M., Laedke, W., Matera, F., Tamburrini, M., Gabitov, I., Haunstein, H., Reid, J., Turitsyn, S.: Role of the *Q* factor estimation in the field trial of 10 Gbit/s transmission at 1300 nm with semiconductor optical amplifiers between Madrid and Merida (460 km). OFC **2**, 325–327 (1999)

Gordon, J.P., Haus, H.A.: Random walk of coherently amplified solitons in optical fiber links. Opt. Lett. **11**, 665–667 (1986)

Gordon, J.P., Mollenauer, L.M.: Phase noise in photonic communications systems using linear amplifiers. Opt. Lett. **15**, 1351–1353 (1990)

Gnauck, A.H., Winzer, P.J.: Optical phase-shift-keyed transmission. IEEE/OSA J. Lightwave Technol. **23**, 115–130 (2005)

Hamacher, M., Troppenz, U., Heidrich, H., Rabus, D.: Active ring resonators based on GaInAsP/InP. Laser Diodes, Optoelectronic Devices and Heterogeneous Integration, Conference on, Proceedings of SPIE, vol 4947, Brugge, Belgium, pp. 212–222 (2002)

Hasegawa, A., Kodama, Y.: Guiding-center solitons in optical fibers. Opt. Lett. **16**, 1385–1387 (1991)

Hasegawa, A., Tapper, F.: Transmission of stationary nonlinear dispersive dielectric fibers I. Anomalous dispersion. Appl. Phys. Lett. **23**, 142–144 (1973)

Heimala, P., Katila, P., Aarnio, J., Heinamaki, A.: Thermally tunable integrated optical ring resonator with poly-Si thermistor. J. Lightwave Technol. **14**(10), 2260–2267 (1996)

Humblet, P.A., Azizoglu, M.: On the bit error rate of lightwave systems with optical amplifiers. J. Lightwave Technol. **9**, 1576–1582 (1991)

Iannone, E., Matera, F., Mecozzi, A., Settembre, M.: Nonlinear Optical Communication Networks. Wiley, New York (1998)

Klein, E.: Densely integrated microring-resonator based components for fibre-to-the-home applications. Ph.D. thesis, University of Twente, AE Enschede (2007)

Lebref, F., Ciani, A., Matera, F., Tamburini, M.: Numerical model for the study of the behaviour of an erbium amplifier for the evaluation of the transmission performance of optical WDM systems. Fiber Integr. Opt. **18**, 245–254 (1999)

Lichtman, E., Evangelides, S.G.: Reduction of the nonlinear impairment in ultralong lightwave systems by tailoring the fiber dispersion. Electron. Lett. **30**, 346–348 (1994)

Maier, M., Reisslein, M.: AWG-based metro WDM networking. IEEE Commun. Mag. **42**(11), S19–S26 (2004)

Marcuse, D.: Single-channel operation in very long nonlinear fibers with optical amplifier and zero dispersion. J. Lightwave Technol. **9**, 356–361 (1991a)

Marcuse, D.: Derivation of analytical expression for the bit-error probability in lightwave systems with optical amplifiers. J. Lightwave Technol. **9**, 1576–1582 (1991b)

Martelli, F., Greco, C.M., D'Ottavi, A., Mecozzi, A., Spano, P., Dall'Ara, R.: Frequency-conversion efficiency independent of signal-polarization and conversion-interval using four-wave mixing in semiconductor optical amplifiers. IEEE Photonics Technol. Lett. **11**, 656–658 (1999)

Matera, F., Settembre, M.: Compensation of the polarization mode dispersion by means of the Kerr effect for non return-to-zero signals. Optic. Lett. **20**(1), 28–30 (1995)

Matera, F., Settembre, M.: Comparison of the performance of optical systems in fiber links 4000 km long. Fiber Integr. Opt. **14**, 109–119 (1995)
Matera, F., Settembre, M.: Comparison of the performance of optically amplified transmission systems. J. Lightwave Technol. **14**, 1–12 (1996)
Matera, F., Settembre, M.: Role of *Q*-factor and of time jitter in the performance evaluation of optically amplified transmission systems. IEEE J. Sel. Top. Quantum Electron. **6**, 308–316 (2000)
Matera, F., Matteotti, F., Forin, D., Tosi Beleffi, G.: Numerical Investigation of wide geographical transport networks based on 40 Gb/s transmission with all optical wavelength conversion. Opt. Commun. **247**, 341–351 (2005)
Mecozzi, A., Balslev, C., Shtaif, M.: System impact of intra-channel nonlinear effects in highly dispersed optical pulse transmission. Photonics Technol. Lett. **12**, 1633–1635 (2000)
Menyuk, C.R.: Stability of solitons in birefringent optical fibers. II Arbitrary amplitude. J. Opt. Soc. Am. B **5**, 392–402 (1988)
Mollenauer, L.F., Stolen, R.H., Gordon, J.P.: Experimental observation of picosecond pulse narrowing and solitons in optical fibers. Phys. Rev. Lett. **45**, 1095–1098 (1980)
Mollenauer, L.F., Smith, K., Gordon, J.P., Menyuk, C.R.: Resistence of solitons to the effect of polarization mode dispersion in optical fibers. Opt. Lett. **14**, 1219–1221 (1989)
Mollenauer, L.F., Evangelides, S.G., Haus, H.A.: Long distance soliton propagation using lumped amplifiers and dispersion shifted fibers. J. Lightwave Technol. **9**, 194–196 (1991)
Mollenauer, L.F., Gordon, J.P., Evangelides, S.G.: The sliding frequency guiding filters: an improvement form of soliton jitter control. Opt. Lett. **17**, 1575–1577 (1992)
Munoz, P., Pastor, D., Capmany, J.: Modeling and design of arrayed waveguide gratings. J. Lightwave Technol. **20**(4), 661–674 (2002)
Poggiolini, P., Bosco Prat, J., Killey, R., Savory, S.: Branch metrics for effective long-houl MLSE IMDD receivers, ECOC 2006, Cannes, 24–28 Sept 2006
Poole, C.D., Wagner, R.E.: Phenomenological approach to polarization dispersion in long single-mode fibers. Electron. Lett. **22**, 1029–1030 (1986)
Settembre, M., Matera, F., Hagele, V., Gabitov, I., Mattheus, A.W., Turitsyn, S.K.: Cascaded optical communication systems with in-line semiconductor optical amplifiers. J. Lightwave Technol. **15**, 962–967 (1997)
Qiao, C., Yoo, M.: Optical burst switching (OBS) – a new paradigm for an optical internet. J. High Speed Netw. **8**, 69–84 (1999)
Sabella, R., Iannone, E., Pagano, E.: Optical transport networks employing all-optical wavelength conversion: limits and features. IEEE J. Sel. Areas Commun. **14**(5), 968–978 (1996)
Shen, Y.R.: Principle of Nonlinear Optics. Wiley, New York (1984)
Smit, M., Van Dam, C.: PHASAR-based WDM-devices: principles, design and applications. IEEE J. Sel. Top. Quantum Electron. **2**(2), 236–250 (1996)
Soref, R., Bennett, B.: Electrooptical effects in silicon. IEEE J. Quantum Electron. **23**(1), 123–129 (1987)
Srivastava, A.K., Sun, Y., Zyskind, J.L., Sulhoff, J.W., Wolf, C., Tkach, R.W.: Fast gain control in an erbium-doped fiber amplifier. In: Proceedings of the Optical Amplifier and Their Applications 1996:PD-P4, Monterey (1996)
Stampoulidis, L., Petrantonakis, D., Stamatiadis, C., Kehayas, E., Bakopoulos, P., Kouloumentas, C., et al.: Microring-resonator-assisted, all-optical wavelength conversion using a single SOA and a second-order Si3 N4 SiO2 ROADM. J. Lightwave Technol. **28**(4), 476–483 (2010)
Tabacchiera, M., Matera, F., Mecozzi, A., Settembre, M.: Dispersion management in phase modulated optical transmission systems. In: Proceedings of the ECOC 2010, Mo2.C.2, Torino, 20–23 Sept 2010
Wang, C., Tucker, R., Summerfield, M.: Tunable optical wavelength converter with reconfigurable functionality. In: Proceedings of the Optical Fiber Communication (OFC 97) Conference, pp. 76–77, Dallas (1997)
Way, P.K.A., Menyuk, C.R.: Polarization mode dispersion. Decorrelation, and diffusion in optical fibers with randomly varying birefringence. J. Lightwave Technol. **14**, 148–157 (1996)

Winzer, P.J., Essiambre, R.J.: Advanced optical modulation formats. Proc. IEEE **94**(5), 952 (2006)
Yariv, A.: Quantum Electronics, 3rd edn. Wiley, New York (1989)
Yu, X., Li, J., Chen, Y., Qiao, C.: Traffic statistics and performance evaluation in optical burst switching networks. J. Lightwave Technol. **22**, 2722–2738 (2004)
Zirngibl, M.: Gain control in erbium-doped fiber amplifiers by an all-optical feedback loop. Electron. Lett. **27**, 560–561 (1991)
Zitelli, M., Matera, F., Settembre, M.: Single-channel transmission in dispersion managed links in condition of very strong pulse broadening: application to 40 Gbit/s signals on step-index fibers. J. Lightwave Technol. **17**, 2498–2505 (1999)

Chapter 4
Experiments on Long-Haul High-Capacity Transmission Systems

Gabriella Bosco, Francesco Matera, Karin Ennser, Morten Ibsen, Lucia Marazzi, Francesca Parmigiani, Periklis Petropoulos, Pierluigi Poggiolini, Marco Tabacchiera, and Marcelo Zannin

4.1 Introduction

Progress in optical communications has been one of the key factors for the enormous growth of the ICT sector, and, in particular, of the Internet phenomenon. Such a progress has been driven by experimental successes that have been obtained in the last three decades in several laboratories all over the world, and we have witnessed a fantastic challenge among such labs to reach record targets such as maximum bit rate, maximum propagation distance, higher performance, maximum efficiency, and, recently, minimum energy consumption. The search for the *maximum bit rate* $\times$ *distance* was based on the principle of *infinite* bandwidth of the optical fiber that let us to imagine transmission of enormous capacities over transoceanic distances, especially after the invention of the optical amplifier.

G. Bosco (✉) • P. Poggiolini
Electronic Department, Politecnico di Torino, Corso Duca degli Abruzzi, 24, 10129 Torino, Italy
e-mail: gabriella.bosco@polito.it; pierluigi.poggiolini@polito.it

F. Matera • M. Tabacchiera
Fondazione Ugo Bordoni, Viale America, 201, 00144 Roma, Italy
e-mail: mat@fub.it; marco.tabacchiera@gmail.com

K. Ennser • M. Zannin
College of Engineering, Swansea University, Singleton Park, SA2 8PP Swansea, West Glamorgan, UK
e-mail: k.ennser@swansea.ac.uk; marcelo.z.rosa@gmail.com

M. Ibsen • F. Parmigiani • P. Petropoulos
Optoelectronics Research Centre (ORC), University of Southampton, Highfield, Southampton, Hampshire, SO17 1BJ
e-mail: mi@orc.soton.ac.uk; frp@orc.soton.ac.uk; pp@orc.soton.ac.uk

L. Marazzi
Corecom, Via G.Colombo 81, 20133 Milano, Italy
e-mail: marazzi@corecom.it

A. Teixeira and G.M.T. Beleffi (eds.), *Optical Transmission: The FP7 BONE Project Experience*, Signals and Communication Technology,
DOI 10.1007/978-94-007-1767-1_4,

Another physical principle which drove the experiments for many years was the soliton transmission. In fact, the soliton, with its characteristics of pulse that does not change its shape along the propagation, was the candidate for the ideal signal, to achieve the "ideal channel." Soliton was introduced in the optical fiber by Hasegawa and Tapper in 1970 (Hasegawa 1973) and was experimentally demonstrated by Mollenauer (1980). After this experimental demonstration, we observed fantastic experiments of very high bit rate transmission systems based on solitons in different propagation conditions.

Fostered by such achievements, many optical devices were invented, fabricated, and experimented for transmission, detection, amplification, and processing. Many such devices have been used also in other technological fields as in medicine, hi-fi equipments, and so on. Furthermore, while the labs were involved in efforts to reach the highest capacities, optical communications was introduced in short distance applications with relatively low capacity, changing the transmissions in the access networks and permitting the replacement of copper cables with optical fibers. Concerning long distance systems, soliton pulses were more useful to understand the nonlinear propagation in optical fibers than in the implementation of networks based on soliton transmission; as a result, thanks to studies on solitons, novel and simpler transmission regimes were proposed, for example, the periodically chromatic dispersion compensation. In addition, coherent systems, that seemed useless for the diffusion of optical amplifiers at the end of the 1990s, have been widely exploited in the last decade since they permit an efficient bandwidth treatment.

Experiments drove the commercial diffusion of optical transmission. In particular, we can say that the ratio in terms of maximum capacity between the experimental environment and the commercial one is about 10. So, if in the 1990s we were in a phase of systems installation based on a channel bit rate of 2.5 Gbit/s, the experiments already aimed at 40-Gbit/s transmission. Similarly, if today the installation mainly regards systems with a channel bit rate at 10 Gbit/s (and some at 40 Gbit/s), the target channel bit rate for experimental investigation is currently 100 Gbit/s and beyond.

Based on these considerations, in this chapter, we report some experiments on 100-Gbit/s transmission that have been achieved in the labs of some partners (Politecnico di Torino and Politecnico di Milano) that took part in the FP7 BONE project. In particular, the authors report some method to achieve fluxes at 100 Gbit/s, using QPSK modulation with PM-QPSK and DQPSK modulation with polarization multiplexing (PM-DQPSK or POLMUX). Furthermore, different propagation conditions are experimented as the case over installed fibers, transoceanic distances, and also by adopting the wavelength conversion.

To make clear the status of the current experiments in the labs all over the world, we also report a short overview of results that were presented at the last European Conference on Optical Communication (ECOC) that was held in Turin in September 2010.

We conclude this chapter with some results on transmission in the presence of a burst traffic, which is related to diffusion of Internet data, traffic that could be degraded by the behavior of erbium optical amplifiers.

4.2 100-Gbit/s PM-QPSK WDM Transmission Experiments Over Installed Fiber

Coherent polarization-multiplexed quadrature-phase-shift keying (PM-QPSK) has recently emerged as the forerunner for practical 100 Gbit/s per channel wavelength division multiplexing (WDM) ultra-long-haul transmission, with dense 50-GHz channel spacing (Fludger 2007; Salsi et al. 2009). A requirement that many carriers demand is to maximize compatibility with existing plants, which are built for 10-Gbit/s IM-DD systems and have DCUs along the line for dispersion management. However, the presence of DCUs has been predicted to impact system performance quite substantially, especially when transmitting over SSMF (Curri et al. 2008; Grellier et al. 2009). In addition, due to splices, jumpers, and bends, actual installed fibers are far less ideal than spooled lab fibers.

Here, the performance of a 16-channel WDM 100 Gbit/s per channel NRZ PM-QPSK system, with 50-GHz channel spacing, is tested over SSMF installed within the operational Metro Network of the operator Fastweb in Torino (Italy) (Poggiolini et al. 2006). In particular, the impact of different dispersion maps on the link span budget is assessed in a 16-span, 1,000-km configuration, comparing the results with simulations (Gavioli et al. 2010). The system maximum reach is measured, as well.

4.2.1 *System Setup*

The setup is shown in Fig. 4.1. The WDM source had 16 DFB lasers spaced 50 GHz. A tunable laser with a linewidth of 100 kHz was used to replace a DFB at the comb center for the channel under test. QPSK modulation was generated using NMZ low-$V_{ß}$ (3 V) modulators. Odd and even channels had separate modulators. The used test sequence was $(2^{15}-1)$, fed to the electrodes of the two NMZs with different electrical delays so that the data encoded over odd and even channels were different. The modulation rate was 25 GBaud. An overall 1.6-Tbit/s WDM NRZ-PM-QPSK signal was assembled by first combining the odd and even QPSK channels, decorrelated by hundreds of symbols, through a 50-GHz interleaver and then splitting and combining over orthogonal polarizations with a 915 symbol delay to achieve multiplexing and decorrelation of the two polarizations.

The overall signal was then launched into a recirculating loop consisting of 63.6 km of installed SSMF (D = 16.3 ps/nm/km). Even if the span length was somewhat short, the total fiber loss was 16.5 dB, mimicking the attenuation of

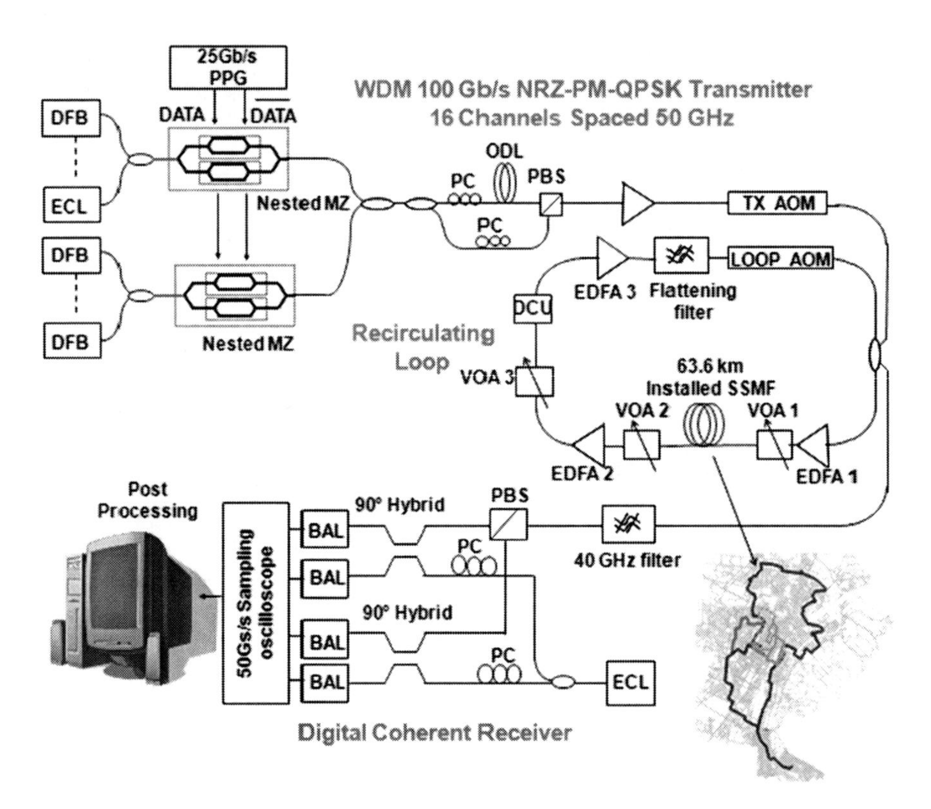

Fig. 4.1 Setup of the transmission loop experiment over installed SSMF in the city of Torino. *ECL* external cavity laser, *PPG* pulse pattern generator, *ODL* optical delay line, *PBS* polarization beam splitter, *AOM* acousto-optic switch, *BAL* balanced photodetector, *PC* polarization controller

a longer and more realistic span, for example, 82.5 km with loss of 0.2 dB/km. The rather large average loss value (0.26 dB/km) was due to the many splices and connectorized jumpers located along the installed link (Poggiolini et al. 2006). The fiber was preceded by a VOA1 to set the launched power and it was followed by VOA2 to vary the span loss. The loop included three low-noise EDFAs and an MZ-based gain-flattening filter. The DCU was inserted between EDFA2 and EDFA3 and was preceded by VOA3 which set the DCU input power to −10 dBm per channel.

This value of DCU input power realized the best trade-off between DCU nonlinearity (if power too high) and excess noise from EDFA3 (if power too low). The effective noise figure of the loop amplification of Fig. 4.1 was about 6.5 dB, and it decreased to 5.35 dB when EDFA3 is removed.

Two different DCUs were used: DCU1 with −1,165 ps/nm and DCU2 with −985 ps/nm. Since the per-span accumulated CD of the SSMF is 1,037 ps/nm,

DCU1 and DCU2 created an over- and an under-compensated map, with span residues −128 ps/nm and +48 ps/nm, respectively. Precompensation was also considered and is discussed later.

At the receiver (Rx), there was an optical filter with 40-GHz bandwidth. Then, the signal was passed through a PBS and combined with the local oscillator (LO) via two optical 90° hybrids to detect the inphase and quadrature components of the two polarizations. The eight outputs of the two hybrids were detected using amplified linear dual-balanced photoreceivers with 30-GHz bandwidth. The signal was digitized at 50 GSa/s using a real-time scope with 13-GHz analog bandwidth and 5 bit resolution, with the waveforms then processed off-line in a conventional PC. The Rx was tuned to the WDM comb center channel. The samples went through a first DSP stage consisting of two complex FIR filters, one for each polarization, set to fully compensate for the nominal amount of accumulated dispersion (taking DCUs into account, when present). Then, an FFT-based algorithm estimated the transmitter (Tx) laser vs. LO frequency shift and compensated for it. Following, there was a second stage equalizer consisting of a fractionally spaced ($T_S/2$) MIMO 4-by-4 structure, made up of 16 real FIR filters with 11 taps each. This equalizer is initially driven to convergence through a training-sequence-based LMS algorithm. Then it moves into a decision-directed tracking mode. The second stage was followed by a decision-driven carrier-phase estimation stage, windowing over 20 symbols, and a value which is close to the optimum for all analyzed cases. Symbol decision and error counting were performed over 1.6 million bits, or 400,000 symbols.

4.2.2 Results

Figure 4.2 shows the back-to-back BER performance measured on the center channel. The required optical signal to noise ratio (OSNR) at $\text{BER} = 10^{-3}$ is 15 dB (0.1 nm), 2.2 dB from the quantum limit.

In a first set of tests, the signal was recirculated 16 times for a total fiber length of 1,018 km. Figures 4.3 and 4.4 show the resulting tolerable span loss (A_{span}) vs. per channel fiber launch power (P_{Tx}), measured at the SSMF fiber input. Note that all points of all curves in figure correspond to $\text{BER} = 10^{-3}$. This gives a more accurate assessment of system performance limits than probing the system behavior at a nonconstant operating BER, such as in Q vs. P_{Tx} plots. The dashed straight line represents linear propagation and therefore has a slope of exactly one. It is shown for convenience to help to visually detect when curves start bending due to the onset of nonlinearity. Figures 4.3 and 4.4 allow to clearly identify the maximum A_{span} and the corresponding optimum launch power, at the reference $\text{BER} = 10^{-3}$.

For instance, Fig. 4.3 shows that the no-DCU system allows to sustain at most 24.1 dB of span loss, for an optimum $P_{\text{Tx}} = 0.8$ dBm. Inserting DCUs deteriorates performance. The degradation is about 2 dB of span loss for DCU1 and 3 dB for

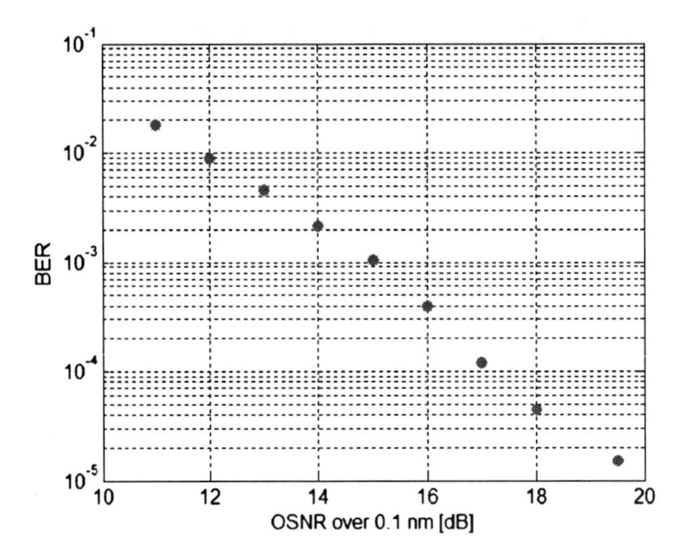

Fig. 4.2 BER vs. OSNR (0.1 nm) in back-to-back

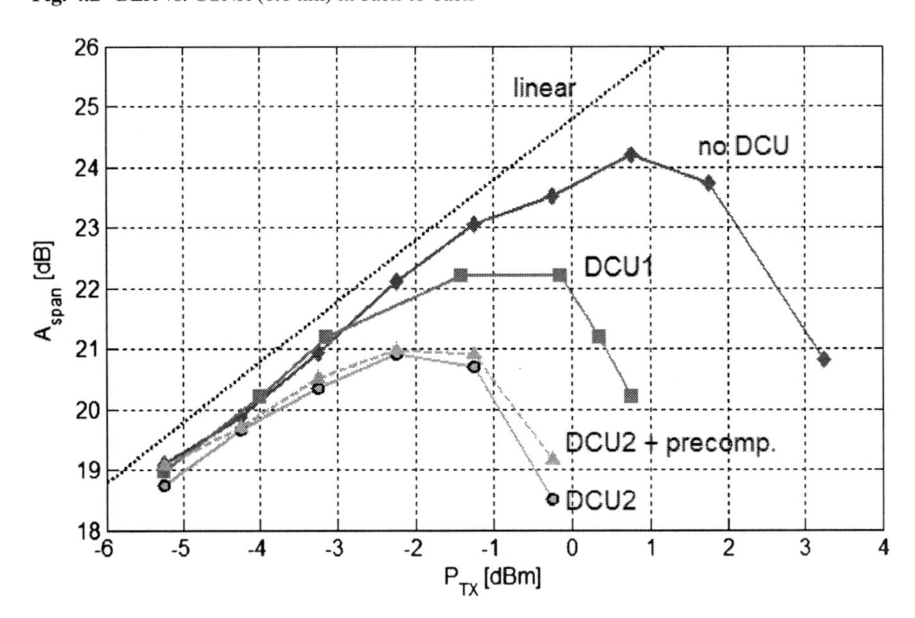

Fig. 4.3 Experimental results of max span loss A_{span} vs. Tx power per channel P_{Tx} at BER = 10^{-3}, 16 spans. *Dashed lines*: linearity fitting. DCU1: −1165 ps/nm (span residue −128 ps/nm), DCU2: −985 ps/nm (span residue +48 ps/nm). Precompensation: −700 ps/nm

DCU2. Note that no-DCU results were obtained by replacing the DCU with a linear loss equivalent to the DCU loss. This way, the power into EDFA3 was kept constant so that the linearity performance was similar with and without DCU. The computer

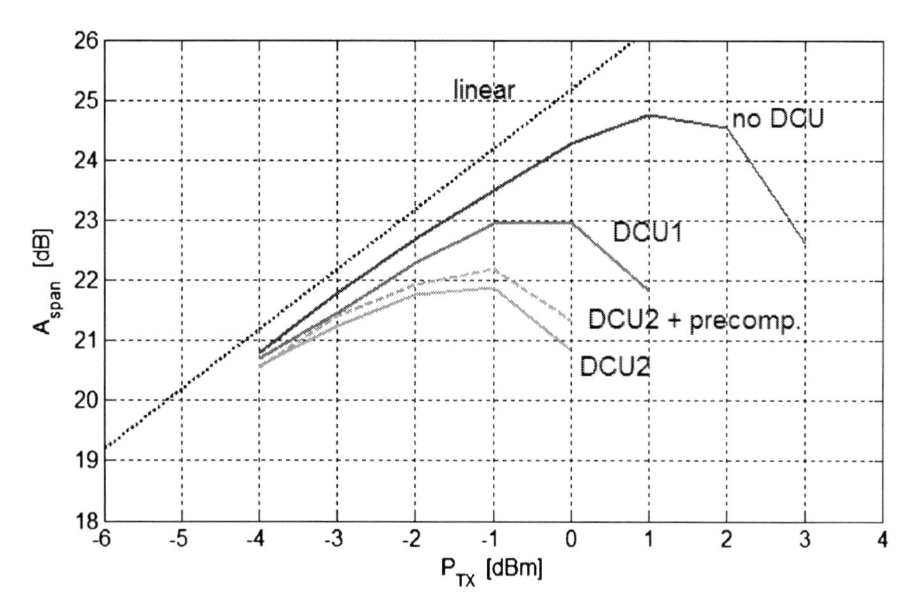

Fig. 4.4 Simulation results of max span loss A_{span} vs. Tx power per channel P_{Tx} at BER $= 10^{-3}$, 16 spans. *Dashed lines*: linear fitting. DCU1: −1165 ps/nm (span residue −128 ps/nm), DCU2: −985 ps/nm (span residue +48 ps/nm). Precompensation: −700 ps/nm

simulations shown in Fig. 4.4 are in good agreement with the experimental results. They are also in general agreement with the simulated results in Curri et al. (2008), Van Den Borne et al. (2009), and Grellier et al. (2009).

With DCU2, we tried precompensation. Due to DCU availability, we could use −150, −550, −1,165 ps/nm, or any sum thereof. The computer simulation optimum was close to −700 ps/nm so we used this value. However, the predicted advantage was small (Fig. 4.4) and the experiment confirmed such forecast (Fig. 4.3).

The fact that computer simulations appeared quite reliable allowed us to exploit them to explore a much larger dispersion map domain than reachable through experiments. In Fig. 4.5, a contour plot is shown of max A_{span} vs. in-line dispersion residue $D_{\mathrm{res,IL}}$ and Tx power per channel P_{Tx}. The vertical dashed lines correspond to the $D_{\mathrm{res,IL}}$ of no-DCU, DCU1, and DCU2 (+1,037, −128, and +48 ps/nm, respectively). The plot shows that typical legacy compensation is located near the plot center, where a performance dip is present. Instead, large values, both positive and negative, of $D_{\mathrm{res,IL}}$ yield a much better performance. However, large negative $D_{\mathrm{res,IL}}$ values are unrealistic, as they would require too much DCU. As a result, the best practically implementable setup seems no-DCU.

The suboptimality of the links with DCUs is even more evident when looking at the maximum achievable range, or number of spans (Fig. 4.6). To carry out this test, VOA2 was removed from the loop so that the span loss was fixed at its minimum value of 16.5 dB. Then, for each number of recirculations, we found the value of P_{Tx} granting BER $= 10^{-3}$. This time, the no-DCU system was measured without

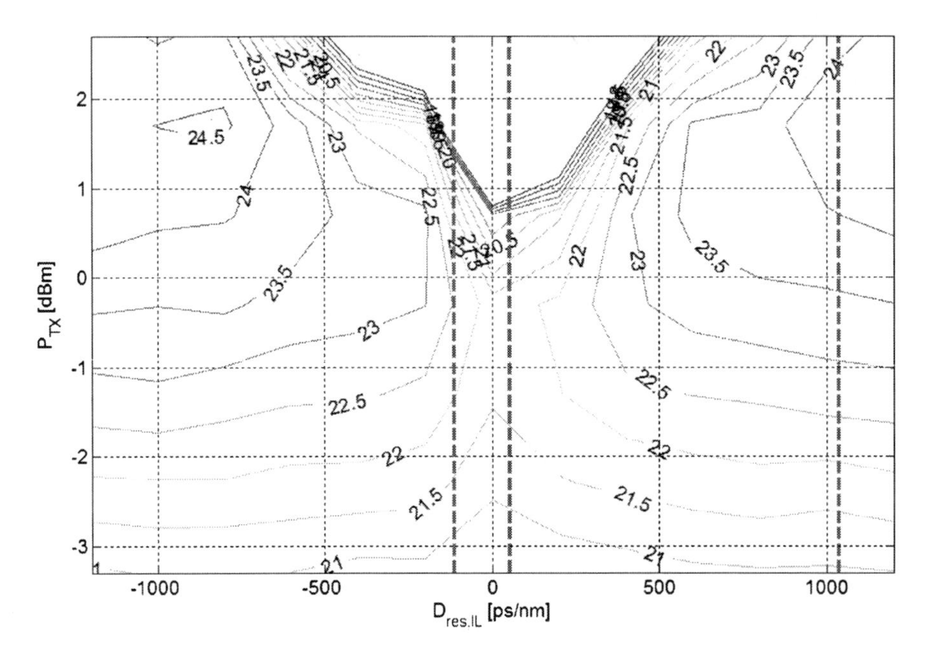

Fig. 4.5 Simulated contour plot of max span loss in [dB] for BER $= 10^{-3}$ vs. in-line dispersion residue $D_{\mathrm{res,IL}}$ and Tx power per channel. Vertical dashed lines: loci of points corresponding to no-DCU, DCU1, and DCU2

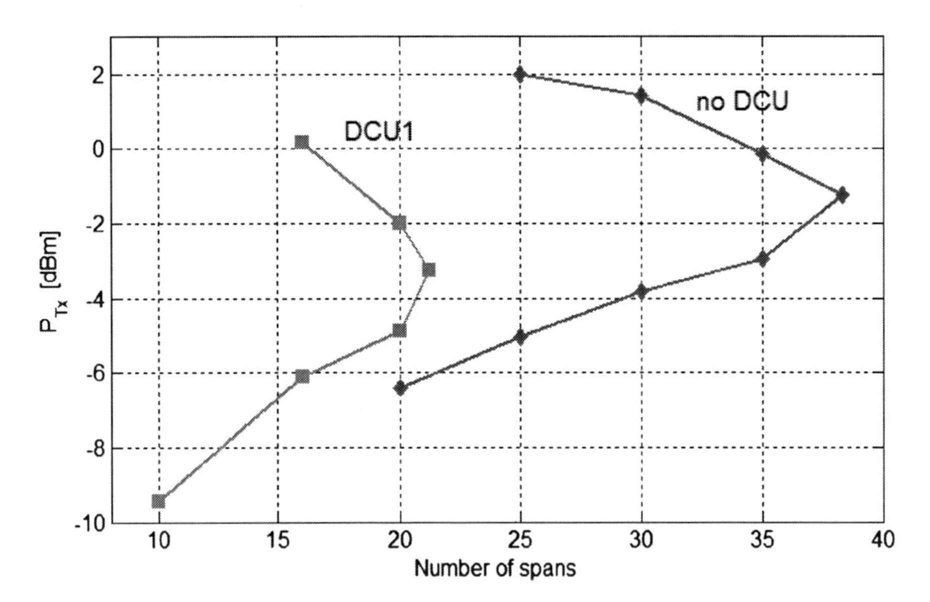

Fig. 4.6 Number of spans vs. Tx power per channel at BER $= 10^{-3}$

VOA3 and EDFA3. This explains why the linearity regions of the two curves (the lower branches in Fig. 4.6) are not aligned: the shift exactly matches the decrease in the effective noise figure (1.25 dB). This was done on purpose, to fully exploit the advantage of completely removing DCUs. Figure 4.6 shows that the presence of DCU1 almost halves the max span number, with respect to no-DCU: from 38 to 21 spans only. The penalty on the maximum reach for compensated transmission is due to the enhancement of fiber nonlinear effects induced by the presence of in-line DCUs (Curri et al. 2008): in linear regime the penalty is 1.25 dB only due to the different amplification setup (Fig. 4.7).

In conclusion, our installed-fiber tests confirmed the good resilience of PM-QPSK to nonlinear effects, showing its viability also over highly lossy installed plants. However, even with 10-G-IM-DD channel removal, the use of PM-QPSK over long-haul legacy links with DCUs might be problematic.

With no-DCUs, despite the impairments typical of a metro installed cable, transmission was possible with an ample span loss budget of 24 dB. However, when DCUs were inserted, mimicking legacy plants originally designed for IM-DD systems, performance was seriously impaired, causing a span budget reduction of 2–3 dB. Precompensation was attempted but its impact was essentially negligible. Also, using computer simulations, we first matched the experimental results and then extended penalty predictions over a large variety of single periodicity maps. The no-DCU configuration still emerged as the best one, even across such wider range of maps.

Finally, we carried out a maximum reach experiment over the same installed fiber: a compensated system achieved 21 spans (1,335 km) whereas a no-DCU system reached 38 spans (2,417 km). In conclusion, NRZ-PM-QPSK yields the best performance without any in-line DCUs. It can also be used over legacy links with DCUs but span budget, launch power, and reach may get substantially curtailed.

4.3 Transoceanic PM-QPSK Terabit Superchannel Transmission Experiments

Recently, there has been an increasing interest in the investigation of 1-Tbit/s "superchannels" (Ma et al. 2009; Chandrasekhar et al. 2009) in support of an eventual Terabit Ethernet Standard. According to this technique, a number of "subcarriers" are seamlessly aggregated to form individual superchannels which would be routed optically through the network as a single channel.

In Ma et al. (2009) and Dischler et al. (2009), the subcarriers were electrically OFDM modulated and the superchannel reached 600 and 400 km, respectively. In Chandrasekhar et al. (2009), 24 subcarriers were modulated using polarization multiplexed (PM) QPSK at 12.5 GBaud each. The subcarriers were spaced exactly at the Baud rate and each subcarrier was frequency-locked and symbol transition-aligned to all others, thus realizing coherent optical OFDM (Co-OFDM). Using

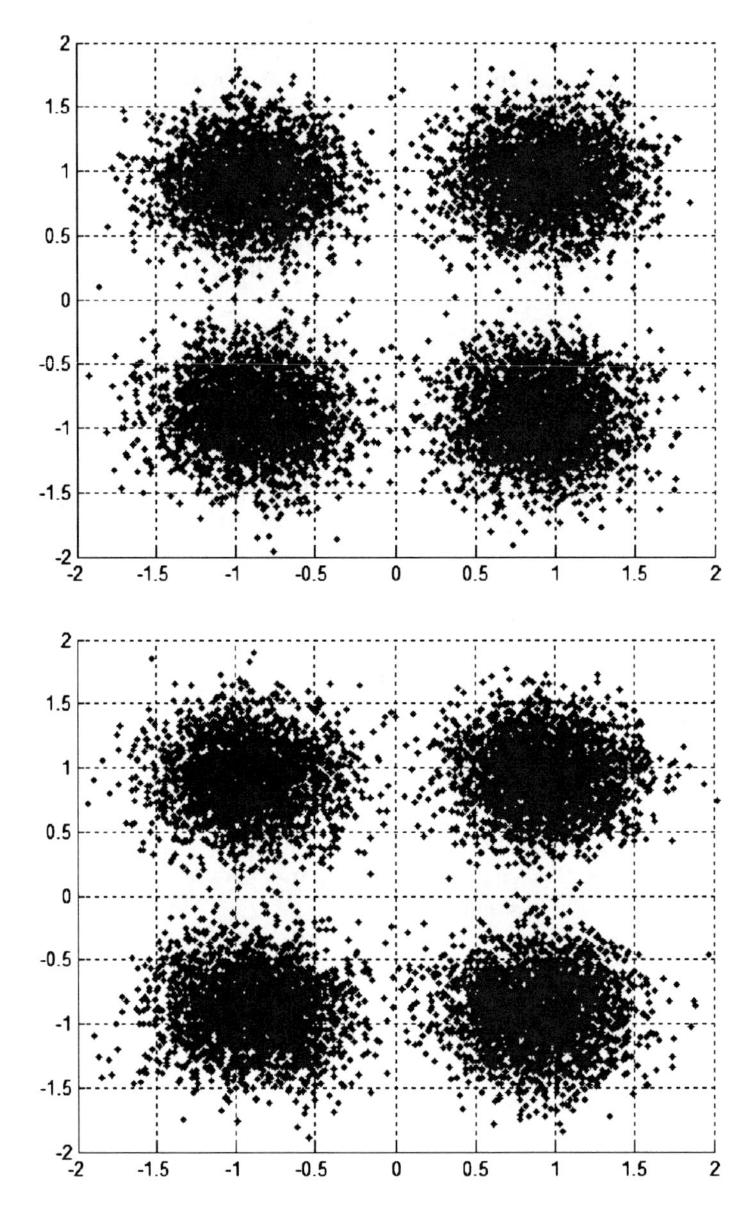

Fig. 4.7 Scattering diagrams at farthest reach (38 spans, 2,400 km)

low-loss and ULAF in combination with Raman amplification, Chandrasekhar et al. (2009) reached a transmission distance of 7,200 km.

A possible alternative approach to Co-OFDM, to realize near-Baud-rate or Baud-rate spacing of subcarriers, is that of creating a superchannel by tightly packing conventional WDM subcarriers, while achieving low crosstalk by using subcarrier narrow optical filtering at the transmitter (Tx). This technique has been widely used in radio links for decades and has been recently proposed for optical links too (Gnauck et al. 2009; Cai et al. 2010). This concept is sometimes called "Nyquist WDM" (Bosco et al. 2010).

In this section, we show experimental results on the long-haul transmission of a Terabit superchannel using 10× (30 GBaud) PM-QPSK subcarriers (1.2 Tbit/s), with off-line processing (Torrengo et al. 2010). The spacing was set at either 33 or 30 GHz, that is, either 1.1 times the Baud rate or exactly the Baud rate. An experiment was also performed using three Terabit superchannels together. Overall, these different setups always reached at least 8,000 km, and in one case, up to 10,000 km. Note that we decided to push the Baud rate to 30 GBaud, despite component bandwidth and sampling rate limitations, to be able to assume 20% FEC overhead, which appears to be the target overhead for second-generation PM-QPSK transponders. Advanced hard-FECs with 20% overhead, with a threshold $\mathrm{BER} = 10^{-2}$, are currently being engineered. With the same overhead, soft-FECs could reach a substantially higher BER (210^{-2} or greater). In this work, we aimed at the hard-FEC threshold of $\mathrm{BER} = 10^{-2}$.

4.3.1 Description of the Experiment

The Tx schematic for the generation of the three superchannels with 33-GHz spacing is shown in Fig. 4.8. Fifteen CW wavelengths spaced 66 GHz were modulated using a NMZM to generate a 15-subcarrier signal, with each subcarrier carrying 60 Gbit/s. This WDM signal was then narrow filtered using a reconfigurable Finisar optical Waveshaper™ filter, with −3-dB bandwidth and 33 GHz. This filter is capable of generating all passbands at once, so only one filter unit was used. In addition to narrow filtering, the Waveshaper filter profile was set to enhance the high-frequency components of each QPSK subcarrier (see Fig. 4.9) to precompensate for electrical bandwidth limitations both at the Tx and Rx. All 15 66-GHz spaced QPSK subcarriers were then launched into an optical frequency doubler, comprising a pass-through branch and a FS branch. The latter included an NMZ modulator operated as an FS, configured to shift the 15 input subcarriers by 33 GHz. The pass-through and FS branches were delayed for decorrelation and then combined to form a 30-subcarrier, 33-GHz spaced QPSK-modulated signal. This signal was then polarization multiplexed to form a 30-subcarrier PM-QPSK signal, with 33-GHz spacing. Each PM-QPSK subcarrier carries 120 Gbit/s. The Tx signal spectrum is shown in Fig. 4.10.

The signal was then launched into a recirculating fiber loop (see Fig. 4.11) consisting of 98.1 km of uncompensated Z-PLUS® PSCF. The fiber loss is 17.6 dB. Nominal dispersion at 1,550 nm is 20.6 ps/nm/km and a slope of 0.06 ps/nm^2/km. The effective area is about 110 μm^2. Backward Raman amplification was used with a net gain of 9.2 dB. A three-pump Raman source with a total power of 800 mW was employed at 1,425, 1,436, and 1,459 nm. The loop included a dual-stage EDFA

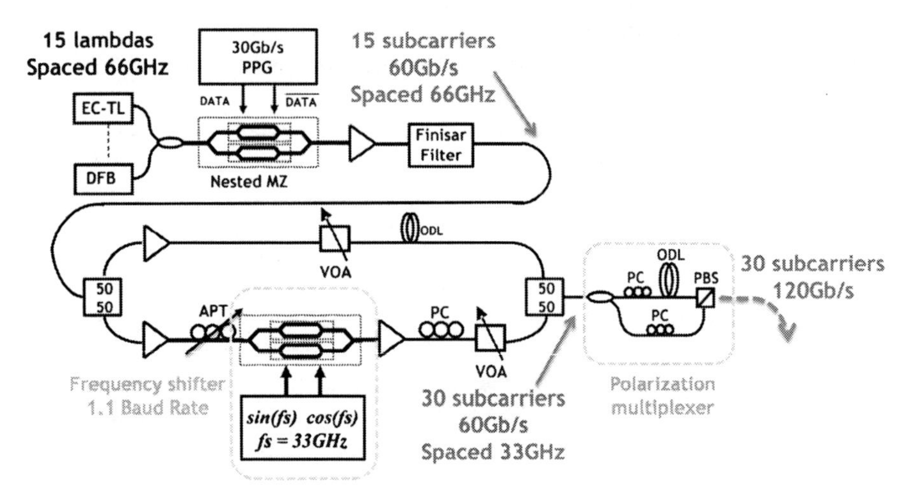

Fig. 4.8 Transmitter schematic for the 3 × 1.2 Tbit/s superchannels. *DFB* distributed-feedback laser, *MZ* Mach–Zehnder, *PPG* pulse pattern generator, *ODL* optical delay line, *PBS* polarization beam splitter, *PC* polarization controller, *APT* automatic polarization controller, *ECL* external cavity laser

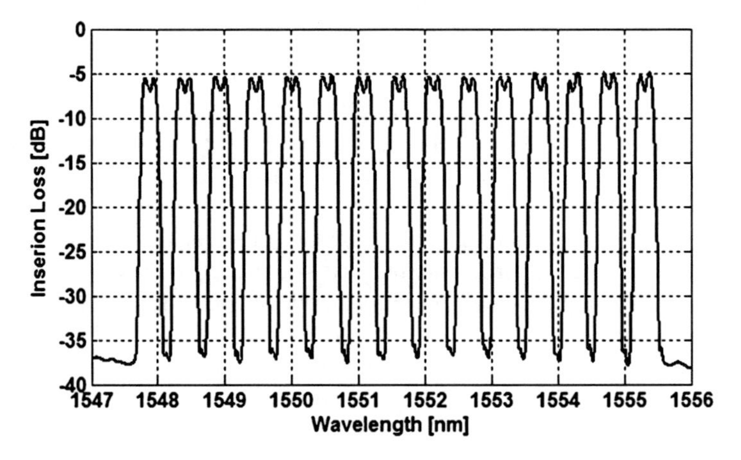

Fig. 4.9 Waveshaper™ filter profile with 33 GHz bandwidth (high resolution)

as well. A gain-flattening filter, the loop acousto-optic switch, and the 3-dB coupler were inserted between the EDFA first and second stage.

The Rx, shown in Fig. 4.12, had a standard setup for coherent reception, with an LO and two 90° hybrids. The eight outputs of the two hybrids were detected using linear-amplified dual-balanced photodetectors with 30-GHz bandwidth provided by Linkra-Teleoptix. Subcarrier selection was performed exclusively by tuning the LO. The four photodetectors electrical signals were digitized at 50 GSa/s using a Tektronix DPO71604 real-time scope. The sampling rate was 1.66 samples/symbol.

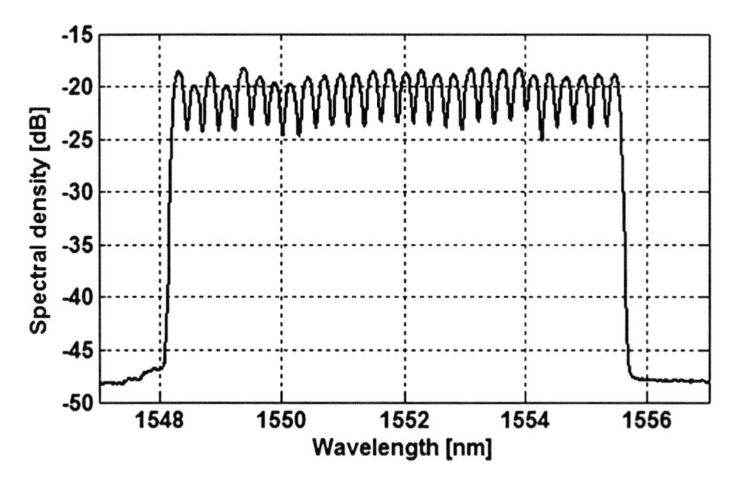

Fig. 4.10 Superchannel spectrum at Tx with 30 subcarriers at 33 GHz spacing (0.06 nm resolution)

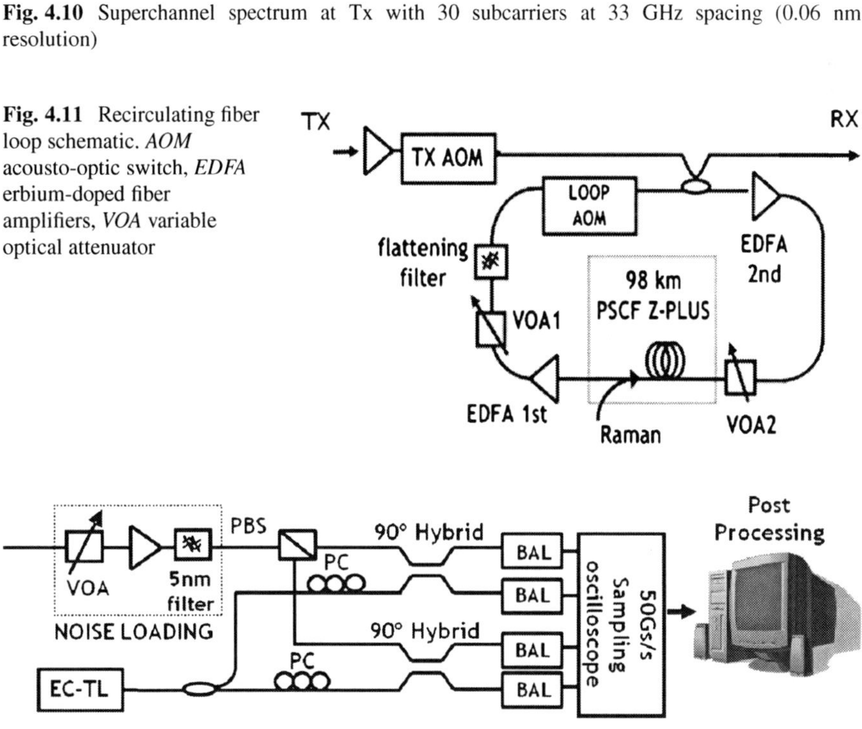

Fig. 4.11 Recirculating fiber loop schematic. *AOM* acousto-optic switch, *EDFA* erbium-doped fiber amplifiers, *VOA* variable optical attenuator

Fig. 4.12 Digital coherent receiver schematic. *BAL* balanced photodetector, *PC* polarization controller, *ECL* external cavity laser

The measured scope analog −3 dB bandwidth was only 12 GHz, i.e., 0.4 times the Baud rate. It was compensated for by Tx optical preemphasis, as mentioned, and by the Rx DSP equalizer.

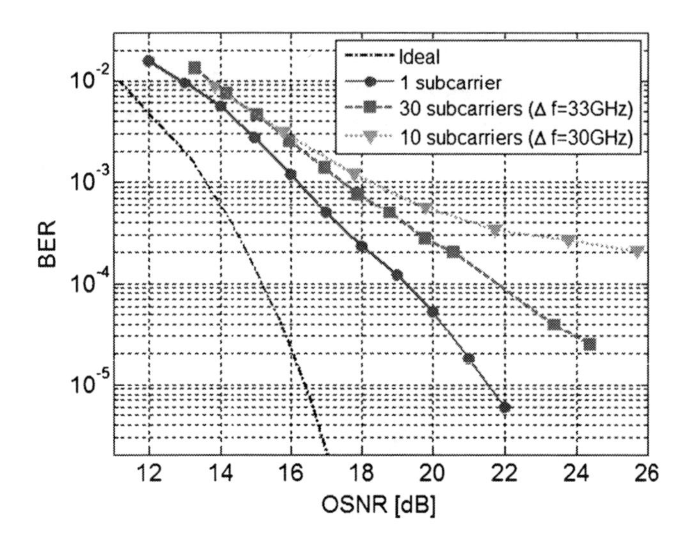

Fig. 4.13 BER vs. OSNR evaluated with an optical bandwidth of 0.1 nm in back-to-back for several combinations of subcarriers in number and spacing

The Rx DSP consisted of a CD-compensation first stage followed by a 25-taps MIMO stage adjusted through a decision-driven CMA algorithm, followed in turn by frequency estimation, and a Viterbi & Viterbi stage. Tx and LO lasers were two distinct, ECL, with linewidth of about 100 kHz. At the Tx, the ECL was tuned to each channel position, replacing each DFB source in turn, for BER measurements.

4.3.2 Experimental Results

Figure 4.13 shows the back-to-back (BTB) BER performance measured on the 16th channel. The required OSNR at BER $= 10^{-2}$ was 13.7 dB, 2.6 dB from ideal. The maximum distance achieved with 30 subcarriers at BER $\leq 10^{-2}$ was 82 spans (8,044 km) at an optimal power per subcarrier of –4 dBm.

Figure 4.14 shows the measured BER on each of the 30 subcarriers. Taking out FEC overhead, the net transmission rate was 3 Tbit/s. We then repeated the experiment with just one 1.2-Tbit/s superchannel (10 subcarriers). We reached 102 recirculations, or 10,006 km. The BER for all subcarriers is shown in Fig. 4.15. The reason for the longer reach of the single superchannel was that the loop EDFA could not be optimized for the low-gain regime of mixed EDFA–Raman amplification. Its NF was about 7.5 dB when transmitting three superchannels and about 5.6 dB with a single superchannel. The longer reach obtained with a single 1-Tbit/s superchannel vs. three (10,006 vs. 8,044 km) was therefore mostly due to lower NF, rather than lower nonlinear crosstalk.

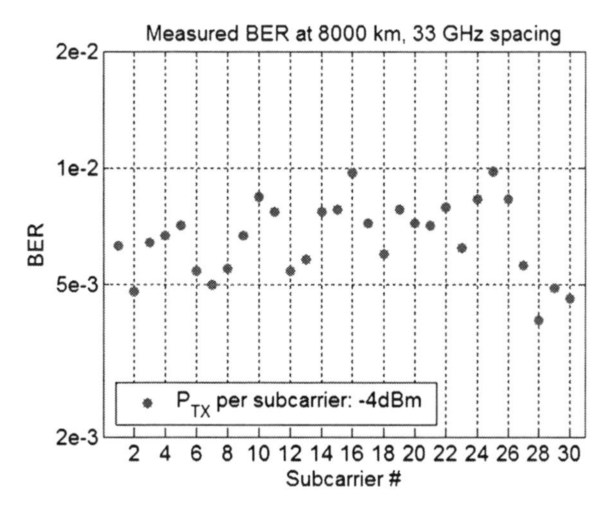

Fig. 4.14 BER of all 30 subcarriers of the $3 \times (1.2\ \text{Tbit/s})$ superchannel at 8,040 km with 33 GHz spacing

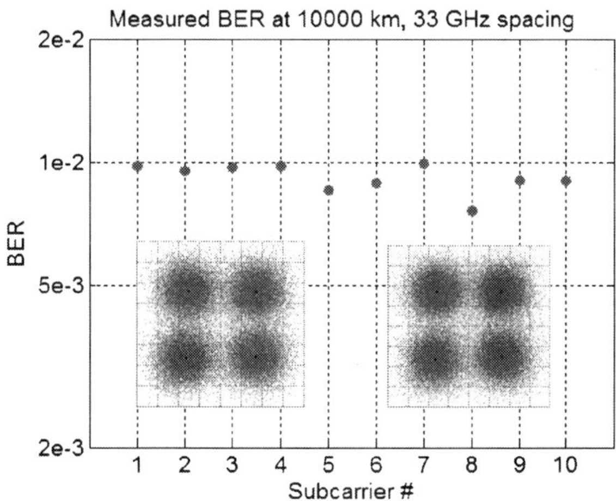

Fig. 4.15 BER of all 10 subcarriers of the single 1.2-Tbit/s superchannel at 10,000 km with 33 GHz spacing

We then lowered the subcarrier spacing to 30 GHz, which is to the Baud rate, and transmitted a single Tbit/s superchannel. The Waveshaper bandwidth was reduced to 30 GHz. The BTB sensitivity measurement over the fifth subcarrier showed a floor, due to crosstalk, of about $\text{BER} = 10^{-4}$ (see Fig. 4.13). Nonetheless, long-haul performance was marginally affected and a max reach curve taken on the fifth subcarrier is shown in Fig. 4.16, for $\text{BER} \leq 10^{-2}$, reaching 9,000 km with an optimum launch power around -4.5 dBm per subcarrier. We measured BER on subcarriers sixth and first as well, which showed same and slightly better performance than the fifth.

Note that the Waveshaper filter was used in these experiments because of its easy reconfigurability. However, its measured profile was not particularly steep (approximately order-2 super-Gaussian). Current technology can achieve steeper filters allowing Baud-rate spacing with lower crosstalk. Note that there is no need to

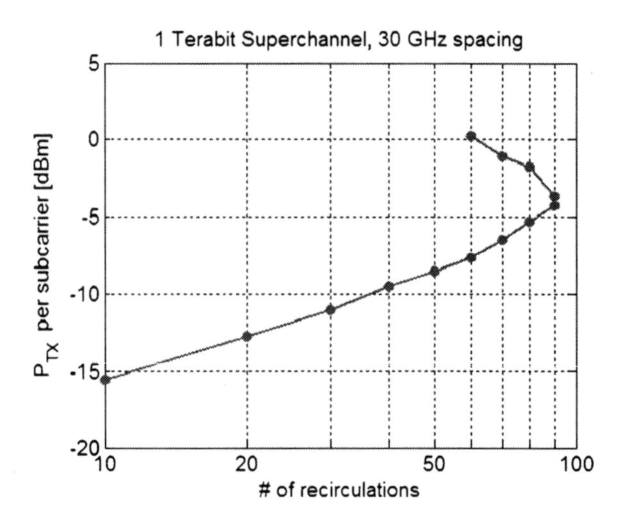

Fig. 4.16 Transmission distance vs. fiber launch power per subcarrier at 30 GHz spacing

use a separate filter for each subcarrier: Tx filtering can be performed, for instance, using interleavers (Gnauck et al. 2009; Cai et al. 2010). In fact, experiment (Cai et al. 2010) reached 10,600 km with spacing 1.2 Baud rate and 4,368 km at an impressive spacing 0.9.

All in all, we believe that the results reported in Cai et al. (2010) and Torrengo et al. (2010) show that the "Nyquist WDM" concept together with PM-QPSK and coherent reception appear to allow extreme subcarrier packing, over ultra-long-haul distances. It is therefore a promising technique for ultra-high-spectral density, ultra-long-haul links.

4.4 Experimental Investigation on 100-Gbit/s POLMUX Including All-Optical Wavelength Conversion

This paragraph is dedicated to the experimental investigation on the performance of the very high bit rate systems at 100 Gbit/s based on POLMUX DQPSK directly detected also by adopting the chromatic dispersion compensation base on the OPC along the line. We first describe the experimental TX-RX setup, then some results on a short fiber link, assessing POLMUX DQPSK chromatic dispersion tolerance and finally the transmission experiments over an uncompensated 500-km SSMF link obtained by exploiting OPC achieved through AOWC.

4.4.1 Experimental Setup

The 100-Gbit/s RZ-DQPSK POLMUX signal used in the experiments is obtained modulating a CW DFB laser with a RZ-carver, i.e., a push–pull Mach–Zehnder

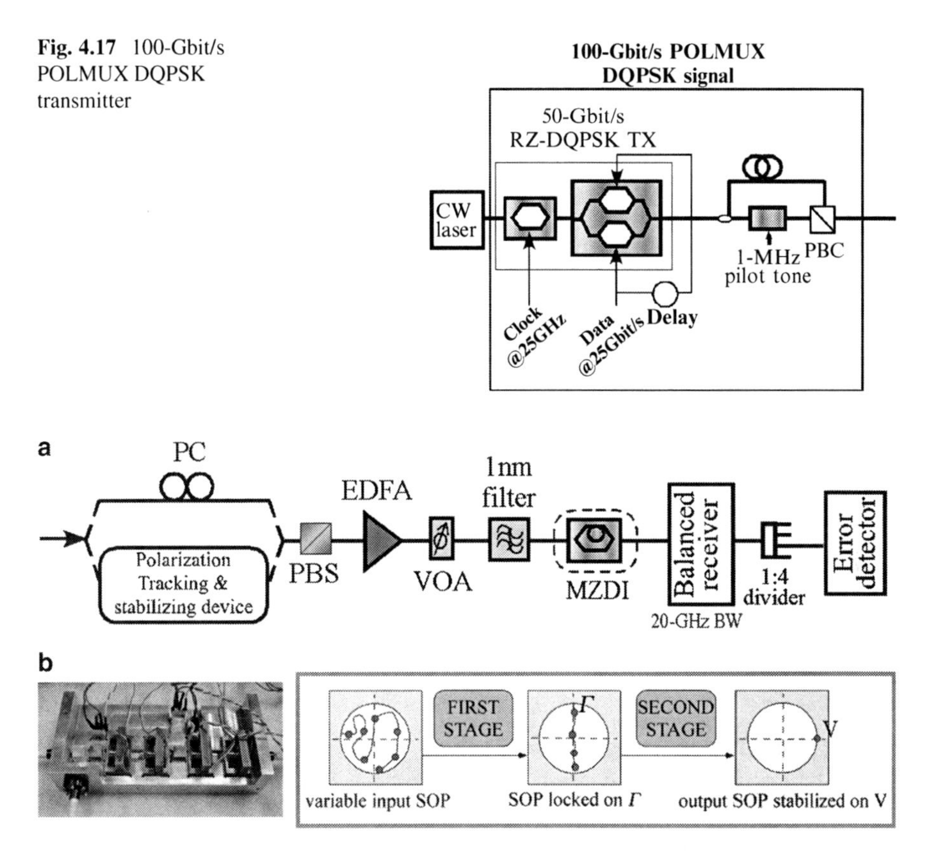

Fig. 4.17 100-Gbit/s POLMUX DQPSK transmitter

Fig. 4.18 100-Gbit/s POLMUX DQPSK receiver

modulator, driven by a 12.5-GHz clock, and then by a dual-parallel Mach–Zehnder modulator driven by two mutually delayed 25-Gbit/s pseudorandom bit sequences (PRBS), as can be seen in Fig. 4.17. The resulting 50-Gbit/s RZ-DQPSK signal is subsequently split into two replicas, which are uncorrelated by a fiber coil. One of the two channels is then labeled by a 1-MHz pilot tone, which is superimposed to the signal as intensity modulation (with a peak-to-peak amplitude corresponding to 10% of the average), in order to identify the channel whose SOP is to be monitored and controlled at the direct-detection receiver. The channels are recombined after their SOPs are made orthogonal, thus reaching an overall capacity of 100 Gbit/s. The uncorrelation between the two channels correctly emulates two different data streams.

This signal can be directly received provided that the two polarizations are properly separated; this is to be done with an automatic polarization stabilizer (Martinelli 2006), which exploits the pilot tone label to stabilize one of the two channels, while the other is subsequently stabilized. At the receiver, shown in Fig. 4.18, the SOPs of the POLMUX channels are properly adjusted by means of a real-time polarization stabilizer, before a PBS demultiplexes the two polarization channels. The exploited endless polarization stabilizer is based on two similar stages, each one including a pair of birefringent elements with fixed eigenstates and controllable phase retardation. These birefringent elements are realized as bulk magneto-optic polarization rotators, which exploit the Faraday effecting bismuth-substituted rare-earth iron garnet. The role of the first stage is to map the input SOP onto a fixed meridian of the Poincaré sphere, GAMA in Fig. 4.18, representing the elliptical SOPs with diagonal axes. Then, the second stage stabilizes the output SOP in the vertical state (point V in Fig. 4.18). The stabilizer tracks the labeled channel through the pilot tone; the orthogonal channel is consequently stabilized as well.

The inphase and quadrature components of each single polarization demultiplexed RZ-DQPSK channel are subsequently detected by a pair of integrated-optic Mach–Zehnder delay interferometers, with 40-ps delay and $a \pm \pi/4$ bias. A 20-GHz balanced differential detector is used. By means of a 1:4 demultiplexer, we extract four 6.25-Gbit/s data sequences, which are fed to the error detector in the BER tester.

In either case, at the receiver, shown in Fig. 4.18, the SOPs of the POLMUX channels are properly adjusted before demultiplexing the two polarization channels by means of a PBS. The inphase and quadrature components of each single polarization demultiplexed RZ-DQPSK channel are detected by a pair of integrated-optic Mach–Zehnder delay interferometers, with 40-ps delay and $a \pm \pi/4$ bias.

4.4.2 Results Without AOWC

The CD robustness of the POLMUX RZ-DQPSK solution is experimentally verified by measuring the BER versus OSNR after propagation in uncompensated SSMF (Boffi 2009). As a reference, POLMUX and single-channel transmission are also evaluated replacing the automatic polarization stabilizer by a manually adjustable fiber PC. The experimental data are shown in Fig. 4.19. At 50 Gbit/s, the automatic SOP tracking, for 10^{-5} BER, introduces a 1-dB penalty with respect to stabilization obtained by manually adjusted PC, as evidenced by comparing black open-triangle with gray open-triangle lines. At 100 Gbit/s, the penalty rises to 2 dB (black full-triangle lines) due to optical crosstalk after polarization demultiplexing. The circle curves describe the performance after propagation in 6-km SSMF (corresponding to 100 ps/nm). The penalty introduced by the SOP tracking at 100 Gbit/s is still 2 dB. The automatically SOP-tracked 100-Gbit/s signal and the 100-Gbit/s signal demultiplexed with manually adjusted PC (respectively, black full circles and gray full circles) suffer nearly the same dispersion penalty of 2 dB, at 10^{-5} BER, with respect to their reference back-to-back (respectively, black full triangles and gray full triangles).

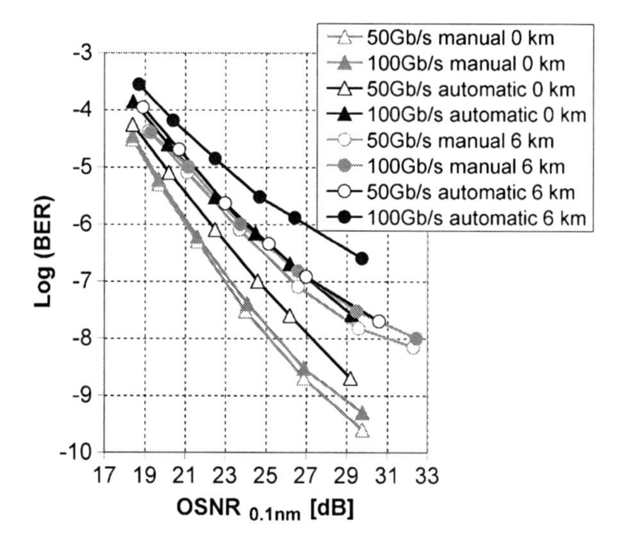

Fig. 4.19 BER vs. OSNR for different transmission conditions

4.4.3 *Results with AOWC*

We finally performed transmission experiments over uncompensated 500-km SSMF by exploiting OPC techniques to overcome propagation impairments. It is noteworthy that the OPC can compensate for nonlinear and linear effects (Marazzi 2009) along the fiber. OPC can be obtained as a by-product through coherent AOWC, as the converted output has an optical spectrum conjugated with respect to the input one. It is also important to underline that, when dealing with polarization multiplexed signals, such as POLMUX DQPSK, the constraints on the polarization independence of the setup for OPC and AOWC are much tighter than those required for use with single-polarization signals, and thus the use of configurations avoiding any polarization crosstalk is mandatory.

The available technologies for coherent AOWC can be usually divided into three groups, depending on the exploited nonlinear effect: FWM process (FWM) in $\chi^{(3)}$ media, SOA, and "cascading" in media with high second order nonlinearity. The third group, currently achieving the best performance, is based on the use of the so-called "cascading technique" in $\chi^{(2)}$ materials. This technique allows obtaining the equivalent effect of a FWM process by combining two $\chi^{(2)}$ effects: a CW "fundamental radiation" is used to produce, by means of the SHG a "pump beam" that is then combined with the original signal to produce, by DFG a wavelength converted copy of the signal. One material suitable for this kind of devices is LN. LN is a ferroelectric material, widely used for optical applications because it is possible to grow large homogeneous crystals, it has large electro-optic coefficients, and it is possible to realize high-quality waveguides (by Ti-indiffusion or proton exchange). Moreover, LN is particularly suitable for the realization of an AOWC device because LN has high $\chi^{(2)}$ nonlinear coefficients ($d_{33} \approx 27$ pm/V), and because the nonlinear

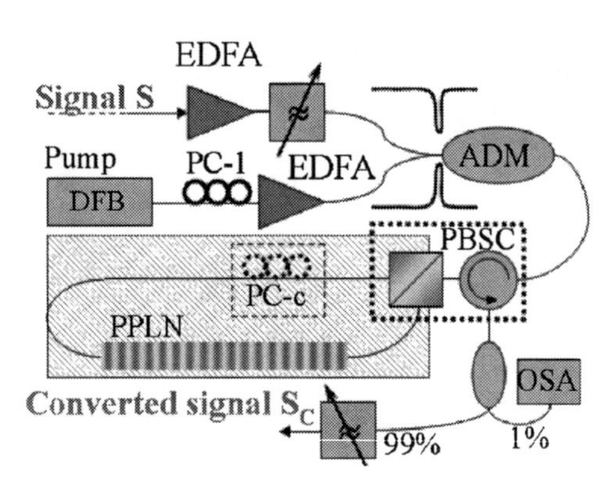

Fig. 4.20 All-optical wavelength converter setup

interactions can be phase matched by using the quasi-phase matching technique, thanks to the periodic poling technology. Moreover, the degradation introduced on the signal by SPM is negligible, as the LN $\chi^{(3)}$ coefficient is relatively low and the interaction length is in the range of 5–10 cm, instead of a few meters as happens with fiber-based AOWCs. These properties, combined with the possibility to develop long and homogeneous waveguides allow the implementation of devices with very high conversion efficiencies. The main problem associated to the use of LN technologies is due to photorefractive effect: the photons from the SHG beam can excite electrons from the shallow traps into the conduction band, allowing them to drift along the crystal optical axis (because of the internal electric field of LN), before being trapped at the borders of the illuminated region. The produced space-charge field, highly inhomogeneous, affects the material refractive indices because of the electro-optic effect, thus yielding to a reduction of the device efficiency and causing beam distortions.

The AOWC device used in the experiments to achieve OPC is based on a 67-mm-long PPLN waveguide fabricated by means of the RPE technique (Parameswaran et al. 2002), and its operating bandwidth is about 60 nm (measured at −3 dB with respect to the maximum efficiency). The end faces of the waveguide are cut at an angle of 6° to avoid back reflections, and are fiber-coupled to PM pigtails by means of UV-curing epoxy glue with a good resistance to high temperatures. Indeed, the device needs to be maintained at a temperature above 100°C in order to avoid the photorefractive damage. The RPE-PPLN waveguide propagates only the TM mode with a measured polarization extinction ratio higher than 40 dB. Polarization independence is obtained by using an AOWC scheme, shown in Fig. 4.20, based on a ring configuration (Chou 2000). The incoming signal (at the wavelength of 1550.12 nm) is combined with the pump (1,552.52 nm) using an ADM. Then a port-4 optical component (PBSC) combining a circulator and a polarizing beam splitter splits both the pump and the signal in two orthogonal polarization components, coupled to the slow axis of two polarization-maintaining fibers, which counter propagate through the ring. Thanks to the manually adjustable fiber polarization controller PC-1, the pump splitting ratio can be adjusted in order to achieve the same conversion efficiency for both polarization components of the signal. TM input

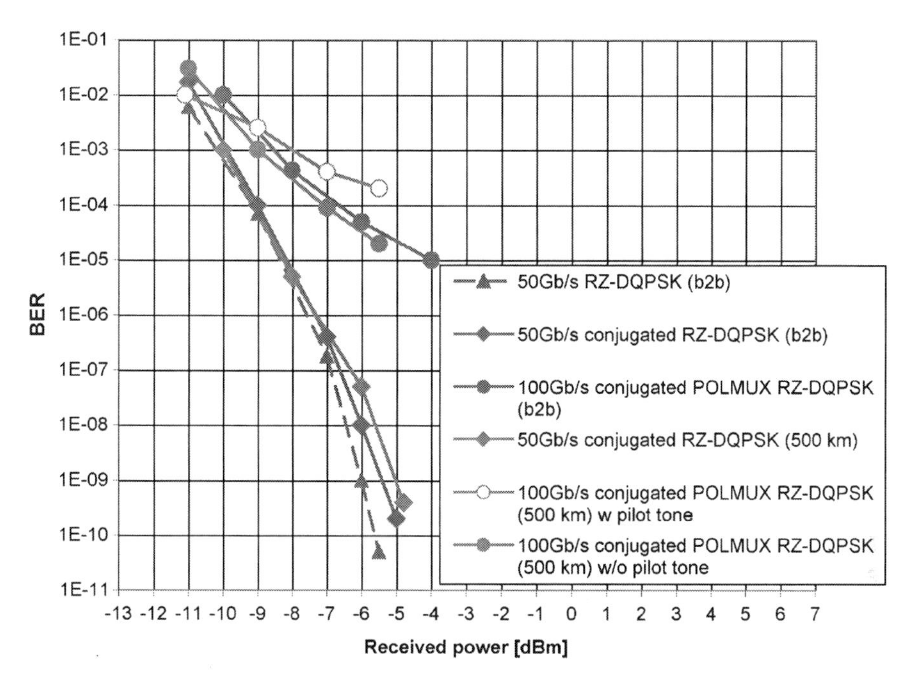

Fig. 4.21 BER vs. received power for different transmission configurations after 500 km

at both sides of the PPLN waveguide is ensured by PBSC. After the wavelength-converted signal has been generated inside the PPLN waveguide, its orthogonal polarization components are recombined at the PBSC, then a tunable filter separates both the pump and the original signal from the converted signal (1,554.94 nm).

Using this scheme, a polarization-independent AOWC can be obtained, provided that PM fiber is used in the ring configuration, and that the PBSC has perfect polarization discrimination. The insertion losses, measured from amplifier outputs to PBSC, are lower than 9 dB and 8 dB for the signal and the pump, respectively.

The OPC device has been exploited in our experimentation as a cost-effective solution to allow propagation of the 100-Gbit/s PDM RZ-QPSK signal over uncompensated 500-km-long SSMF by employing direct detection, avoiding expensive optical dispersion compensation, or coherent detection. In this case, placing OPC in the middle of the transmission link, a perfect compensation for all the even orders of fiber chromatic dispersion is provided. The transmission line is composed of six spans of SSMF, whose length varies between 77 and 105 km. The signal is amplified using double-stage EDFAs with an output power set to +3 dBm, in order to guarantee operation in the linear propagation regime. At the receiver, we employed the above described polarization stabilizer.

The BER curves versus received power, shown in Fig. 4.21, demonstrate the successful propagation both of 50 Gbit/s RZ-DQPSK and of 100 Gbit/s PDM RZ-DQPSK over the 500-km-long SSMF link employing the OPC technique. By comparing the blue triangle and the blue diamond curves, it is clear that the phase conjugation process does not introduce signal penalties at 50 Gbit/s. Conversely, as the blue full circle curve demonstrates, optical phase conjugation does introduce some penalties on the 100-Gbit/s PDM signal (2 dB at 10^{-4} BER). The significant measured penalty at 100 Gbit/s is due to a polarization-dependent crosstalk between the PDM channels in the OPC. Such a behavior is induced by the nonideal features of the OPC polarization diversity scheme described before. The 50-Gbit/s RZ-DQPSK propagation over the 500-km-long link does not present any penalty down to 10^{-6} BER, while at 10^{-9} the resulting penalty is less than 1 dB. Such a penalty has to be ascribed to OSNR reduction produced by the signal amplifier present in the OPC setup. At 100 Gbit/s, PDM RZ-DQPSK exhibits a 2-dB penalty on the unlabeled channel (the one without the pilot tone indicated with full red circles), whereas the orthogonal channel (with the pilot tone, open red circles) is slightly more penalized after 500 km at 10^{-4} BER. No significant penalty is thus found, after 500 km, with respect to the 100-Gbit/s conjugated back-to-back signal.

These sets of BER measurements nevertheless unexpectedly present a marked dependence of the conversion performance on the SOP of the signal. Moreover, a polarization fluctuation of the SOP of the wavelength-converted signal, both in case of single polarization and POLMUX conversion, has been found over a time scale of about 0.1 s, which prevents manual polarization demultiplexing. These results suggest that the AOWC setup induces some interplay between the two POLMUX channels and consequently, when a POLMUX format is considered, it is not possible to obtain a good BER performance for both channels simultaneously (OFC_Marazzi and OFC_Pusino). A careful analysis carried out by means of a polarimeter verifies that the AOWC causes a reduction of the two SOP orthogonality, which is the origin of a significant crosstalk between the POLMUX channels. The angular separation between the SOPs, measured onto the Poincaré sphere, reduces from 180° at the AOWC input to about 175° at its output. Measuring the polarization extinction ratio after a PM-pigtailed PBS, the orthogonality loss is due to a very small rotational misalignment ($\Delta\theta$) between the PPLN waveguide and the PM-fiber pigtails. In particular, one pigtail is well aligned ($\Delta\theta < 0.5°$), conversely, the other is slightly misaligned of $\Delta\theta \approx 3°$. Such a value, that is usually negligible for single polarization formats, is sufficient to cause a significant AOWC performance degradation due to the polarization crosstalk.

In order to compensate for the 3°-angular misalignment and to verify its impact on the conversion performance, a manually adjustable polarization controller (PC-c in Fig. 4.20) has been between place between the PPLN and the PBSC. By properly adjusting PC-c, so as to compensate for the pigtail misalignment, we were able to eliminate fluctuations in the converted signal SOP (Martelli et al. 2009). In order to determine the penalty strictly associated with the AOWC impact on the POLMUX signal, the automatic polarization stabilizer has been from the setup. Polarization demultiplexing is then performed with a PBS, by properly orienting the SOPs of the POLMUX channels with a manually adjusted fiber polarization controller. BER measurements as a function of the received power are presented in Fig. 4.22. The performance of both 50-Gbit/s RZ-DQPSK polarization channels are checked separately. The full blue-triangle BER curve refers to the BTB of the

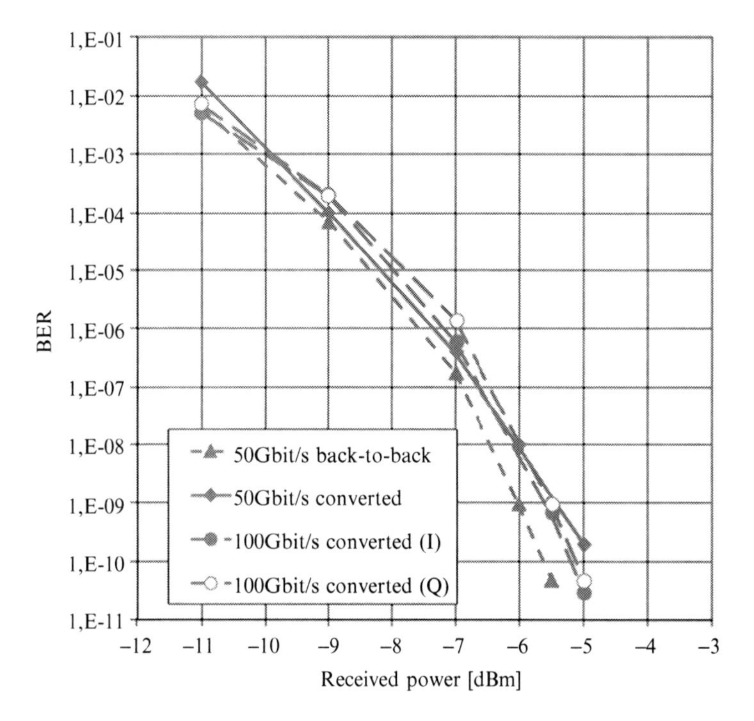

Fig. 4.22 BER vs. receiver power for different propagation conditions

labeled single-polarization channel, while the blue-diamond BER curves refer to the separate conversion of the single-polarization channel, showing that when a single-polarization channel was converted, the AOWC penalty was less than 0.5 dB with respect to the BTB.

The BER for the overall 100-Gbit/s POLMUX signal is subsequently measured. If PC-c is properly adjusted, a variation of the input signal SOPs produces only a small variation of the AOWC performance, as it can be seen comparing open and full blue circles curves in Fig. 4.21. In the case of full circles curves, almost an identical performance is obtained for both POLMUX channels, and the penalty with respect to the single-channel BTB is less than 0.5 dB.

By varying the signal SOP at the AOWC input through the polarization controller, a performance difference lower than 0.2 dB between channels A and B (open and full green circles in Fig. 4.22) is still present. This indicates that the setup still presented a small polarization dependence that, anyway, can be considered negligible. This result demonstrates that PPLN setup allows obtaining excellent performance in 100-Gbit/s POLMUX formats. Even in the worst case, the penalty with respect to the single polarization BTB is less than 1 dB. Crosstalk between the two polarization channels is avoided if a fine alignment of the waveguide pigtails is provided. Such a requirement can be easily achieved by means of the current packaging technology.

As demonstrated by our experiments, the POLMUX format imposes very strict requirements in polarization independent schemes. Indeed, even a small nonideal polarization behavior of the AOWC components degrades the orthogonality of the polarization channels, thus causing a significant optical crosstalk at the receiver. Indeed, a slight rotational misalignment ($\approx 3°$) between the TM mode of the PPLN waveguide and the slow axis of one PM-fiber pigtail causes a residual polarization dependence producing a relevant degradation of the performance in case of POLMUX signals, due to the loss of orthogonality between the polarization channels and the consequent polarization crosstalk. By compensating the rotational misalignment, it is possible to achieve all-optical wavelength conversion of a 100-Gbit/s POLMUX RZ-DQPSK signal with a "worst-case penalty" of less than 1 dB with respect to back-to-back operation.

4.5 The Current Status of Long-Haul High Bit Rate Transmission Experiments: Results from ECOC 2010

ECOC is one of the main events where the most important experiments on optical systems and networks are shown every year. Therefore, looking at the results reported in such a conference can be very useful to understand the state of the optical communications from the experimental point of view.

Looking at the topic of the long-haul high bit rate transmission systems, during the last years, we have witnessed a continuous challenge among the main research institutes to demonstrate how to transmit tens of Tbit/s over thousands of kilometers. As an example, we report in Table 4.1 a list of some experiments of ECOC 2001, where we illustrate the total capacity, the number of channels, the distance, and some hints on the transmission techniques.

We can see how the capacity is mainly obtained with WDM systems based on nx40 Gbit/s, by adopting mainly special fibers with IM-DD technique.

In the last years, the wider and wider bandwidth required from the users has stimulated the increase of the bit rate transmission for long-haul optical fiber systems and one of the aim is the channel transmission at 100 Gbit/s. To reach such capacities, several technological approaches are required both in terms of advanced modulation formats and methods to mitigate the impairments due to the in-line degradation effects as the ASE of optical amplifiers and the dispersive and nonlinear fiber effects. Concerning advanced transmission and detection formats, the combination of amplitude and phase modulation allows us to use higher bit rate reducing the signal bandwidth. Therefore, we have seen several experiments of transmission using phase-, amplitude-, and polarization-modulated signal, also in links with high chromatic dispersion. In particular, the DQPSK, POL DQPSK, and POL QPSK are assumed fundamental techniques, since, thanks to the reduction of the bandwidth, the impact of the dispersive impairments is strongly mitigated. Nonetheless, in long-haul fiber links dispersive and nonlinear effects had to be

Table 4.1 Main experiments presented in ECOC 2001

Total capacity (Tbit/s)	$N \times$ channel bit rate (Gbit/s)	Total distance (km)	Transmission technique	Kind of fiber, amplifier spacing	Authors
5	125 × 42.7	1,200	NRZ ETDM	TeraLight™	S. Bigo ALCATEL
3.2	80 × 40	300	PSBT format 50 GHz Polar. mux	TeraLight™	H. Bissessur ALCATEL
2.4	240 × 10	7,400	Raman ampl.	G.652, comp. 50 km	N. Shimojoh Fujitsu
1.6	40 × 42.7	2,000	NRZ EMUX	Particular fiber 100 km	B. Zhu Lucent
1.6	160 × 10.66	380	NRZ, Raman ampl	Particular fiber Unrepeated	P. Le Reux ALCATEL
1.28	64 × 20	4,200	Raman ampl.	Particular disp. man. 100 km	K. Ishida Mitsubishi
1.28	32 × 40	1,704	ETDM NRZ	Particular disp. man.	A. Hugbart ALCATEL
1	25 × 40	306	OTDM	Particular fibers Unrepeated	K. Tanaka KDDI
0.24	3 × 80	600	OTDM	G.652 50 km	H. Murai OKI

controlled to reduce their impact on the propagation of wide signal bandwidth: the use of a local high chromatic dispersion could strongly limit the impact of Kerr impairments (especially FWM effects in WDM systems), and the chromatic dispersion accumulation could be avoided by the use of compensating devices as fiber grating and dispersion compensating fiber. Several works have shown that the periodic chromatic dispersion compensation, also known as dispersion management, is the method that also limits the accumulation of self-phase modulation effects.

The comeback of coherent detection has somewhat changed this scenario, since in this class of systems, linear transmission impairments (like chromatic dispersion or PMD) can be perfectly compensated by the receiver DSP. In particular, no in-line dispersion compensation is needed, and in many cases, the best performance is achieved with optically uncompensated transmission lines. The main transmission impairments is thus the Kerr fiber nonlinearity. Several studies have been performed on the possibility of compensating such nonlinear effects in the electrical domain. Using back propagation (i.e., inverting the Manakov equation at the receiver), a good improvement has been demonstrated in single-channel systems, but the compensation of interchannel nonlinearity in WDM scenarios still seems to have a prohibitive complexity for the state-of-the-art technology.

In fact, recorded experiments (see the list below) have been performed without any electrical compensation of nonlinearities, but using special fibers, like PSCF of ULAF, which are designed in order to have a large effective area (and consequently a low nonlinearity coefficient) and a high chromatic dispersion coefficient (to locally mitigate the impact of Kerr nonlinearity).

In Chap. 3, by means of numerical simulation, and in Sects. 4.1–4.3, we have shown the advantages of the phase-modulated signal to reach the highest capacity. Here, we illustrate some details of the experiments reported at the last ECOC 2010.

112 × 112-Gbit/s Transmission Over 9,360 km with Channel Spacing Set to the Baud Rate (360% Spectral Efficiency). (Jin-Xing Cai)

A transmission of 112 × 112-Gbit/s prefiltered PDM RZ QPSK channels with 360% spectral efficiency over 9,360 km was demonstrated with the channel spacing set to the Baud rate. This results in a record spectral efficiency for transpacific distance.

100 × 120-Gbit/s PDM 64-QAM Transmission over 160 km Using Linewidth-Tolerant Pilotless Digital Coherent Detection. (Akihide Sano, Japan)

11.2-Tbit/s transmission of 12.5-GHz spaced 120-Gbit/s PDM 64-QAM signals over 160 km was demonstrated by using a digital coherent receiver with pilotless demodulation algorithms. The spectral efficiency of 9.0 b/s/Hz is, in for now, the highest reported for 100-Gbit/s/ch-class transmission.

7 × 224-Gbit/s WDM Transmission of Reduced-Guard-Interval CO-OFDM with 16-QAM Subcarrier Modulation on a 50-GHz Grid over 2,000 km of ULAF and Five ROADM Passes. (Xiang Liu)

Wavelength-division multiplexed transmission of seven 224-Gbit/s reduced-guard-interval CO-OFDM channels was demonstrated with 16-QAM subcarrier modulation on a 50-GHz grid, resulting in a net spectral efficiency of 4 b/s/Hz. Transmission over 20 100-km ultra-large-area fiber spans and five wavelength-selective-switch passes is achieved

4 × 170-Gbit/s DWDM/OTDM Transmission Using Only One Quantum Dash Fabry–Perot Mode-Locked Laser. (Marcia Costa)

A 4 × 170-Gbit/s DWDM/OTDM transmission experiment was performed using only one quantum dash Fabry–Perot mode-locked laser. BER measurements show a penalty of 1 dB at BER = 10–9 for back-to-back and error floor for BER = 10–8 for transmission over 100 km

1.28-Tbit/s/channel Single-Polarization DQPSK Transmission Over 525 km Using Ultrafast Time-domain Optical Fourier Transformation. (Pengyu Guan)

A single-channel 1.28-Tbit/s transmission over 525 km is demonstrated for the first time with a single-polarization DQPSK signal. Ultrafast time-domain optical Fourier transformation is successfully applied to DQPSK signals and results in improved performance and increased system margin.

Transoceanic PM-QPSK Terabit Superchannel Transmission Experiments at Baud-Rate Subcarrier Spacing. (Enrico Torrengo et al.)

We show experimental results of transmission of a Terabit superchannel, consisting of 10×120-Gbit/s PM-QPSK densely packed subcarriers. We reached 10,000 km with $1.1 \times$ Baud-rate subcarrier spacing and 9,000 km with Baud-rate spacing. We also reached 8,000 km with three superchannels at 1.1 spacing.

80 × 100-Gbit/s Transmission Over 9,000 km Using Erbium-Doped Fiber Repeaters Only. (Massimiliano Salsi)

A transoceanic transmission at 100 Gbit/s is demonstrated with EDFA repeaters only. The result was obtained with coherent detection, ultra-low-loss large-effective-area fiber, and PDM-QPSK modulation format.

40 × 112-Gbit/s Transmission over an Unrepeatered 365 km Effective Area-Managed Span Comprised of Ultra-Low-Loss Optical Fiber. (John D Downie)

We experimentally demonstrate transmission of 40×112-Gbit/s PM-QPSK channels over a 365 km unrepeatered span enabled by ultra-low-loss fibers in an effective area-managed configuration using only backward-pumped Raman with 25-dB gain and EDFA amplification.

Transmission of a 213.7-Gbit/s Single-Polarization Direct-Detection Optical OFDM Superchannel over 720-km Standard Single-Mode Fiber with EDFA-only Amplification. (Wei-Ren Peng et al.)

An optical multiband receiving method was proposed and demonstrated which supports transmission of a 213.7-Gbit/s (189.7-Gbit/s) DDO-OFDM superchannel over 720-km SSMF. After transmission, all the nine OFDM bands exhibit BERs lower than the FEC threshold (BER = 3.8e–3).

Looking at these results, we can conclude that, currently, the experimental investigation is mainly dedicated to reach the highest capacity exploiting the modulation property of the signal and hence obtaining the maximum bit rate on each channel. The conclusion that can be obtained from the pages of this chapter is that the technology is now mature to transmit tenth of 100-Gbit/s channels, for a total capacity in the C-band as high as 10 Tbit/s, over thousands of kilometers.

4.6 Experimental Investigation of WDM Burst Traffic Amplification

The experimental work describes a thorough investigation of all-optical gain clamped amplifier dynamics for channels add/drop operation (on–off cases) and typical OBS traffic. Main advantages of the optical gain control technique are the use of passive components, low cost, and easy to upgrade standard amplifiers. However, relaxation oscillations frequencies may accumulate spurious power oscillations along the amplifier chain and careful amplifier design is necessary.

Analyses performed are qualitative and quantitative, with measurement of the power fluctuations of the amplified signals, eye diagrams, and BER. We discuss the characteristics and statistics of burst traffic and describe a tailor-made OBS

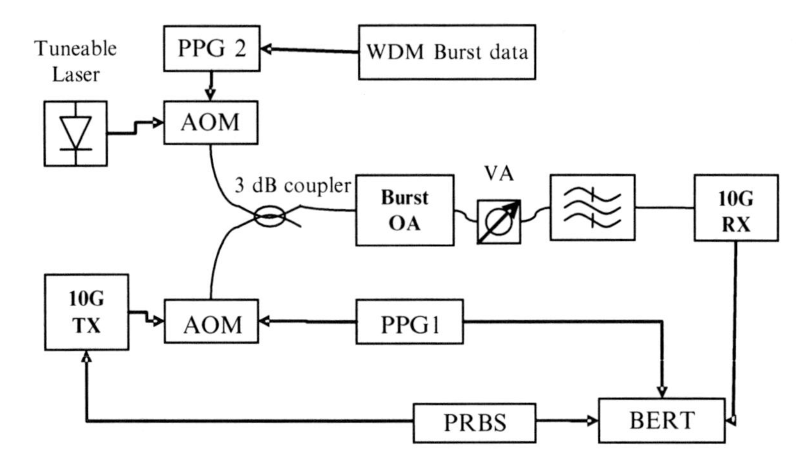

Fig. 4.23 Experimental setup for WDM burst data amplification with gain clamping configuration and BER measurement

platform. The benefit of the clamping technique is experimentally demonstrated for on–off cases and typical WDM burst traffic (Zannin et al. 2009).

4.6.1 Experimental Setup

The setup implemented for the experiments shown in Fig. 4.23 represents a scenario of burst traffic with 16 WDM channels, which is optically amplified. The experiments focus on the degradation in transmission caused by the variation of power at the amplifier input. The constant adding and dropping of channels is perceived by the amplifier as a total input power variation, regardless of which channels are added or dropped. This leads to the assumption that the traffic can be represented by a single laser source, instead of 16, provided that the traffic behavior is known. The laser source here contains the traffic total power concentrated at a single wavelength. This is actually a worse scenario than distributing the power into several channels spread over different wavelengths across the C-band, as it happens in a real system. In fact, in the latter case, the negative effect of spectral hole burning (SHB) (Ennser et al. 2005) is reduced.

The traffic is simulated here with a tunable laser set to −1 dBm and 1,550 nm. The signal is modulated according to the PPG2 connected to the AOM. The PPG2 provides the AOM with the number of channels being transmitted simultaneously over time. This means that the power level of the tunable laser varies with the number of channels being transmitted.

Signal amplification is achieved with a standard commercial EDFA whose pump power is set to 155 mW, which corresponds to a 17-dB gain with −1-dBm input power. In addition, the amplifier is also used as active medium in a laser cavity resonator configuration, which stabilizes its gain by providing optical feedback. Cavity losses are determined by a variable attenuator, coupling losses of the ring, and the insertion losses of the optical devices used. For the sake of simplicity, the ring cavity configuration used allows the lasing wavelength to be varied with a tunable narrow-band filter (0.2-nm FWHM) present in the cavity loop. During the experiment, the tunable narrow-band filter was centered at 1,548 nm, which determines the resonating wavelength. This choice reduces the SHB effect due to the use of a single 16-channel source, with the lasing and signal wavelengths quite close to each other (Ennser et al. 2006). To guarantee the clamping robustness, the pump power is increased by a factor $x = 1.30$ with respect to the pump power needed to start lasing action. The pump power is defined as $P_{\mathrm{Pump}} = x.P_{\mathrm{Pump,th}}$, where $P_{\mathrm{Pump,th}}$ is the pump power at the lasing threshold, and it is about the same power needed to drive the OA to the same gain level without clamping. Note that the cavity passive length is minimized (around 5 m) in order to enhance the performance of the clamping method (Ennser et al. 2005).

In a laser system, there are intrinsic dynamics related to the relaxation oscillations – i.e., the natural frequency at which the laser power (hence the amplifier gain) oscillates once the system is perturbed. These gain oscillations may impact the quality of the signal amplified, therefore an investigation is carried out.

In order to test the signal quality, the experiment uses a probe channel with average power of -15 dBm at 1556.5 nm. The probe channel is configured either as a continuous wave or as a binary sequence, depending on the experiment. In the latter, a PRBS provides a Mach–Zehnder modulator with a $2^{31}-1$ sequence at 10 Gbit/s. After this transmitter block (10 G TX), an AOM acts as a switch for simulating burst transmission with pattern determined by the PPG1. A BERT compares the data generated with the one identified by the receiving block (10 G RX). The OBPF is tuned to the probe channel wavelength while the power received is controlled by a VA.

4.6.2 Investigation of On–Off Cases

The first set of experiments is devoted to evaluate the gain excursion induced by large and abrupt (less than 1 μs rise/fall time) change in the input power. The first approach simulates a worst case scenario, where the maximum variation in power occurs. This represents the case where the system undergoes sudden drop of all the channels transmitted and, analogously, all the channels are simultaneously switched on, due to channel reconfiguration.

The tunable laser that represents all the channels is modulated by a square wave at 1 kHz, with a maximum power equal to -1 dBm. The value is set higher on this experiment in order to stress the worst case of operation of the EDFA, which guarantees that it operates in the saturation regime.

The gain variation of the surviving channel represented by the probe channel is analyzed both with and without gain clamping. The gain excursion of the probe

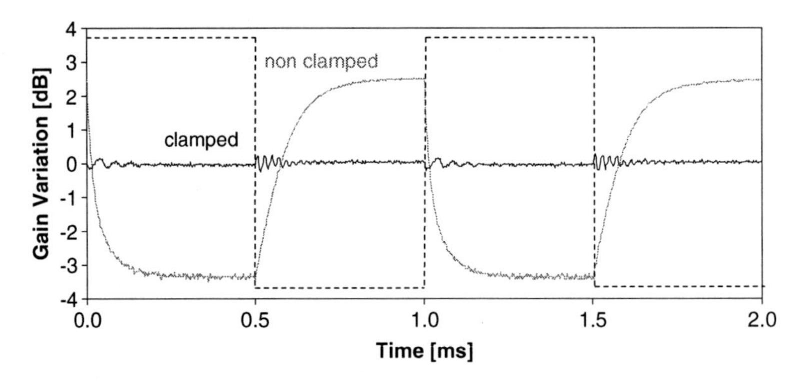

Fig. 4.24 Gain variation of the surviving channel due to simultaneous drop and addition of 16 channels at 1 kHz on a WDM system with both clamped and nonclamped optical amplification

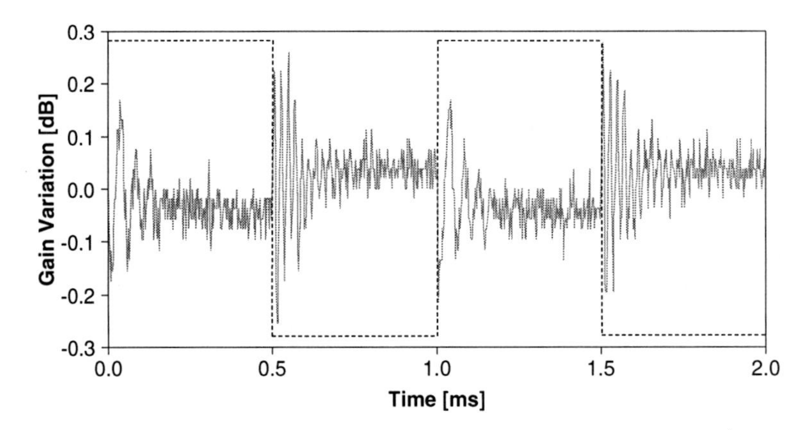

Fig. 4.25 Expanded details of probe channel gain variation with optical clamped gain

channel is shown in Fig. 4.24. We note an offset of nearly 6 dB for unclamped condition while the power excursion is limited to 0.5 dB when the OGC-OA configuration is used. Expanded details of OGC-OA measurement are shown in Fig. 4.25.

The experiment is repeated with lower signal power variation. The maximum power is now set to −7 dBm, which represents a four-channel WDM system. Figure 4.26 shows the gain variation for the optical amplifier without and with gain stabilization. The offset for the first case is around 3.1 dB, while clamping the gain limits the power excursions to less than 0.2 dB (shown in detail on Fig. 4.27).

Both cases showcase significant reduction in the maximum gain excursion when OGC-OA is used. Also note that even the worst case indicates that the clamped gain amplifier reduces considerably the ripple to a value of 0.5 dB, which is quite reasonable.

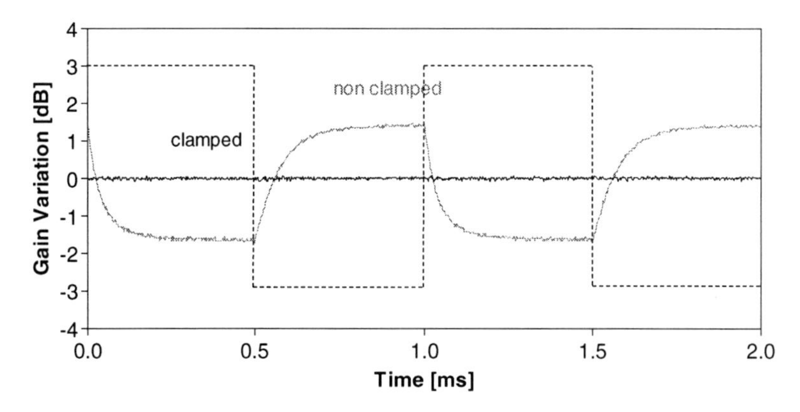

Fig. 4.26 Gain variation of the surviving channel due to simultaneous drop and addition of 4 WDM channels with unclamped and clamped amplifier

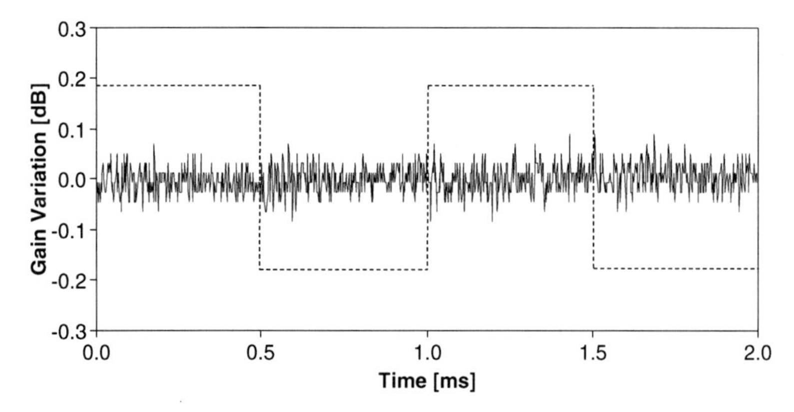

Fig. 4.27 Gain variation of the surviving channel due to simultaneous drop and addition of 4 WDM channels at 1 kHz with optical clamped amplifier

A second set of experiments evaluates the impact of gain excursion on eye-diagram quality and BER measurements. For this purpose, the continuous probe channel is replaced by a 10-Gbit/s PRBS.

Figure 4.28 shows the 16-channel on–off scenario, with the unclamped (a) and the clamped gain amplifier (b). The scenario representing on–off of four channels is shown in Fig. 4.29, unclamped EDFA (a) and clamped gain amplifier (b).

Figure 4.28a shows a noisy eye diagram with three-level signals. The lower level corresponds to the transmission of logic level 0's. The intermediate level corresponds to logic level 1's when all channels are present (simultaneously transmitted). The higher level corresponds to logic level 1's when the probe channel is the only channel transmitted. If the input power of the amplifier is reduced, the amplifier operation is closer to linear regime, which increases its gain. This is the

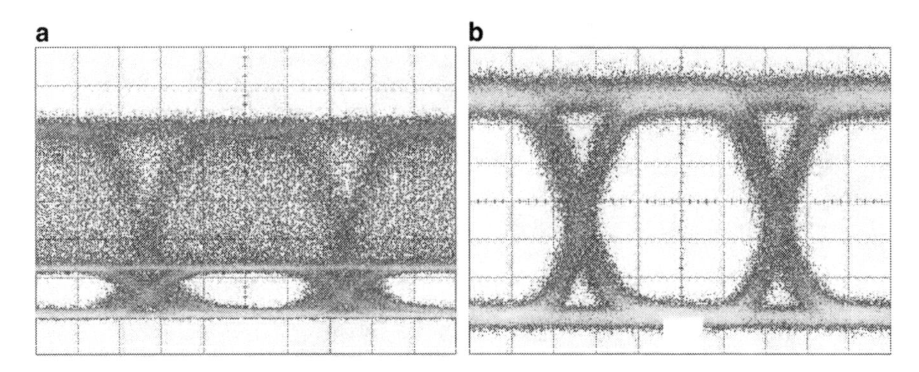

Fig. 4.28 Eye diagram of the probe channel under simultaneous drop and addition of 16 WDM channels with both nonclamped (**a**) and clamped (**b**) amplification

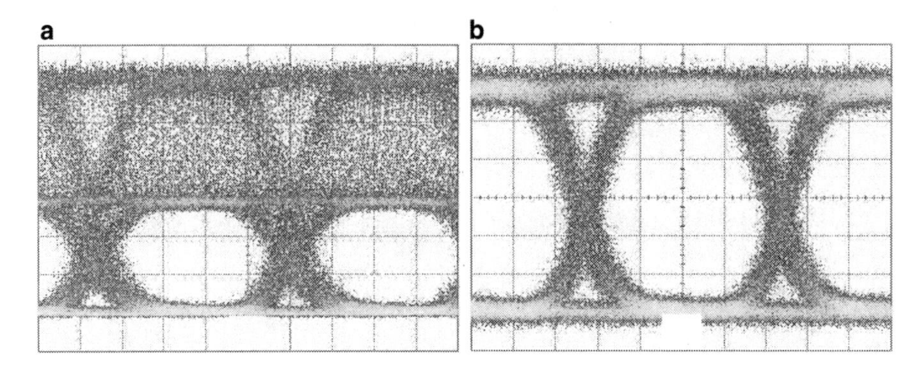

Fig. 4.29 Eye diagram of the probe channel under simultaneous drop and addition of 4 WDM channels with both unclamped (**a**) and clamped (**b**) amplification

reason that the eye opening height is higher in Fig. 4.29a than in Fig. 4.29. Note that the two figures are in the same scale.

To quantify the evidence of eye-diagrams differences, we measured BER values as shown in Fig. 4.30. The BER curves of the two clamped cases are similar. fact, comparing once again the cases with 16 and 4 channels with clamped gain, the slight difference in ripples shown in Figs. 4.25 and 4.27 is not noticeable in the eye diagrams of Figs. 4.28b and 4.29b. If instead the BER curve comparison is made among the unclamped cases the penalty induced with large number of channels is higher, more than 1.5 dB at $\text{BER} = 10^{-9}$, which is in accordance to the difference observed in the eye diagrams in Figs. 4.28a and 4.29.

The curves demonstrate that the OGC amplifier (OGC-OA) is robust in terms of BER for both on–off cases at a 1-kHz rate, with 3-dB improvement for the worst case (16 channels). Since these simultaneous addition and drop of a large channel number is unlikely to occur, the results indicate that input power variation at this given rate do not affect the robustness of the clamping configuration (Fig. 4.30).

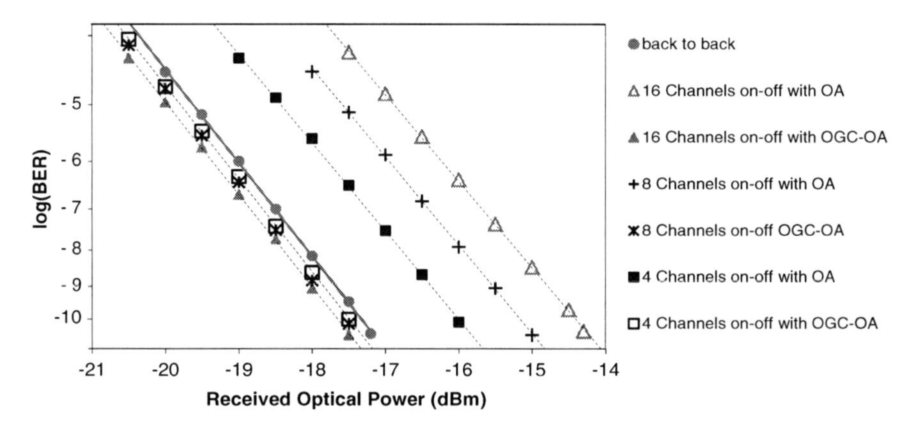

Fig. 4.30 Bit error rate for the probe channel under simultaneous drop and addition of 16, 8, and 4 WDM channels at 1 kHz with both clamped and nonclamped optical amplification

Further understanding of the behavior of the amplifier is obtained by varying the rate of the input power steps. In fact, as previously mentioned, the OGC-OA has some intrinsic dynamics, related to the ROF, while the unclamped OA has a rise/fall dynamic with time constant related to erbium $I_{13/2}$ lifetime of about 8 ms in silica fibers. The worst case of 16 channels on–off is implemented and the channels add/drop frequency is varied. An increase of the on–off frequency has not a trivial meaning: for OGC-OA, it implies that the OA has less time for stabilizing its gain but also that the perturbation lasts less time; for unclamped OA, it implies less time for gain rising and when the frequency is faster than the erbium lifetime, an equivalent average input power is seen by the unclamped amplifier.

Figure 4.31 shows the case where the frequency is 4 kHz. This gain variation should be compared with Fig. 4.24, where the frequency is 1 kHz for the same input power. For the case where the gain is not clamped, it is possible to observe that the faster change of input power does not allow enough time for the gain to stabilize.

Increasing the frequency to 10 kHz, the effect is even clearer, as shown in Fig. 4.32. The response of the amplifier is slow if compared to the on–off frequency and this may generate interplay between oscillations in the on and off time slots (Della Valle et al. 2007).

Figure 4.33 shows that at a frequency equal to 100 kHz, the amplifier is nearly insensitive to the changes in the input power of the amplifier. Therefore, there is no significant difference in clamping or not the amplifier gain.

By comparing the cases with clamped gain in figure, Figs. 4.31 and 4.32, one observes that the initial ripples behave the same way – apart from an interplay behavior that is observed in Fig. 4.32 because the add/drop frequency approaches the resonator cavity relaxation oscillation frequency, which could induce chaotic behavior as reported in Della Valle et al. (2007). Nevertheless, these ripples are not enough to affect the BER. This may be observed in Fig. 4.33, where the cases

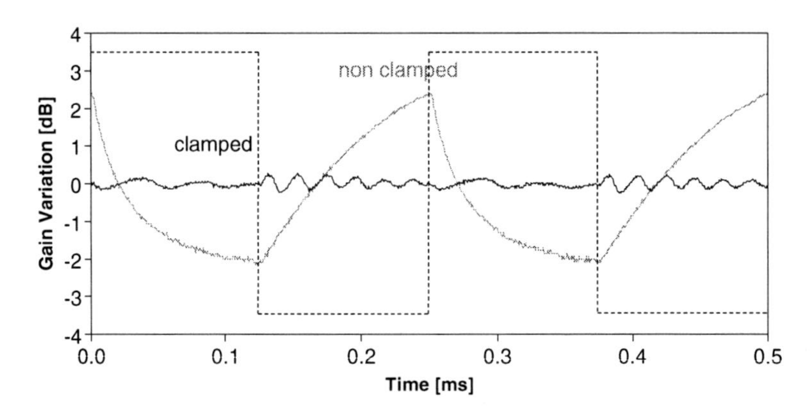

Fig. 4.31 Gain variation of the surviving channel due to simultaneous drop and addition of 16 WDM channels with both clamped and unclamped optical amplification

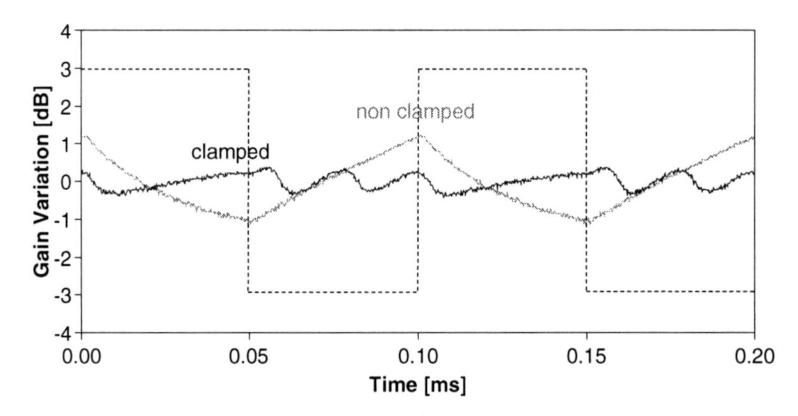

Fig. 4.32 Gain variation of the surviving channel due to simultaneous drop and addition of 16 WDM channels at 10 kHz with both clamped and unclamped optical amplification

with clamped gain are roughly equal within uncertainties. Figure 4.34 illustrates the BER curves for the probe channel under simultaneous drop and addition of 16 WDM channels at rates of 1, 4, 10, and 100 kHz.

On the other hand, as expected, gain ripples in Figs. 4.24, 4.31–4.33 decrease for unclamped OA. Therefore, unclamped OA looks more stable for very fast sustained modulations in the range of 10 kHz to about 100 kHz. However we note here that, as discussed about the OBS traffic, it is very unlikely to have more than two on/off modulation at the same frequencies. Therefore, the ripples of OGC-OA, originated by long sequences of on/off modulation, are likely to be much smaller with optical burst traffic (Taccheo et al. 2007; Zannin et al. 2009).

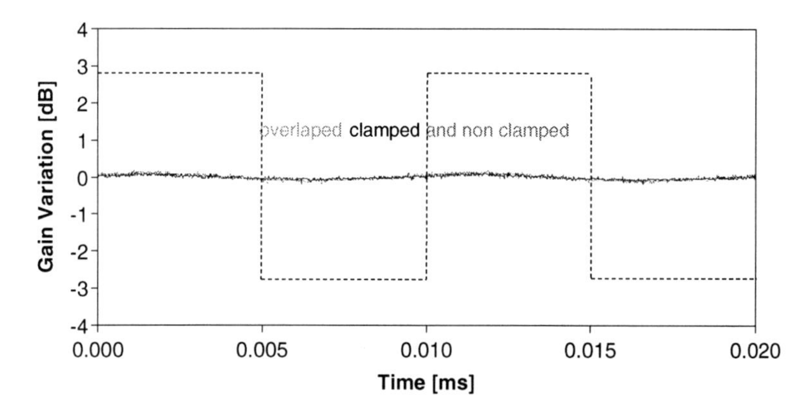

Fig. 4.33 Gain variation of the surviving channel due to simultaneous drop and addition of 16 WDM channels at 100 kHz with both clamped and unclamped optical amplification

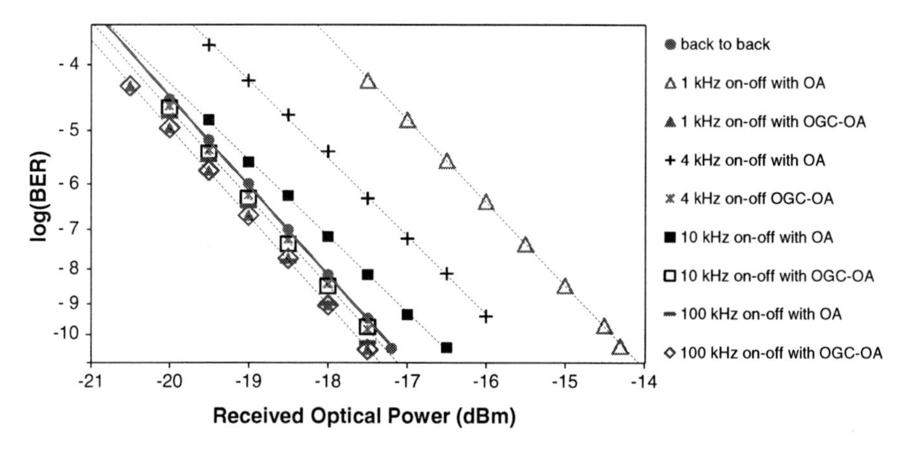

Fig. 4.34 Bit error rate curves for the probe channel under simultaneous drop and addition of 16 WDM channels at rates of 1 kHz, 4 kHz, 10 kHz, and 100 kHz with both clamped and unclamped optical amplification

4.6.3 Investigation of WDM Burst Traffic

The power variation on the input of the amplifier is now determined by the burst traffic data discussed in the previous subsection. The ad-hoc measurement platform described in Fig. 3.32 (see Chap. 3) provides 14-s traces composed by 95,000 bursts. A 1-s long WDM burst stream is assembled from uncorrelated data samples from different time slots. Burst lengths are in the order of tens of microseconds. A sample of the burst optical signal is shown in Fig. 4.35 with the power scaled to the number of channels. The average power at the input of the amplifier is –4 dBm and corresponds to the level of eight channels.

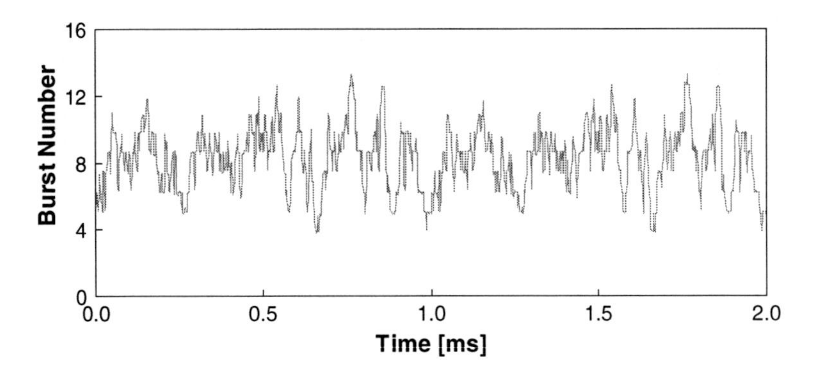

Fig. 4.35 Optical burst data at the amplifier input

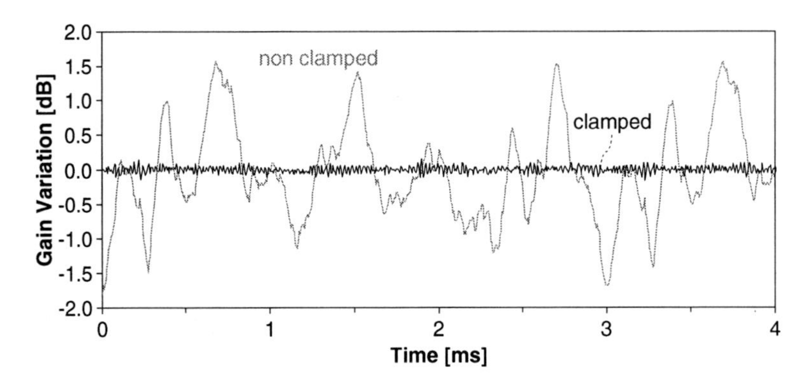

Fig. 4.36 Gain variation of the surviving channel in a 16-channel WDM system with burst traffic with both clamped and unclamped optical amplification

The gain variation for the both clamped and unclamped amplifier is shown in Fig. 4.36. In the case of unclamped amplifier, 3.3-dB peak-to-peak ripples are observed when using a clamped amplifier, the fluctuations has a maximum 0.3 dB peak-to-peak.

The eye diagram of the unclamped amplified burst traffic data in Fig. 4.37a is less noisy than the cases discussed in the previous section (Figs. 4.28a and 4.29). This phenomenon is observed in the case of burst traffic because the power variation steps are smaller than the ones simulated in the previous cases. Another factor that contributes to this improvement is the fact that the steps occur at a faster rate than the one in the previous eye diagrams.

For completion of the experimental investigation, Fig. 4.38 shows that with OCG-OA, the performance of the system is improved by 1.5 dB at a BER equal to 10^{-10}.

As conclusion, the effectiveness of optical gain clamping technique to stabilize amplifier at the presence of OBS traffic is thoroughly investigated. Abrupt 1 kHz

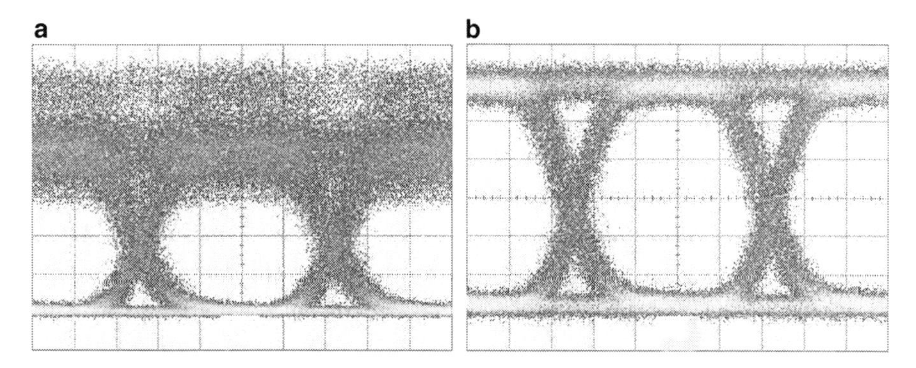

Fig. 4.37 Eye diagram of the probe channel in a 16 channel-WDM system with burst traffic with both unclamped (**a**) and clamped (**b**) optical amplification

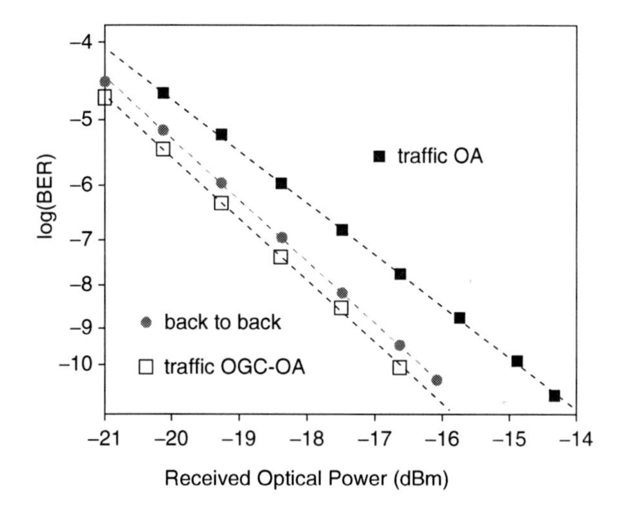

Fig. 4.38 Bit error rate curves for the probe channel in a 16-channel WDM system with burst traffic with both clamped and unclamped optical amplification

add/drop of 16 channels results in power excursions of 6 dB for the surviving channel when no gain stabilization technique is implemented. The passive all-optical clamping method limits power excursions to 0.5 dB. The gain control reduces transmission penalty by 3 dB when BER = 10^{-10}. Cutting the number of channels from 16 to 4 reduces the unclamped power excursions by 3 dB, which translates into 1.5-dB improvement for BER = 10^{-10}. Applying the clamping techniques still improves the BER by additional 1.5 dB while power excursions are limited to 0.2 dB. The 0.3-dB difference between clamped amplifiers with 16 and 4 channels are not sensed by the system in terms of performance: the BER curves are similar.

When typical OBS traffic is used, the average input power remains the same, corresponding to 8 channels. Nevertheless, the variation steps are smaller and therefore the nonstabilized gain amplifier causes the output power to fluctuate within 3.3 dB. Applying the clamping technique described here limits these power excursions to 0.3 dB. This 3-dB improvement increases the system performance by 1.5 dB with BER $= 10^{-10}$.

4.7 Trends on Regeneration

Due to the recent developments in the modulation formats and the evolution of the electrical signal processing, the regeneration based on all-optical functionalities has been a little bit out of the highlights and research efforts. However it is clear, as it happened with the EDFA, that if a regeneration optical device is found with inherent capabilities, simplicity and transparency, this will represent a breakthrough in the present state-of-the-art since it will impact twofold the optical networking: energy and capacity wise.

In this section, some advanced techniques aiming the ultimate achievement of all-optical regeneration of any signal are presented in order not to forget such enhanced evolutions with such a drastic potential.

4.7.1 Use of FBGs in Regeneration Applications

As the capacity and speed requirements in optical communications systems continue to increase every year, the need for sensible use of all-optical switching to enhance the performance and energy efficiency of optical networks becomes ever more imminent. FBG technology can play a much more central role in switching and signal processing applications than the simple route to high-quality bandpass filtering. For example, FBGs have successfully used as nonlinear switching devices in the past, where Kerr nonlinearities are excited within the grating length itself. However, the short fiber lengths associated with these filters imply that extremely high peak powers are required for the operation of these devices, rendering them less attractive for telecommunications applications. In this section, we discuss the use of FBGs for precision shaping of short optical pulses, and demonstrate through examples. Herein, the principles and use of tailored optical pulses to extend and improve the performance of Kerr nonlinearity based optical signal processing devices are discussed for telecom applications.

The optical Kerr effect is a change in the refractive index of a nonlinear material in response to an applied optical signal field, and, in particular, its induced index change is directly proportional to its square. This optical Kerr effect manifests itself through nonlinear phenomena, such as SPM and XPM, depending if the intensity dependence of the refractive index is due to the signal itself or a second one, which is usually referred to as the pump, and leads to the spectral broadening of the optical signal itself. This nonlinear effect has been exploited for many years, especially in fiber, to develop various nonlinear optical signal processing devices in telecom. In more detail, if we ignore the effects of dispersion and higher order nonlinearity, as is appropriate in many instances for state-of-the-art highly nonlinear fibers and telecoms signals, an optical pulse propagating in a nonlinear optical fiber generates a nonlinear phase shift, $\phi_{NL}(t)$, and an associated instantaneous frequency shift, $\delta\omega(t)$, across the pulse envelope, usually referred to as frequency chirp, which can be mathematically described by the following equation:

$$\delta\omega(t) = -\frac{\partial\varphi_{NL}(t)}{\partial t} \propto -\gamma P_0 L_{eff} \frac{\partial |U(t)|^2}{\partial t}, \tag{4.1}$$

where γ is the nonlinear coefficient, P_0 is the pulse peak power, L_{eff} is the fiber effective length, $U(t)$ is the normalized electric field of the optical signal (in the case of SPM) or the optical pump (in the case of XPM), which is used to switch the signal with, and t is the time variable.

In a conventional Kerr-based all-optical nonlinear switch, one usually exploits the fiber nonlinearity, the fiber length, or the signal peak power in order to enhance the corresponding specific function, while assuming the interacting signals (signal and/or pump) constant. Referring to Eq. 4.1, this implies that the term $|U(t)|^2$ is seen as a constant or, in other words, the corresponding optical pulse is represented as a delta Dirac function with value equal to its peak power. However, this is only a very simple approximation of the real situation and, moreover, properly tailoring the optical pulse shape of the signals in combination with the nonlinear material can give higher flexibility and improved performances of the system under investigation. This basic idea is sketched in Fig. 4.39, which schematically shows what is happening, in an all-optical switching device in the instance that no preshaping (left), or a preshaping scheme, is applied to the signals prior the switch. As can be seen in the first case, large impairments of the signal pulses can result in significant signal distortion at the output of the switch when the signal is switched by a temporally finite optical pulse. This could be avoided incorporating an intermediate stage, within which the signal and/or the pump pulses are reshaped into the optimal shape for the specific telecom application prior to being injected into the switch.

More specifically, in order to use the Kerr effect for switching one generally exploits either the induced phase change within some form of interferometer, or the induced frequency chirp through some form of filtering that discriminates between new frequencies generated in the nonlinear process and the original signal spectrum. Appreciating this to be the case it is clear that certain simple forms of pulse shape offer distinct benefits and opportunities within each class of device. For example, from Eq. 4.1, it is evident that flat-top pulses provide the possibility to generate a uniform phase-shift across the full pulse envelope and thus provide for efficient, timing-jitter tolerant switching within interferometric (Parmigiani et al. 2006a) or in-line devices (Parmigiani et al. 2008a). Similarly, it is seen that parabolic pulses can be used to generate a pure linear chirp across the signal and thus provide for efficient retiming (Parmigiani et al. 2006b), optimal pulse compression

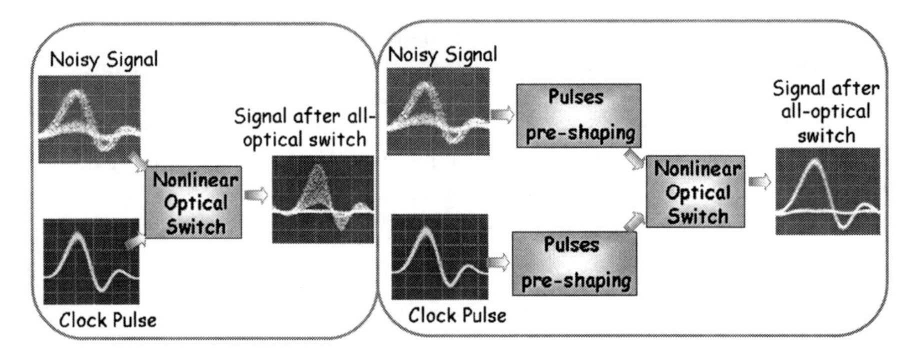

Fig. 4.39 All-optical nonlinear switching without (*left*) and with (*right*) the use of all-optical pulse pre- or post-shaping

(Parmigiani et al. 2006c) or linear compensation distortion (Ng et al. 2008), by further propagating them in a linear dispersive medium with suitably matched net dispersion. Parabolic pulses are also the best shapes for the generation of flat SPM-induced spectra and optimized spectral density, which are ideal for spectrally sliced source applications (Parmigiani et al. 2006c). Equally, from Eq. 4.1, it is also evident that triangular pulses, with their linear slope, provide a means of imparting a uniform frequency shift with respect to the signal initial central wavelength, with minimum additional spectral broadening and thus provide doubling optical pulses in both frequency and time domains, by further propagating them in a linear dispersive medium (Latkin et al. 2009). Furthermore, if an asymmetric triangular shape, i.e., saw-tooth wave, is considered, its spectral intensity evolution is also asymmetric, with most of the energy associated to the smoother leading edge and thus provide for efficient wavelength conversion (Parmigiani et al. 2008b; 2009a), impairment-robust, and high performance time-domain add–drop multiplexing (Parmigiani et al. 2008c; 2009b), when followed by a proper offset filter. The mathematical expressions, target temporal shapes, and corresponding induced chirps of the shapes discussed above are sketched in columns 1–3 in Fig. 4.40.

The preshaping of the signal and/or pump can be achieved using accurate optical filtering in amplitude and phase, applying simple Fourier theory. This can be achieved using tailored SSFBGs (Petropoulos et al. 2001), long period gratings (Krcmarík et al. 2009), and two-dimensional liquid crystal on silicon (LCOS) spatial light modulator (Roelens et al. 2009). The measured temporal traces shown in the last column in Fig. 4.40 were generated using SSFBG technology, which is described in the first part of the chapter.

4.7.2 Regenerators for Phase Encoded Signals

All-optical signal regeneration is an effective approach to eliminate accumulated signal impairments in high-performance optical communication systems without the

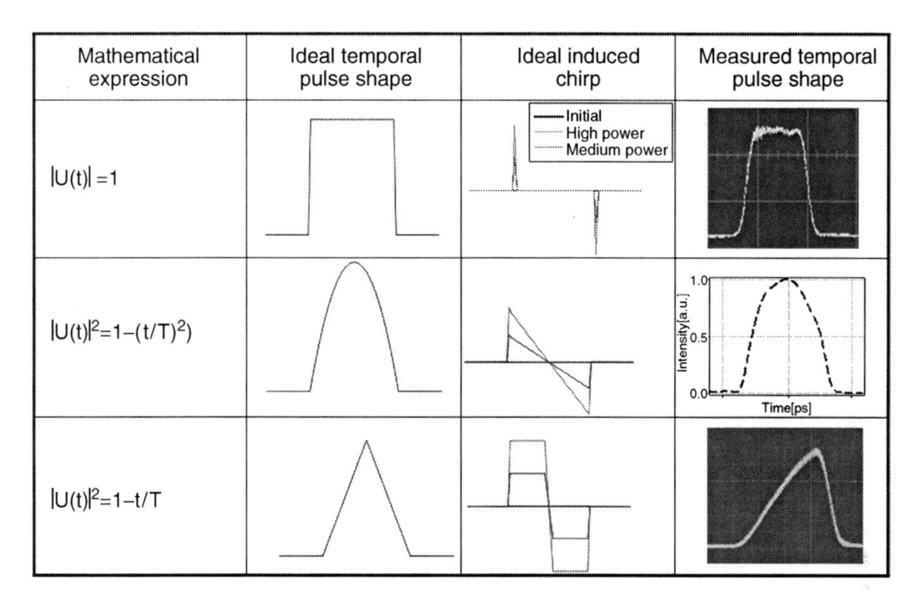

Fig. 4.40 Examples of mathematical expressions, ideal temporal pulse shapes, corresponding induced chirp, and experimentally measured shape generated using SSFBG technology

need of O/E/O conversion. Until recently, the seemingly abundant fiber bandwidth has allowed engineers to follow the relatively simplistic approach of switching light on and off to encode the information to be transmitted, i.e., transmitting the information using OOK modulation formats. However, the bandwidth demands are now such that this approach is no longer sustainable. Cutting edge research in optical transmission adopts techniques which have been originally developed for RF communications and involve manipulation of both the intensity and phase of the transmitted field. Advanced multilevel modulation formats, MPSK and approaches such as DPSK, DQPSK, and QAM, have emerged as an attractive vehicle for optical transmission. As well as offering high spectral efficiency, these approaches also offer considerable advantages in terms of optical signal to noise ratio and resilience to the various transmission impairments. However, the use of the complex optical-field gives rise to a new dominant limitation to system performance, namely nonlinear phase noise. Nonlinear phase noise arises from nonlinear interactions mainly due to Kerr nonlinearity (Gordon and Mollenauer 1990; Demir 2007) and they depend on the amplitude and phase noise evolution of all signals transmitted within the fiber as many individual wavelength channels are usually transmitted simultaneously. Each channel is also degraded by other sources of amplitude/phase noise generated within the transmission line including amplified spontaneous emission from optical amplifiers and quantum noise, to name but two. A good regenerator should then be capable of suppressing the amplitude and phase noise of the signals to allow further transmission and to suppress the seed noise that causes nonlinear signal phase noise to build up in the first place.

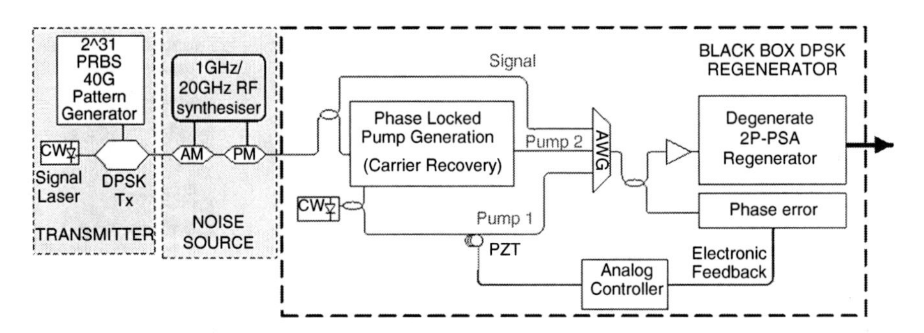

Fig. 4.41 Experimental setup of the 40-Gbit/s phase and amplitude regenerator (Parmigiani et al. 2010a)

To date, most work on all-optical regeneration of DPSK signals has focused on the elimination of amplitude noise only, which can be achieved using cross-phase modulation (XPM) in a nonlinear amplifying loop mirror (NALM) (Cvecek et al. 2008) or using saturation of FWM interaction caused by pump depletion in a nonlinear fiber (Croussore and Li 2007a; Matsumoto and Sanuki 2007) A different approach to elimination of amplitude and phase noise simultaneously is to adopt some forms of phase-to-amplitude format conversion and then to apply amplitude regeneration on the amplitude encoded signal (Matsumoto and Sakaguchi 2008). However, a simpler solution is to directly eliminate the phase noise.

Phase and amplitude regeneration can be achieved directly by exploiting the phase squeezing capability and the saturated regime of phase-sensitive amplifiers (PSAs). Contrary to a phase-insensitive amplifier, where the inphase and quadrature components of the electric field of the signal experience the same gain, in a PSA, the two quadrature intensity components are amplified differently: the inphase component experiences a gain g, while the quadrature component experiences a deamplification $1/g$, so that phase squeezing can be achieved.

Initial proof of concept results have recently been demonstrated by Croussore et al. (Croussore et al. 2006, 2008; Croussore and Li 2007b) in either single- or dual-pump configurations using either interferometric configurations based on nonlinear optical loop mirrors (Croussore et al. 2006), or using FWM processes (Croussore et al. 2008; Croussore and Li 2007b) in fiber. However, due to the technological challenges to recover the carrier and phase-lock, the newly generated pump(s) to the DPSK signal itself, their phase relationship was, so far, artificially guaranteed using a single (master) CW laser (Croussore et al. 2008; Croussore and Li 2007b), making the regenerator impractical for use in a real transmission system. However, technologies move on and a FWM-based all-optical phase and amplitude regenerator setup in a "black box" was recently proposed and experimentally demonstrated. The demonstration was based on a dual-pump degenerate PSA configuration and acted on a 40-Gbit/s DPSK signal (Parmigiani et al. 2010a, b; Slavik et al. 2010). In this section, these results are reported and discussed.

The experimental setup is shown in Fig. 4.41. The data signal was a 40 Gbit/s NRZ-DPSK, 2^{31}–1 PRBS. The effects of phase and amplitude noise were emulated by modulating the signal phase and amplitude in a deterministic fashion using two additional phase and amplitude modulators, driven sinusoidally at frequencies of $\sim$20 GHz and $\sim$1 GHz, respectively. The distorted signal was then launched into the black-box regenerator. Part of the signal was tapped off to facilitate the frequency and phase locking of the two local pumps used in the degenerate PSA configuration, as follows. The tapped signal and a narrow linewidth CW laser (pump 1), detuned from the signal by 200 GHz, were first combined in a germano-silicate highly nonlinear fiber (HNLF) (length, dispersion, nonlinear coefficient, and attenuation of 500 m, -0.09 ps/nm/km, 11.5 W^{-1} km^{-1}, and 0.83 dB/km, respectively) to parametrically generate an idler wave that served as the seed for the second pump. Note that due to the phase erasure process, the binary data modulation was not transferred to the idler (Lu and Miyazaki 2009). Then, the weak idler wave was filtered and injected into a (slave) semiconductor laser to generate the second phase-locked pump (pump 2) by means of a multiplexer (Weerasuriya et al. 2010). As the injection locking is a much slower process than FWM, any high frequency fluctuations (e.g., bit-to-bit phase variations) present on the original data signal were not transferred onto the output of the slave laser. Pump 1 and pump 2 were then coupled together with the data signal, amplified up to $\sim$34 dBm, and launched into an alumino-silicate strained HNLF which was used for the PSA regeneration. This fiber exhibited an increased SBS threshold (Nielsen et al. 2010) which allowed its use without the need for any active SBS suppression scheme. The length, dispersion, polarization mode dispersion, nonlinear coefficient, and attenuation of this fiber are 177 m, -0.13 ps/nm/km, 0.11 ps $km^{-0.5}$, 7.1 W^{-1} km^{-1}, and 15 dB/km, respectively. The relative powers of the pumps and signal were adjusted for optimal regeneration performance. Any slow (sub kHz) relative phase drifts between the interacting waves picked up due to acoustic and thermal effects present prior to the PSA fiber were compensated for by an electrical phase-locked loop that controlled a piezoelectric-based fiber stretcher in the pump path.

The performance of the regenerator was first studied using a constellation analyzer based on a homodyne coherent receiver and offline DSP operating at 10 Gbit/s (Sjödin et al. 2009; Sköld et al. 2008). In this case, phase noise only was added to the signal. The local oscillator used in the measurement was obtained by tapping off the signal laser before data encoding. From these measurements, amplitude and bit-to-bit phase changes were calculated and the corresponding DPSK constellation diagrams are shown in Fig. 4.42. Note that for these measurements, the data rate was also adjusted to 10 Gbit/s and the phase modulator that distorted the signal was subsequently driven at 5 GHz (Parmigiani et al. 2010a). The results show that the phase noise can be squeezed by the regenerator to the back-to-back level and almost negligible amplitude noise is induced even for as extreme peak-to-peak values of phase distortion as $\pm 80°$ (Fig. 4.42) confirmed error-free operation with negligible power penalty as compared to the back-to-back even for the most extreme case of added noise reported in Fig. 4.42.

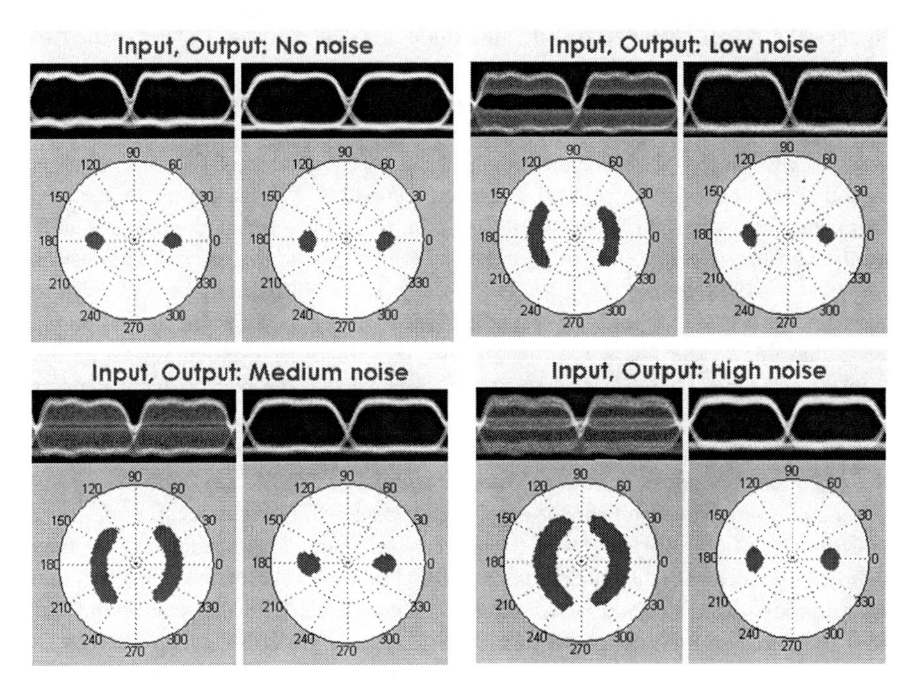

Fig. 4.42 Demodulated eye diagrams after balanced detection and differential constellation diagrams (showing bit-to-bit phase changes) for different levels of phase noise only (Slavik et al. 2010)

The noise performance of the regenerator was also studied at 40 Gbit/s using eye diagrams and BER measured for various levels of phase-only, amplitude-only, and simultaneous amplitude/phase perturbations. Following demodulation in a delay line interferometer and balanced detection, the eyes measured at the input and output of the regenerator for no noise and for the highest level of the added phase noise only are shown in Fig. 4.43a, highlighting that the regenerator was capable of "reopening" a completely closed eye. Corresponding BER curves are shown in Fig. 4.43a: at the input of the regenerator, the added phase noise caused error flooring in the BER curves from 10^{-6} to 10^{-3}. At the output of the regenerator error-free performance was achieved for all the noise levels with power penalties below 3.5 dB as compared to the back-to-back case. For the highest level of noise (corresponding to a completely closed eye at the input of the regenerator), the regenerator output eye diagrams were widely open; however, the corresponding BER curve started showing a slight error floor, showing the PSA limitations.

Figure 4.43b shows similar noise performance assessment of the regenerator when different levels of amplitude noise only were added to the signal (similar signal degradations were chosen). Note that the same experimental conditions (signal/pumps powers) were preserved at the input of the PSA in this case. The demodulated eyes for no noise and for the highest level of the added amplitude noise

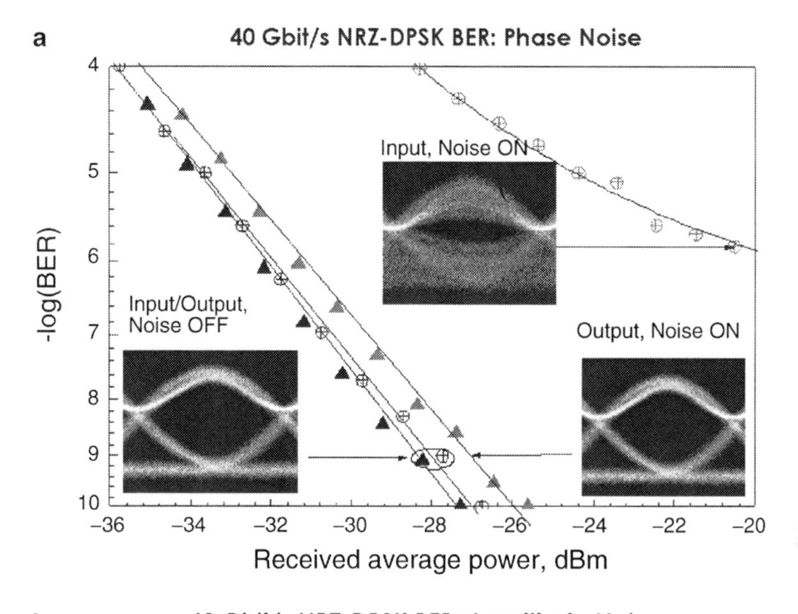

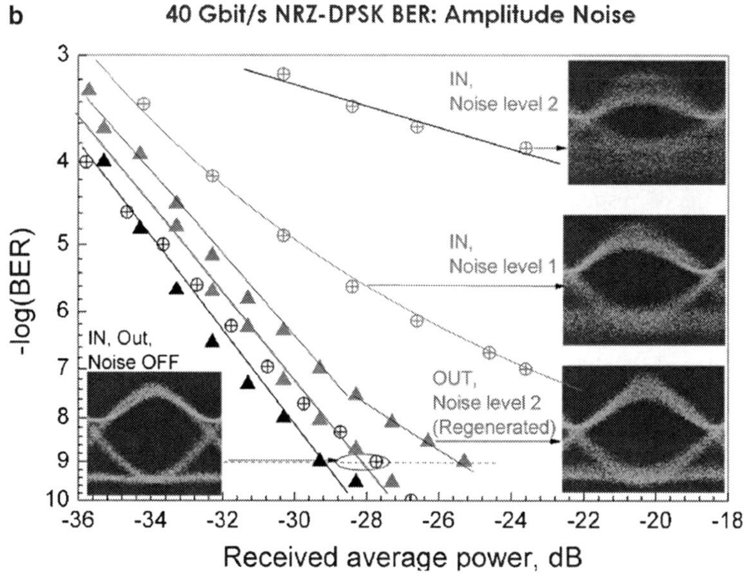

Fig. 4.43 BER curves (measured using single-sided receiver detection after a 1-bit delay line interferometer) and corresponding eye diagrams (measured using a dual-port optical sampling oscilloscope after the 1-bit delay line interferometer) when phase-only (**a**), amplitude-only (**b**), and both amplitude and phase (**c**) noise are added at the input of the system. The performance at the regenerator input and output is shown as *circles* and *triangles*, respectively (no noise: *black*; lower level of amplitude or phase noise: *red*; higher level of amplitude or phase noise: *green*; combined (lower level) amplitude and phase noise: *blue*) (Slavik et al. 2010)

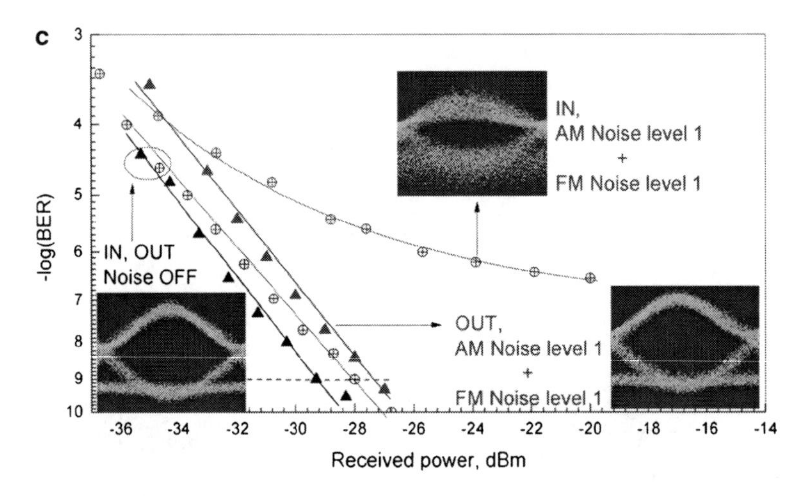

Fig. 4.43 (continued)

are shown at the input/output of the regenerator, highlighting the good amplitude noise regenerative capability. For more quantitative analysis of its performance, BER curves are also reported: while at the input of the regenerator, the added amplitude noise caused the BER curves to vary from 10^{-6} to 10^{-4}, at the output of the regenerator error-free performance was achieved for all of the noise levels considered. For the highest level of noise (corresponding to an almost closed eye at the input of the regenerator), the regenerator output eye diagrams were widely open, however the corresponding BER curve started showing a slight error floor, similar to the previous case.

The regenerator performance was finally assessed using eye diagrams and BER measurements for various levels of simultaneous amplitude/phase perturbations. In this case, the lower levels of amplitude and phase perturbations were applied, see Fig. 4.43c. Once more, the good quality of the demodulated eyes at the input/output of the regenerator highlights the excellent phase and amplitude regenerative quality of the system, which is also confirmed by the corresponding BER measurements.

These experimental demonstrations confirm that nonlinear fiber optics have the potential to provide the solutions required for the processing of even complex modulation formats, and significant research efforts are currently devoted towards achieving this goal, both in Europe and elsewhere. The system presented in this section allows for the correction of even large amounts of phase and amplitude noise, and is scalable to far higher repetition rates than 40 Gbit/s. Furthermore, it is claimed that the technique can be used as the basic building block for the regeneration of even more complex modulation formats.

References

Boffi, P.: Stable 100-Gb/s POLMUX-DQPSK transmission with automatic polarization stabilization. IEEE Photon. Technol. Lett. **21**, 745–747 (2009)

Bosco, G., Carena, A., Curri, V., Poggiolini, P., Forghieri, F.: Performance limits of Nyquist-WDM and CO-OFDM in high-speed PM-QPSK systems. IEEE Photon. Technol. Lett. **22**(15), 1129–1131 (2010). 1 Aug 2010

Cai, J.-X., et al.: Transmission of 96x100G pre-filtered PDM-RZ-QPSK channels with 300% spectral efficiency over 10,608 km and 400% spectral efficiency over 4,368 km. In: Proceedings of the OFC 2010, post-deadline paper PDPB10, San Diego, 21–25 Mar 2010

Chandrasekhar, S., Liu, X., Zhu B., Peckham, D.W.: Transmission of a 1.2 Tb/s 24-carrier no-guard-interval coherent OFDM superchannel over 7200-km of ultra-large-area fiber. In: Proceedings of the ECOC 2009, paper PD2.6, Vienna, 20–24 Sept 2009

Chou, M.H.: Optical signal processing and switching with second-order nonlinearities in waveguides. IEICE Trans. Electron. **E83-C**, 869–874 (2000)

Croussore, K., Li, G.: Amplitude regeneration of RZ-DPSK signals based on four-wave mixing in fibre. Electron. Lett. **43**(3), 177–178 (2007a)

Croussore, K., Li, G.: Phase regeneration of NRZ-DPSK signals based on symmetric-pump phase-sensitive amplification. IEEE Photon. Technol. Lett. **19**(11), 864–866 (2007b)

Croussore, K., Li, G.: BPSK phase and amplitude regeneration using a traveling-wave phase-sensitive amplifier. IEEE/LEOS Winter Topical Meeting Series, paper MB1.3, Sorrento, 14–16 Jan 2008, pp. 45–46

Croussore, K., Kim, I., Kim, Ch, Han, Y., Li, G.: Phase-and-amplitude regeneration of differential phase-shift keyed signals using a phase-sensitive amplifier. Opt. Expr. **14**, 2085–2094 (2006)

Curri, V., Poggiolini, P., Carena, A., Forghieri, F.: Dispersion compensation and mitigation of nonlinear effects in 111 Gb/s WDM coherent PM-QPSK systems. IEEE Photon. Technol. Lett. **20**(17), 1473–1475 (2008). 1st Sept 2008

Cvecek, K., Sponsel, K., Stephan, C., Onishchukov, G., Ludwig, R., Schubert, C., Schmauss, B., Leuchs, G.: Phase-preserving amplitude regeneration for a WDM RZ-DPSK signal using a nonlinear amplifying loop mirror. Opt. Expr. **16**, 1923–1928 (2008)

Della Valle, G., Festa, A., Taccheo, S., Ennser, K., Aracil, J.: Nonlinear dynamics induced by burst amplification in optically gain-stabilized erbium-doped amplifiers. Opt. Lett. **32**, 903–905 (2007)

Demir, A.: Nonlinear phase noise in optical-fiber-communication systems. J. Light. Technol. **25**(8), 2002–2031 (2007)

Dischler, R., Buchali, F.: Transmission of 1.2 Tb/s continuous waveband PDM-OFDM-QPSK signal with spectral efficiency of 3.3bit/s/Hz over 400 km of SSMF. In: Proceedings of the OFC 2009, paper PDPC2, San Diego, 22–26 Mar 2009

Ennser, K., Della Valle, G., Ibsen, M., Shmulovich, J., Taccheo, S.: Erbium-doped waveguide amplifier for reconfigurable WDM metro networks. IEEE Photon. Technol. Lett. **17**, 1468–1470 (2005a)

Ennser, K., Della Valle, G., Mariani, D., Ibsen, M., Shmulovich, J., Taccheo S., Laporta P.: Erbium-Doped waveguide amplifier insensitive to channel transient and to spectral-hole-burning offset. In: Proceedings of the Advanced Solid-State Photonics 2005:MB32 (2005b)

Ennser, K., Taccheo, S., Rogowski, T., Shmulovich, J.: Efficient erbium-doped waveguide amplifier insensitive to power fluctuations. Opt. Expr. **14**, 10307–10312 (2006)

Fludger, C.R.S.: 10x111 Gbit/s, 50 GHz spaced, POLMUX-RZ-DQPSK transmission over 2375 km employing coherent equalisation. In: Proceedings of the OFC/NFOEC 2007, post-deadline paper PDP-22, Anaheim, Feb 2007

Gavioli, G., Torrengo, E., Bosco, G., Carena, A., Curri, V., Miot, V., Poggiolini, P., Forghieri, F., Savory, S.J., Molle, L., Freund, R.: NRZ-PM-QPSK 16 × 100 Gb/s transmission over installed fiber with different dispersion maps. IEEE Photon. Technol. L. **22**(6), 371–373 (2010). 1 Mar 2010

Gnauck, H., Winzer, P.J., Doerr, C.R., Buhl, L.L.: 10 × 112-Gb/s PDM 16-QAM transmission over 630 km of fiber with 6.2-b/s/Hz spectral efficiency. In: Proceedings of the OFC 2009, San Diego, Mar 2009

Gordon, J.P., Mollenauer, L.F.: Phase noise in photonic communications systems using linear amplifiers. Opt. Lett. **15**(23), 1351–1353 (1990)

Grellier, E., Antona, J.-C., Bigo, S.: Revisiting the evaluation of non-linear propagation impairments in highly dispersive systems. In: Proceedings of the ECOC 2009, paper 10.4.2, Vienna, 20–24 Sept 2009

Hasegawa, A., Tapper, F.: Transmission of stationary nonlinear dispersive dielectric fibers I Anomalous dispersion. Appl. Phys. Lett. **23**, 142–144 (1973)

Krcmarík, D., Slavík, R., Park, Y., Azaña, J.: Nonlinear pulse compression of picosecond parabolic-like pulses synthesized with a long period fiber grating filter. Opt. Expr. **17**, 7074–7087 (2009)

Latkin, A.I., Boscolo, S., Bhamber, R.S., Turitsyn, S.K.: Doubling of optical signals using triangular pulses. J. Opt. Soc. Am. **B 26**, 1492–1496 (2009)

Lu, G.-W., Miyazaki, T.: Optical phase add/drop for format conversion between DQPSK and DPSK and its application in optical label switching systems. IEEE Photon. Technol. Lett. **21**, 322–324 (2009)

Ma, Y., Yang, Q., Tang, Y., Chen, S., Shieh, W.: 1-Tb/s per channel coherent optical OFDM transmission with subwavelength bandwidth access. In: Proceedings of the OFC 2009, paper PDPC1, San Diego, 22–26 Mar 2009

Marazzi, L.: Real-time 100-Gb/s POLMUX RZ-DQPSK transmission over uncompensated 500 km of SSMF by optical phase conjugation. In: Proceedings of the OFC '09, Optical Fiber Communication Conference, paper JWA44, San Diego, 22–26 Mar 2009

Martelli, P., et al.: All-optical wavelength conversion of a 100-Gb/s polarization-multiplexed signal. Opt. Expr. **17**(20), 17758–17763 (2009)

Martinelli, M.: Polarization stabilization in optical communication systems. J. Light. Technol. **24**(1), 4172–4183 (2006)

Matsumoto, M., Sakaguchi, H.: DPSK signal regeneration using a fiber-based amplitude regenerator. Opt. Expr. **16**, 11169–11175 (2008)

Matsumoto, M., Sanuki, K.: Performance improvement of DPSK signal transmission by a phase-preserving amplitude limiter. Opt. Expr. **15**, 8094–8103 (2007)

Mollenauer, L.F., Stolen, R.H., Gordon, J.P.: Experimental observation of picosecond pulse narrowing and solitons in optical fibers. Phys. Rev. Lett. **45**, 1095–1098 (1980)

Ng, T.T., Parmigiani, F., Ibsen, M., Zhange, Z., Petropoulos, P., Richardson, D.J.: Compensation of linear distortions by using XPM with parabolic pulses as a time lens. IEEE Photon. Technol. Lett. **20**(13), 1097–1099 (2008)

Nielsen, L.G., Dasgupta, S., Mermelstein, M.D., Jakobsen, D., Herstrom, S., Pedersen, M.E.V., Lim, E.L., Alam, S.-U., Parmigiani, F., Richardson, D.J., Palsdottir, B.: A silica based highly nonlinear fibre with improved threshold for stimulated Brillouin scattering. ECOC 2010, Turin, 19–23 Sept 2010

Parameswaran, K.R., et al.: Highly efficient second-harmonic generation in buried waveguides formed by annealed and reverse proton exchange in periodically poled lithium niobate. Opt. Lett. **27**(3), 179–181 (2002)

Parmigiani, F., Petropoulos, P., Ibsen, M., Richardson, D.J.: All-optical pulse reshaping and retiming systems incorporating pulse shaping fiber Bragg grating. IEEE J. Light. Technol. **24**(1), 357–364 (2006a)

Parmigiani, F., Petropoulos, P., Ibsen, M., Richardson, D.J.: Pulse retiming based on XPM using parabolic pulses formed in a fiber Bragg grating. IEEE Photon. Technol. Lett. **18**(7), 829–831 (2006b)

Parmigiani, F., Finot, C., Mukasa, K., Ibsen, M., Roelens, M.A.F., Petropoulos, P., Richardson, D.J.: Ultra-flat SPM-broadened spectra in a highly nonlinear fiber using parabolic pulses formed in a fiber Bragg grating. Opt. Expr. **14**(17), 7617–7622 (2006c)

Parmigiani, F., Oxenlowe, L.K., Galili, M., Ibsen, M., Zibar, D., Petropoulos, P., Richardson, D.J., Clausen, A.T., Jeppesen, P.: All-optical 160 Gbit/s retiming system using fiber grating based pulse shaping technology. J. Light. Technol. **27**(9), 1135–1141 (2008a)

Parmigiani, F., Ibsen, M., Ng, T.T., Provost, L., Petropoulos, P., Richardson, D.J.: An efficient wavelength converter exploiting a grating based saw-tooth pulse shaper. Photon. Technol. Lett. **20**(17), 1461–1463 (2008b)

Parmigiani, F., Ng, T.T., Ibsen, M., Petropoulos, P., Richardson, D.J.: Timing jitter tolerant all-optical TDM demultiplexing using a saw-tooth pulse shaper. Photon. Technol. Lett. **20**, 1992–1994 (2008c)

Parmigiani, F., Ibsen, M., Petropoulos, P., Richardson, D.J.: Efficient all-optical wavelength conversion scheme based on a saw-tooth pulse shaper. Photon. Technol. Lett. **21**(24), 1837–1839 (2009a)

Parmigiani, F., Petropoulos, P., Ibsen, M., Almeida, P.J., Ng, T.T., Richardson, D.J.: Time domain add-drop multiplexing scheme enhanced using a saw-tooth pulse shaper. Opt. Expr. **17**, 6562–6567 (2009b)

Parmigiani, F., Slavík, R., Kakande, J., Lundström, C., Sjödin, M., Andrekson, P.A., Weerasuriya, R., Sygletos, S., Ellis, A.D., Grüner-Nielsen, L., Jakobsen, D., Herstrøm, S., Phelan, R., O'Gorman, J., Bogris, A., Syvridis, D., Dasgupta, S., Petropoulos, P., Richardson, D.J.: All-optical phase regeneration of 40 Gbit/s DPSK signals in a black-box phase sensitive amplifier. In: Proceedings of the Optical Fiber Communication Conference (OFC/NFOEC 2010), Paper PDPC3, San Diego (2010a)

Parmigiani, F., Slavik, R., Kakande, J., Gruner-Mielsen, L., Jakobsen, D., Herstrom, S., Weerasuriya, S., Sygletos, S., Ellis, A.D., Petropoulos, P., Richardson, D.J.: All-optical phase and amplitude regeneration properties of a 40 Gbit/s DPSK black-box phase sensitive amplifier, ECOC 2010, Turin, 9–23 Sept 2010b

Petropoulos, P., Ibsen, M., Ellis, A.D., Richardson, D.J.: Rectangular pulse generation based on pulse reshaping using a superstructured fiber Bragg grating. IEEE J. Light. Technol. **19**(5), 746–752 (2001)

Poggiolini, P., Nespola, A., Abrate, S., Ferrero, V., Lezzi, C.: Long-term PMD characterization of a metropolitan G.652 fiber plant. J. Light. Technol. **24**(11), 4022–4029 (2006). Nov 2006

Roelens, M.A.F., Bolger, J.A., Williams, D., Frisken, S.J., Baxter, G.W., Clarke, A.M., Eggleton, B.J.: Flexible and reconfigurable time-domain demultiplexing of optical signals at 160 Gb/s. IEEE Photon. Technol. Lett. **21**(10), 618–620 (2009)

Salsi, M., et al.: 155x100Gbit/s coherent PDM-QPSK transmission over 7,200 km. In: Proceedings of the ECOC 2009, post-deadline paper PD2.5, Vienna, 20–24 Sept 2009

Sjödin, M., Johannisson, P., Sköld, M., Karlsson, M., Andrekson, P.; Cancellation of SPM in self-homodyne coherent systems. ECOC 2009, paper We8.4.5, Vienna, 20–24 Sept 2009

Sköld, M., Yang, J., Sunnerud, H., Karlsson, M., Oda, S., Andrekson, P.A.: Constellation diagram analysis of DPSK signal regeneration in a saturated parametric amplifier. Opt. Expr. **16**, 5974–5982 (2008)

Slavik, R., Parmigiani, F., Kakande, J., Lundstrom, C., Sjodin, M., Andrekson, P., Weerasuriya, R., Sygletos, S., Ellis, A.D., Gruner-Nielsen, L., Jakobsen, D., Herstrom, S., Phelan, R., O'Gorman, J., Bogris, A., Syvridis, D., Dasgupta, S., Petropoulos, P., Richardson, D.J.: All-optical phase and amplitude regenerator for next-generation telecommunications systems. Nat. Photon. **4**, 690–695 (2010)

Taccheo, S., Della Valle, G., Festa, A., Ennser, K., Aracil, J.: Amplification of optical bursts in gain-stabilized erbium-doped optical amplifier. In: Proceedings of the Optical Fiber Communication Conference 2007:OMN3 (2007)

Torrengo, E., et al.: Transoceanic PM-QPSK terabit superchannel transmission experiments at baud-rate subcarrier spacing. In: Proceedings of the ECOC 2010, paper We.7.C.2, Torino, Sept 2010

Van Den Borne, D., Sleiffer, V., Alfiad, M., Jansen, S., Wuth, T.: POLMUX-QPSK modulation and coherent detection: the challenge of long-haul 100 G transmission. In: Proceedings of the ECOC 2009, invited paper 3.4.1, Vienna, 20–24 Sept 2009

Weerasuriya, R., Sygletos, S., Ibrahim, S.K., Phelan, R., O'Carroll, J., Kelly, B., O'Gorman, J., Ellis, A.D.: Generation of frequency symmetric signals from a BPSK input for phase sensitive amplification. OWT6, paper number 1928, OFC 2010

Zannin, M., Ennser, K., Taccheo, S., Careglio, D., Solé-Pareta, J., Aracil, J.: On the benefits of optical gain clamped amplification in optical burst switching networks. IEEE J. Light. Technol. **27**, 5475–5482 (2009)

Chapter 5
Economics of Next-Generation Networks

Marco Forzati, Jiajia Chen, Miroslaw Kantor, Bart Lannoo, Claus Popp Larsen, Christer Mattsson, Attila Mitcsenkov, Giorgia Parca, Elisabeth Pereira, Armando Pinto, António Teixeira, Lena Wosinska, and Muneer Zuhdi

5.1 The Cost Side: Techno-Economics of Future Internet

Telecommunication networks are made of links and nodes. Links and nodes are interconnected through compatible interfaces in order to form the physical topology. A link receives information in one interface and places it in all the other interfaces; a node receives information in one interface and places it in some of the other interfaces. The nodes must be able to take decisions in order to route the

M. Forzati (✉)
Department for Networking and Transmission, Acreo AB, 164 40 Kista, Sweden
e-mail: Marco.Forzati@acreo.se

J. Chen • L. Wosinska
The Royal Institute of Technology (KTH), Electrum 229, 16440 Kista, Stockolm, Sweden
e-mail: jiajiac@kth.se; wosinska@kth.se

M. Kantor
Department of Telecommunications, AGH University of Science and Technology, Al. Mickiewicza 30, 30–059 Kraków, Poland
e-mail: kantor@kt.agh.edu.pl

B. Lannoo
Ghent University - IBBT, Gaston Crommenlaan 8, 102, B-9050 Ghent-Ledeberg
e-mail: bart.lannoo@intec.ugent.be

C.P. Larsen • C. Mattsson
Acreo Netlab, Electrum 236, 164 40 Kista, Sweden
e-mail: claus.popp.larsen@acreo.se; crister.mattsson@acreo.se

A. Mitcsenkov
Department of Telecommunication and Media Informatics, Budapest University of Technology and Economics, Pf. 91., H-1521 Budapest, Hungary
e-mail: mitcsenkov@tmit.bme.hu

G. Parca • A. Pinto • A. Teixeira
DETI/Instituto de Telecomunicações, University of Aveiro, P-3810-193 Aveiro, Portugal
e-mail: anp@det.ua.pt; teixeira@ua.pt

A. Teixeira and G.M.T. Beleffi (eds.), *Optical Transmission: The FP7 BONE Project Experience*, Signals and Communication Technology,
DOI 10.1007/978-94-007-1767-1_5,

information through the appropriate interfaces. This ability to take decisions can be given to the node by some form of manual configuration or can be acquired automatically through routing protocol. The nodes can be themselves source or destination of information.

Being the networks made of nodes and links, we can define the CapEx as the sum of the cost of nodes and links. Hence, the total setup cost C_{T}, i.e. the CapEx, is given by

$$C_{\mathrm{T}} = C_{\mathrm{L}} + C_{\mathrm{N}}, \tag{5.1}$$

where C_{L} and C_{N} are the links' and nodes' setup cost, respectively. In the calculation of the CapEx, we follow closely the approach present in (Korotky 2004). The setup cost of links and nodes can be written as

$$C_{\mathrm{L}} = \sum_{l=1}^{L} c_l, \tag{5.2}$$

$$C_{\mathrm{N}} = \sum_{n=1}^{N} c_n, \tag{5.3}$$

respectively, where L is the number of links, c_l is the cost of link l, N is the number of nodes, and c_n is the cost of node n. From (5.1), (5.2) and (5.3), we obtain

$$C_{\mathrm{T}} = \sum_{l=1}^{L} c_l + \sum_{n=1}^{N} c_n. \tag{5.4}$$

Expression (5.4) can be rewritten as

$$C_{\mathrm{T}} = \frac{L}{L}\sum_{l=1}^{L} c_l + \frac{N}{N}\sum_{n=1}^{N} c_n. \tag{5.5}$$

Considering the definition of average

$$\langle q \rangle = \frac{1}{M}\sum_{m=1}^{M} q_m, \tag{5.6}$$

E. Pereira
Department of Economics, Management and Industrial Engineering, Research Unit on Governance, Competitiveness and Public Policies (GOVCOPP), Universidade de Aveiro, Campus universitário de Santiago, 3810–193 Aveiro, Portugal
e-mail: melisa@ua.pt

M. Zuhdi
University of Aveiro/Etisalat, Campus Universitário de Santiago, P-3810-193 Aveiro, Portugal
e-mail: muneer_zuhdi@hotmail.com

we can obtain

$$C_{\mathrm{T}} = L\langle c_l \rangle + N\langle c_n \rangle, \tag{5.7}$$

being

$$\langle c_l \rangle = \frac{1}{L}\sum_{l=1}^{L} c_l, \tag{5.8}$$

and

$$\langle c_n \rangle = \frac{1}{N}\sum_{n=1}^{N} c_n, \tag{5.9}$$

the average cost of links and nodes, respectively. Equations 5.4 and 5.7 are mathematically identical, therefore the CapEx estimation is reduced to the estimation of four quantities: N, L, $\langle c_l \rangle$ and $\langle c_n \rangle$. The accuracy of the CapEx estimation is only a result of the accuracy of the estimation of these four quantities. We should note that depending on the nature of the problem, some of these quantities can be known in advance. For instance, the number of nodes is usually known, and we only have to estimate the other three quantities. We are going to treat the problem in quite a general way; from this general treatment, a model can, in principle, be developed for any specific case.

5.1.1 Backbone Networks and Cost Analysis Theory

Backbone networks are composed of links and nodes. In transport networks, links are typically point-to-point links, physically supported by fibre-optic transmission systems. Each link usually interconnects two nodes of the network. Links accept traffic in one end, and after a certain amount of time, the propagation and processing time, the traffic appears in the other end, i.e. links accept the traffic from one node and deliver the traffic to the other node. Links have to provide interfaces with a given bit rate, compatible with the interfaces of the adjacent nodes. The particular implementation of a link is done according to number of channels that have to be supported, required bit rate per channel, and quality of service; by quality of service we mean the bit-error-rate and link failure probability. Links can be implemented with one or more transmission systems; ultimately, it is a matter of minimising the costs. The cost of link l is given by

$$c_l = \sum_{t=1}^{T_l} c_{l,t}, \tag{5.10}$$

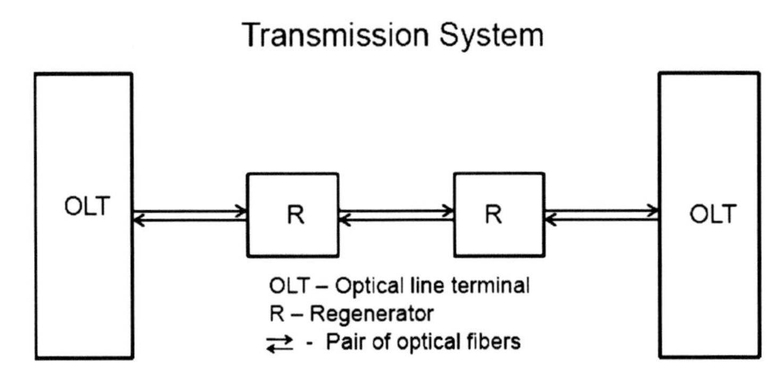

Fig. 5.1 Generic optical transmission system

where $c_{l,t}$ is the cost of the transmission system t in the link l and T_l is the number of transmission systems used to support the link l.

The design of a given link consists in selecting the transmission systems able to support the traffic. This choice is done in order to minimise the costs. In order to perform a link design, we must know the amount of traffic and the cost and capacity of the transmission systems. With that information, the design of a link is in principle a simple task.

5.1.1.1 Link Cost

The choice of the link architecture is done in order to obtain a given transmission capacity, in terms of number of optical channels supported and bit rates, minimising the costs. Although some variations are possible, the typical architecture of an optical communication system (Fig. 5.1) comprises the optical line terminal equipment (OLT – optical line terminal), the regeneration stages (R – regenerators), and the optical fibre.

Usually, a pair of terminal equipments is interconnected through a pair of optical fibres. If the distance between the terminal equipments is large, regenerators' stages may be required in order to guarantee a certain quality of signal at the receiver input. If the optical node accepts a WDM signal, it is possible to avoid the use of a pair of OLTs in each side of the link. This can be considered as a particular case in which the cost of the OLTs is null, as this is not a common situation.

In terms of setup cost for the transmission system t, $c_{l,t}$, three major terms can be considered, one related to the cost of the terminal equipment, $c_{l,t}^{\mathrm{OLT}}$, the other related to the cost of the regeneration stages, $c_{l,t}^{\mathrm{R}}$ and another related to the fibre cost per unit of length $c_{l,t}^{\mathrm{F}}$. We can say that the setup cost of a transmission system is given by

$$c_{l,t} = 2c_{l,t}^{\mathrm{OLT}} + n_{l,t}^{\mathrm{R}} c_{l,t}^{\mathrm{R}} + s_{l,t} c_{l,t}^{\mathrm{F}} + c_{l,t}^{\mathrm{others}}. \tag{5.11}$$

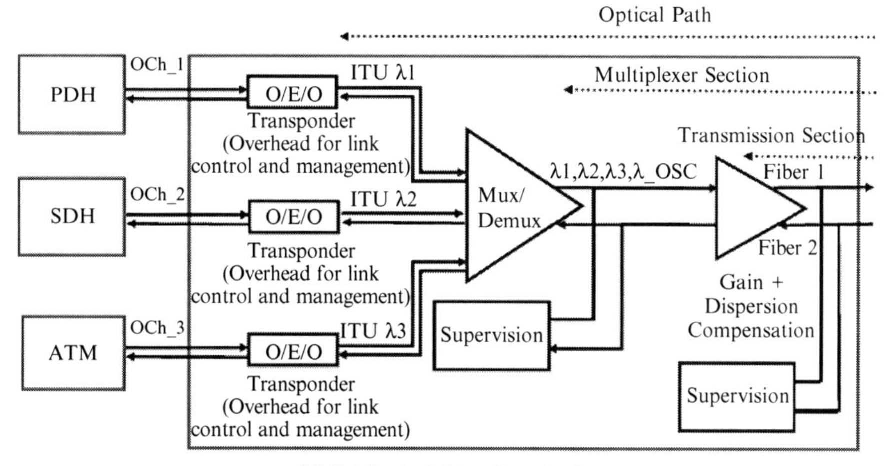

Fig. 5.2 OLT architecture for a WDM transmission system

In expression (5.11), we are considering a pair of terminal equipments, $n^{\mathrm{R}}_{l,t}$ bidirectional regenerators and a pair of fibres with length $s_{l,t}$; $c^{\mathrm{others}}_{l,t}$ takes into consideration costs that cannot be directly related to the terminal equipments, regenerators and fibres. These other costs take into consideration, for instance, administrative costs related to the installation of a new transmission system in link l. Considering that $c^{\mathrm{others}}_{l,t}$ is negligible, we obtain

$$c_{l,t} = 2c^{\mathrm{OLT}}_{l,t} + n^{\mathrm{R}}_{l,t} c^{\mathrm{R}}_{l,t} + s_{l,t} c^{\mathrm{F}}_{l,t}. \tag{5.12}$$

Nowadays, telecommunications operators tend to rent the fibre, which converts this cost to an operation cost; therefore, $c^{\mathrm{F}}_{l,t} = 0$, and

$$c_{l,t} = 2c^{\mathrm{OLT}}_{l,t} + n^{\mathrm{R}}_{l,t} c^{\mathrm{R}}_{l,t}. \tag{5.13}$$

Expression (5.13) gives the setup cost of a transmission system. However, we should note that (5.13) is not suitable for all cases, it is just a plausible model. For any particular case under analysis, a model must be developed in order to reflect the setup costs of that particular transmission system.

The terminal equipment of a system with multiple wavelengths (WDM) is made of a base structure which typically includes an optical multiplexer, a module of optical processing, and a set of transponders. It requires one transponder per optical channel. A transponder has two interfaces, a short-range interface to connect with the local equipment and a long-range interface to connect with another transponder in the other side of the link; see Fig. 5.2.

The cost of the terminal equipment can usually be divided into two major components: one component that is fixed, $\gamma_{l,t,0}^{\text{OLT}}$, and another component that grows linearly with the traffic, i.e. the bit rate, τ, times the number of optical channels, $w_{l,t}$. Therefore, we can write

$$c_{l,t}^{\text{OLT}} = \gamma_{l,t,0}^{\text{OLT}} + \gamma_{l,t,1}^{\text{OLT}} \tau w_{l,t}. \tag{5.14}$$

The regenerators used in optical networks are classified according to the domain of operation. If the regenerator includes opto-electric conversion, it is said to be an electrical regenerator; if it operates entirely in the optical domain, it is named optical regenerator. The regenerators are also classified according to their functionalities. The regenerator 1R only amplifies the signal, the 2R amplifies and restores the signal shape, and the 3R regenerator amplifies, restores and re-times the signal. The regenerators can be placed in the transmission system or in the network nodes. The regenerators placed along the transmission systems, i.e. between a pair of terminal equipments, are usually named "line regenerators." Nowadays, line regenerators tend to be 1R and tend to operate entirely in the optical domain. Line regenerators are designed as optical amplifiers and operate using the light-stimulated emission physical effect. We should note that a system able to emit light in the form of stimulated emission also emits light in the form of spontaneous emission; therefore, optical amplifiers always add noise to the signal, i.e. always degrade the optical signal to noise ratio.

Let us now look at the fibre. After the manufacturing process, fibres are bundled into optical cables. The type of cables depends on the applications. For some applications, soft plastic cables are enough; for others, hard steel-based cables are required. Optical cables are installed in ducts along roads, railways, etc. Nowadays, optical cables are installed by several entities. For instances, local governments can install optical-fibre cables using the pipes of water and sewer networks. The operators of highways and railroads can lay optical-fibre cables along the road. Electrical energy companies can also place optical-fibre cables using their networks. Several other entities are laying optical-fibre cables to support private networks or to lease to telecom operators to support public networks. In this scenario, telecom operators tend to lease fibre-optic cables instead of installing them. Therefore, the cost of the optical-fibre cables tends to appear as an operational expense instead of a capital expenditure.

A pair of fibres is usually used per transmission system. One fibre of the pair is used in one direction, and the other fibre is used to transmit in the opposite direction. In modern WDM transmission systems, several optical channels mapped in different wavelengths are transmitted through a single fibre.

In order to obtain the installation cost of the links, we can use (5.2), (5.10), (5.13) and (5.14) to obtain

$$C_{\text{L}} = \sum_{l=1}^{L} \sum_{t=1}^{T_l} 2\gamma_{l,t,0}^{\text{OLT}} + \sum_{l=1}^{L} \sum_{t=1}^{T_l} 2\gamma_{l,t,1}^{\text{OLT}} \tau w_{l,t} + \sum_{l=1}^{L} \sum_{t=1}^{T_l} n_{l,t}^{\text{R}} c_{l,t}^{\text{R}}. \tag{5.15}$$

The first term of (5.15) can be rewritten as

$$\sum_{l=1}^{L}\sum_{t=1}^{T_l} 2\gamma_{l,t,0}^{\mathrm{OLT}} = 2L\langle\gamma_0^{\mathrm{OLT}}\rangle, \tag{5.16}$$

where $\langle\gamma_0^{\mathrm{OLT}}\rangle$ is the average base cost of the OLTs per link, i.e.

$$\langle\gamma_0^{\mathrm{OLT}}\rangle = \frac{\sum_{l=1}^{L}\sum_{t=1}^{T_l} \gamma_{l,t,0}^{\mathrm{OLT}}}{L}. \tag{5.17}$$

Considering the second term of (5.15), we can write

$$\sum_{l=1}^{L}\sum_{t=1}^{T_l} 2\gamma_{l,t,1}^{\mathrm{OLT}}\tau w_{l,t} = 2\langle\gamma_1^{\mathrm{OLT}}\rangle\tau W, \tag{5.18}$$

where W is the total number of optical channels in the network and

$$\langle\gamma_1^{\mathrm{OLT}}\rangle = \frac{\sum_{l=1}^{L}\sum_{t=1}^{T_l} \gamma_{l,t,1}^{\mathrm{OLT}} w_{l,t}}{W}. \tag{5.19}$$

In the derivation of (5.19), we have assumed that all optical channels support the same amount of traffic, τ, given by

$$\tau = \frac{T}{D}, \tag{5.20}$$

where T is the total bidirectional traffic and D is the number of bidirectional demands. Considering the average number of optical channels per link given by

$$\langle w\rangle = \frac{W}{L}, \tag{5.21}$$

we can rewrite (5.18) as

$$\sum_{l=1}^{L}\sum_{t=1}^{T_l} 2\gamma_{l,t,1}^{\mathrm{OLT}}\tau w_{l,t} = 2L\langle\gamma_1^{\mathrm{OLT}}\rangle\tau\langle w\rangle. \tag{5.22}$$

The third term of (5.15) can be rewritten as

$$\sum_{l=1}^{L}\sum_{t=1}^{T_l} n_{l,t}^{\mathrm{R}} c_{l,t}^{\mathrm{R}} = \langle c_{\mathrm{R}}\rangle N^{\mathrm{R}}, \tag{5.23}$$

where N^{R} is the total number of regeneration stages and $\langle c_{\mathrm{R}} \rangle$ is the average cost of a bidirectional regeneration stage,

$$\langle c_{\mathrm{R}} \rangle = \frac{\sum_{l=1}^{L} \sum_{t=1}^{T_l} n_{l,t}^{\mathrm{R}} c_{l,t}^{\mathrm{R}}}{N^{\mathrm{R}}}. \tag{5.24}$$

The average number of regeneration stages is given by

$$\langle n_{\mathrm{R}} \rangle = \frac{N^{\mathrm{R}}}{L}; \tag{5.25}$$

Therefore, (5.23) can be rewritten as

$$\sum_{l=1}^{L} \sum_{t=1}^{T_l} n_{l,t}^{\mathrm{R}} c_{l,t}^{\mathrm{R}} = L \langle n_{\mathrm{R}} \rangle \langle c_{\mathrm{R}} \rangle. \tag{5.26}$$

Considering (5.16), (5.22) and (5.26), we can rewrite (5.15) as

$$C_{\mathrm{L}} = 2L\langle \gamma_0^{\mathrm{OLT}} \rangle + 2L\langle \gamma_1^{\mathrm{OLT}} \rangle \tau \langle w \rangle + L\langle n_{\mathrm{R}} \rangle \langle c_{\mathrm{R}} \rangle; \tag{5.27}$$

Therefore,

$$\langle c_l \rangle = 2\langle \gamma_0^{\mathrm{OLT}} \rangle + 2\langle \gamma_1^{\mathrm{OLT}} \rangle \tau \langle w \rangle + \langle n_{\mathrm{R}} \rangle \langle c_{\mathrm{R}} \rangle. \tag{5.28}$$

We should note that $\langle \gamma_0^{\mathrm{OLT}} \rangle$, $\langle \gamma_1^{\mathrm{OLT}} \rangle$ and $\langle c_{\mathrm{R}} \rangle$ are given in units of money, and we are assuming that they are input parameters of our problem, or at least they can be obtained from the available information. Consequently, to estimate $\langle c_l \rangle$, we only have to estimate $\langle w \rangle$ and $\langle n_{\mathrm{R}} \rangle$, i.e. the average number of optical channels per link and the average number of regeneration stages per link.

5.1.1.2 Node Cost

In a network, to the edge nodes converge links from the access network and from the transport network. The links from the transport network are high-bit-rate links that interconnect nodes from the transport network. The links from the access networks tend to be low-bit-rate links that transport to/from the edge nodes traffic from/to end users.

One of the major functions of the nodes is to route the traffic. To route the traffic, the nodes must have some sort of "intelligence"; this "knowledge" can be acquired through routing protocols or can be obtained through the configuration process. In order to make the routing process efficient, the traffic is usually aggregated and mapped into some sort of standardised container. After the mapping process, the

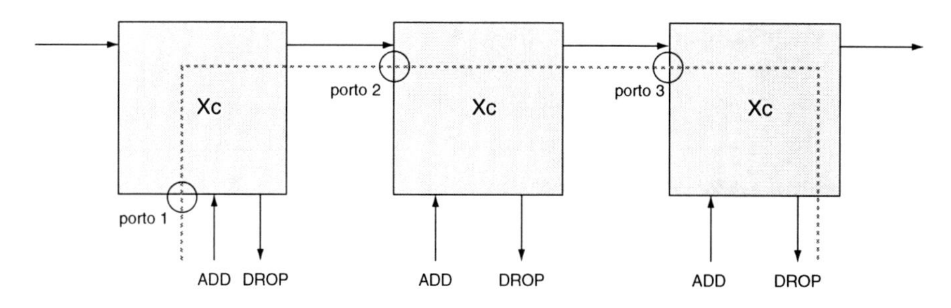

Fig. 5.3 Example network with three add and drop ports

traffic is routed. Therefore, the nodes perform three major functions: aggregation, mapping and routing. These functions are usually designated as bandwidth management. The aggregation and mapping tend to occur only in the edge nodes in order to simplify the nodes' job. Before an advance to the calculation of the nodes' setup cost, we are going to see how to count the nodes' ports.

Let us consider first unidirectional input ports; see Fig. 5.3. An unidirectional demand between node i and node j requires P_{ij} input ports. The calculus of P_{ij} can be done by adding the input port in the first node and the input port in the last node, and adding the number of hops between the node i and the node j, h_{ij}, less one, i.e.

$$P_{ij} = 1 + 1 + \left(h_{ij} - 1\right) = 1 + h_{ij}, \tag{5.29}$$

See example in Fig. 5.3.

Considering all unidirectional demands, D_1, we can calculate the total number of unidirectional input ports, assuming no survivability scheme,

$$P_1 = \sum_{d=1}^{D_1} (1 + h_d) = \sum_{d=1}^{D_1} 1 + \sum_{d=1}^{D_1} h_d, \tag{5.30}$$

h_d being the number of hops associated with the demand d. From (5.30) we obtain

$$P_1 = D_1 + \frac{D_1}{D_1} \sum_{d=1}^{D_1} h_d = D_1 \left(1 + \langle h \rangle\right) = N \langle d \rangle \left(1 + \langle h \rangle\right), \tag{5.31}$$

where the average number of hops $\langle h \rangle$ is given by

$$\langle h \rangle = \frac{1}{D_1} \sum_{d=1}^{D_1} h_d. \tag{5.32}$$

The average number of input ports is given by

$$\langle P_1 \rangle = \frac{P_1}{N} = \frac{N \langle d \rangle (1 + \langle h \rangle)}{N} = \langle d \rangle (1 + \langle h \rangle), \tag{5.33}$$

For uniform traffic, we obtain

$$\langle P_1 \rangle = (N - 1)(1 + \langle h \rangle). \tag{5.34}$$

In order to consider bidirectional ports now, we should note that a bidirectional port is also an input port; therefore, the number of bidirectional ports equals the number of unidirectional input ports

$$P_2 = P_1 = P, \tag{5.35}$$

where P is the number of ports. In a similar way, we obtain for the average values

$$\langle P_2 \rangle = \langle P_1 \rangle = \langle P \rangle. \tag{5.36}$$

Considering (5.35) and (5.34), we obtain

$$P = N \langle d \rangle (1 + \langle h \rangle) \tag{5.37}$$

and

$$\langle P \rangle = \langle d \rangle (1 + \langle h \rangle). \tag{5.38}$$

Considering the example network (see Fig. 5.4), we calculate the average number of ports. This can be done by counting directly in the scheme or using the derived expressions; in both cases, the obtained value should be $P = 76$, and $\langle P \rangle = 12.666$ with $\langle h \rangle = 1.533$.

Another way to count the ports is to consider adding ports, P_{ADD}, dropping ports, P_{DROP}, and through ports, P_{THRU}; see Fig. 5.5. The adding ports are used to insert traffic in the node, the dropping ports are used to extract traffic, and the through ports are used by traffic that just crosses the node toward its final destination.

We should note that all adding ports are also dropping ports because all ports are bidirectional ports; therefore,

$$P_{ADD} = P_{DROP}. \tag{5.39}$$

Any adding or dropping port is going to originate two ports, the tributary (adding and dropping ports) and a line port. Therefore, the total number of bidirectional ports equals two times the number of adding or dropping ports plus the through ports, i.e.

$$P = 2P_{ADD} + P_{THRU} = 2P_{DROP} + P_{THRU}. \tag{5.40}$$

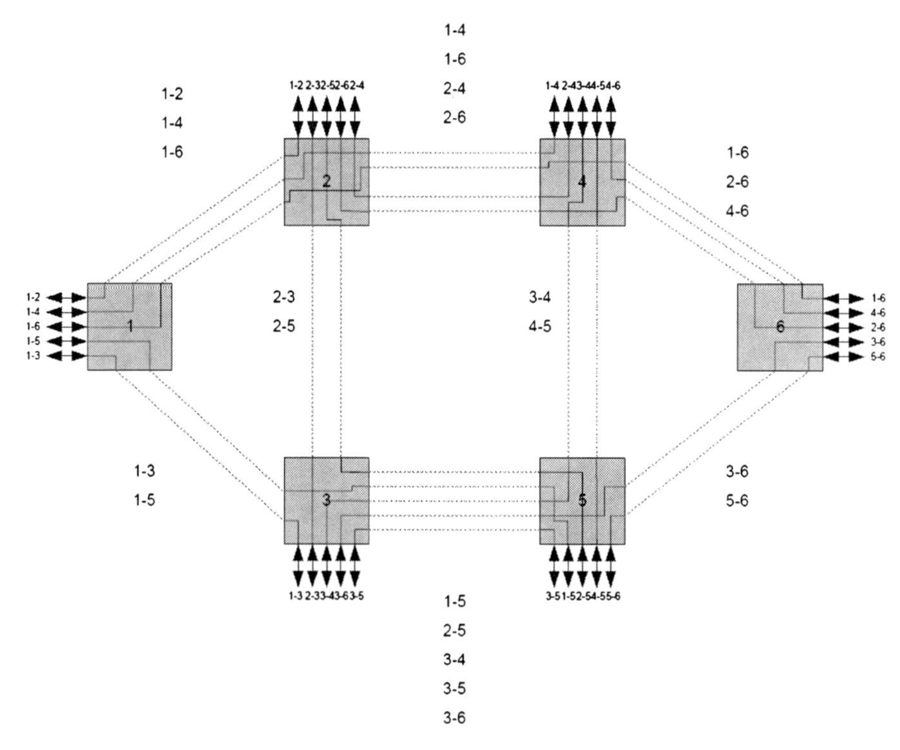

Fig. 5.4 Example network with 76 ports and $< h > = 1.533$ (example from Correia 2008)

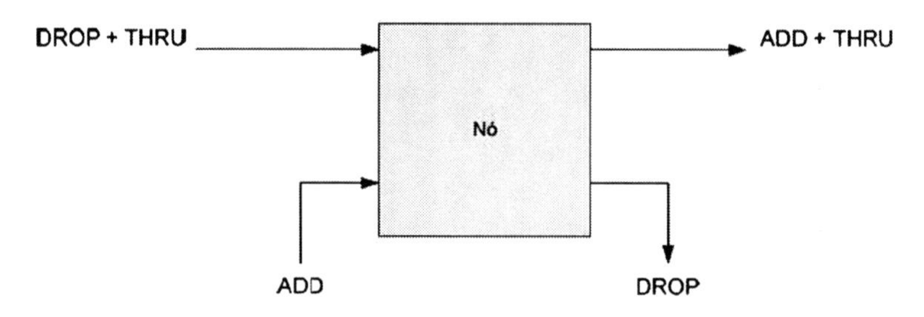

Fig. 5.5 Adding, dropping, and through ports of a node

Considering (5.39) and (5.40), we obtain

$$P = P_{ADD} + P_{DROP} + P_{THRU}. \tag{5.41}$$

Dividing both members of (5.41) by N, we obtain

$$\langle P \rangle = \langle P_{ADD} \rangle + \langle P_{DROP} \rangle + \langle P_{THRU} \rangle. \tag{5.42}$$

We should note that the number of adding or dropping ports must equal the number of unidirectional demands, i.e.

$$P_{ADD} = P_{DROP} = D_1 = N\langle d\rangle; \tag{5.43}$$

therefore, dividing by N,

$$\langle P_{ADD}\rangle = \langle P_{DROP}\rangle = \langle d\rangle. \tag{5.44}$$

Considering (5.41), (5.37) and (5.43), we obtain

$$P_{THRU} = P - P_{ADD} - P_{DROP} = N\langle d\rangle(1 + \langle h\rangle) - 2N\langle d\rangle = N\langle d\rangle(\langle h\rangle - 1), \tag{5.45}$$

and

$$\langle P_{THRU}\rangle = \langle P\rangle - \langle P_{ADD}\rangle - \langle P_{DROP}\rangle = \langle d\rangle(1 + \langle h\rangle) - 2\langle d\rangle = \langle d\rangle(\langle h\rangle - 1). \tag{5.46}$$

Applying these results to the example network (see Fig. 5.4), we obtain $P_{ADD} = P_{DROP} = 30$, $P_{THRU} = 16$, $\langle P_{ADD}\rangle = \langle P_{DROP}\rangle = 5$ and $\langle P_{THRU}\rangle = 2.666$. Note that the same results can be obtained by direct counting in Fig. 5.4.

Another alternative way to group the ports is by considering tributary ports, P_{TRIB}, and line ports, P_{LINE}. The tributary ports make the connection with links of the access network, and the line ports make the connection with links of the core network. We should have

$$P = P_{TERM} + P_{LINE}, \tag{5.47}$$

and

$$\langle P\rangle = \langle P_{TERM}\rangle + \langle P_{LINE}\rangle. \tag{5.48}$$

We must also have

$$P_{TERM} = P_{ADD} = P_{DROP} = D_1, \tag{5.49}$$

and

$$\langle P_{TERM}\rangle = \langle P_{ADD}\rangle = \langle P_{DROP}\rangle = \langle d\rangle. \tag{5.50}$$

From (5.47), (5.43) and (5.45), we obtain

$$P_{LINE} = P_{ADD} + P_{THRU} = P_{DROP} + P_{THRU} = D_1 + D_1(\langle h\rangle - 1) = N\langle d\rangle\langle h\rangle, \tag{5.51}$$

and

$$\langle P_{LINE}\rangle=\langle P_{ADD}\rangle + \langle P_{THRU}\rangle=\langle P_{DROP}\rangle + \langle P_{THRU}\rangle=\langle d\rangle + \langle d\rangle\,(\langle h\rangle - 1)=\langle d\rangle\langle h\rangle. \tag{5.52}$$

Considering again the example network (see Fig. 5.4), we obtain $P_{\mathrm{TRI}} = 30$, $P_{LINE} = 46$, $\langle P_{TRIB}\rangle = 5$ and $\langle P_{LINE}\rangle = 7.666$. We can verify these results using the developed formulation or by counting directly the tributary and line ports and dividing by the number of nodes.

The implementation of a survivability strategy is going to increase the number of line ports by a factor $1 + \langle k\rangle$, the number of tributary ports remaining constant; therefore, we have

$$P_{LINE}^{k} = P_{LINE}\,(1 + \langle k\rangle) = N\,\langle d\rangle\langle h\rangle\,(1 + \langle k\rangle)\,, \tag{5.53}$$

and

$$\langle P_{LINE}^{k}\rangle = \langle P_{LINE}\rangle\,(1 + \langle k\rangle) = \langle d\rangle\langle h\rangle\,(1 + \langle k\rangle)\,. \tag{5.54}$$

Consequently, the total number of ports is given by

$$P^{k} = P_{TERM} + P_{LINE}^{k} = N\,\langle d\rangle + N\,\langle d\rangle\langle h\rangle\,(1 + \langle k\rangle) = N\,\langle d\rangle\,[1 + \langle h\rangle\,(1 + \langle k\rangle)]\,, \tag{5.55}$$

and the average value by

$$\langle P^{k}\rangle = \langle P_{TERM}\rangle + \langle P_{LINE}^{k}\rangle = \langle d\rangle + \langle d\rangle\langle h\rangle\,(1 + \langle k\rangle) = \langle d\rangle\,[1 + (1 + \langle k\rangle)\,\langle h\rangle]\,. \tag{5.56}$$

We should note that the number of adding and dropping ports also remains unchanged with the implementation of the survivability strategy. Considering the through ports, we have to distinguish between different survivability schemes. If restoration is used, the extra resources are shared and it is not possible to count in advance the through ports. If $1+1$ dedicated protection is implemented, there are always two disjunct paths to support a given demand. In this case, the number of through ports equals the number of line ports minus the number of ports that transport traffic local to the node. As the number of local ports is two times the number of adding ports, we have

$$P_{THRU}^{k_p} = P_{LINE}^{k} - 2P_{ADD} = N\,\langle d\rangle\,[\langle h\rangle\,(1 + \langle k\rangle) - 2]\,, \tag{5.57}$$

and the average value is given by

$$\langle P_{THRU}^{k_p}\rangle = \langle d\rangle\,[\langle h\rangle\,(1 + \langle k\rangle) - 2]\,. \tag{5.58}$$

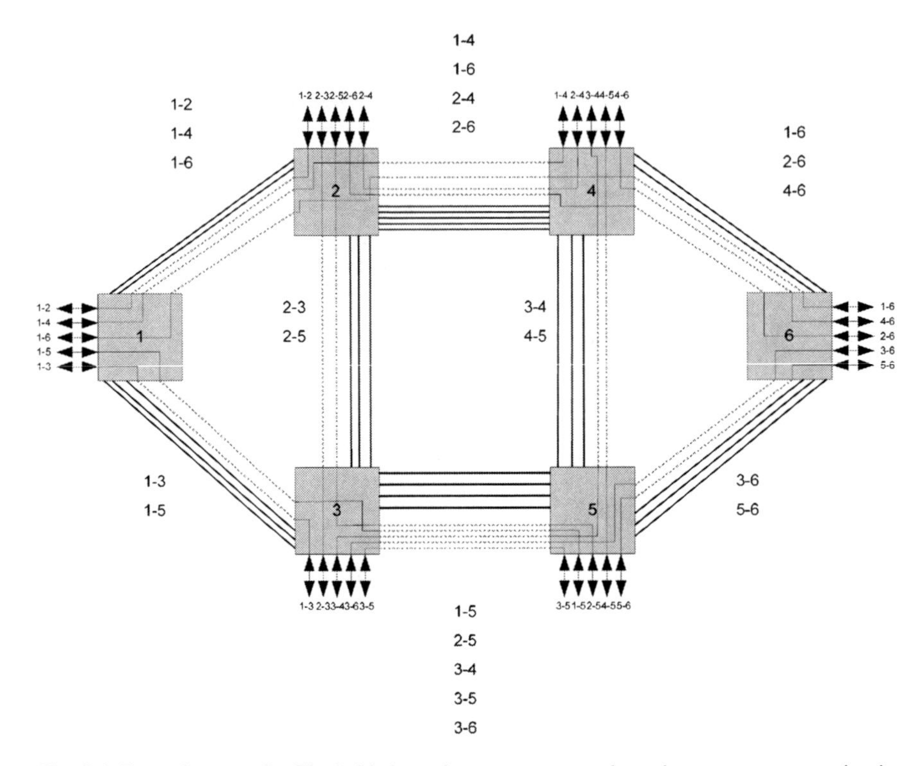

Fig. 5.6 Example networks. The bold channels represent extra channels to support restoration in case of link failure

Considering the network presented in Fig. 5.6, by counting and dividing by the number of nodes or by applying the developed formulation, we can obtain $P_{TRIB} = P_{ADD} = P_{DROP} = 30$, $P^k_{LINE} = 96$, $P^k = 126$, $\langle P_{TRIB} \rangle = \langle P_{ADD} \rangle = \langle P_{DROP} \rangle = 5$, $\langle k_r \rangle = 1.087$, $\langle P^k_{LINE} \rangle = 16$ and $\langle P^k \rangle = 21$.

Let us now look at the problem of bandwidth management. This can be performed entirely in the electrical domain, or can also be performed partially in the optical domain. Therefore, the setup cost of a node can be written as

$$c_n = c_{exc,n} + c_{oxc,n}, \tag{5.59}$$

where $c_{exc,n}$ and $c_{oxc,n}$ are the setup costs of the electrical and optical bandwidth management systems, respectively, of the node n. We should note that if the bandwidth management is performed only in the electrical domain, we have $c_{oxc,n} = 0$.

Using (5.3) and (5.59), we obtain

$$C_{\mathrm{N}} = \sum_{n=1}^{N} c_n = \sum_{n=1}^{N} (c_{exc,n} + c_{oxc,n}) = C_{exc} + C_{oxc}, \tag{5.60}$$

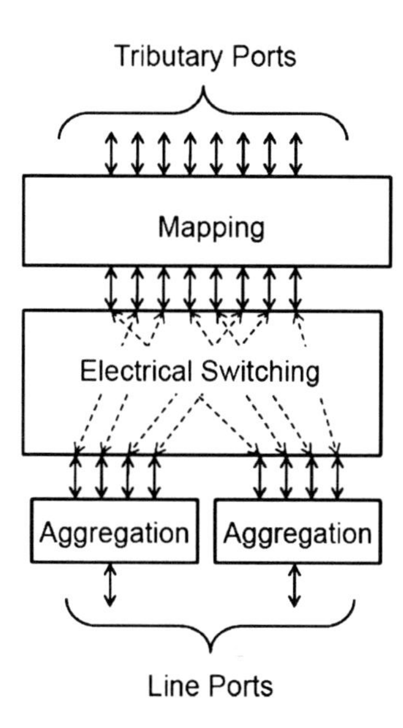

Fig. 5.7 Electrical bandwidth management system

with

$$C_{exc} = \sum_{n=1}^{N} c_{exc,n} = N \langle c_{exc} \rangle, \tag{5.61}$$

and

$$C_{oxc} = \sum_{n=1}^{N} c_{oxc,n} = N \langle c_{oxc} \rangle. \tag{5.62}$$

In Fig. 5.7, an electrical system for bandwidth management is presented. The traffic from the tributary ports is mapped, switched and aggregated before being sent through the line ports. We should note that the counting ports must be performed considering the elementary unit of traffic. In this case we can see that the elementary unit of traffic allows us to map four tributary channels; therefore, in our counting system, we will say that we have two tributary ports. Physically there are eight, but as we can map four tributary channels into one elementary unit of traffic, these eight tributary channels are equivalent to only two tributary channels in our elementary unit of traffic. In the example of Fig. 5.7, we have two line ports, two tributary ports, two adding ports, two dropping ports and zero through ports. The total number of ports is four.

A plausible model for the setup cost of an electrical bandwidth management system is to consider a fixed base cost and another cost term proportional to the traffic that has to be switched, i.e.

$$c_{exc,n} = \gamma_{e0} + \gamma_{e1} \tau P_n^k, \tag{5.63}$$

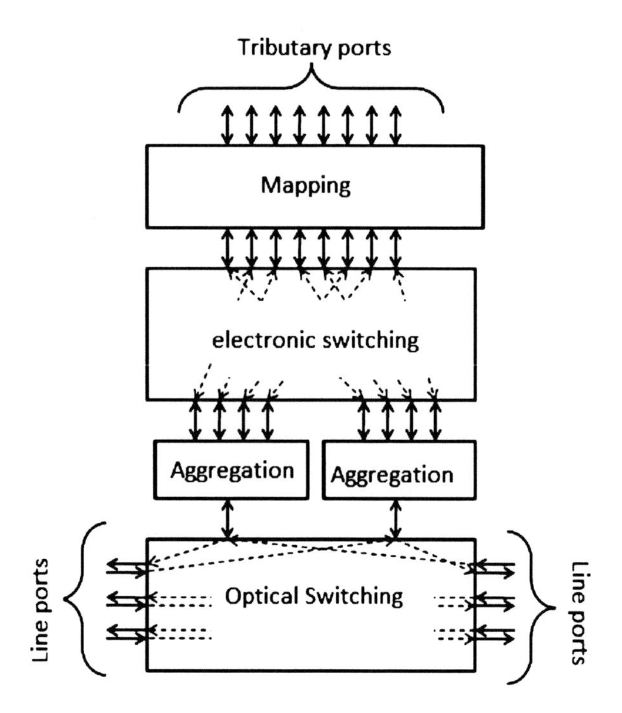

Fig. 5.8 Electrical and optical bandwidth management system

where γ_{e0} is the base cost, γ_{e1} is a cost per unit of processed traffic, and P_n^k is the total number of ports of the electrical node.

The average setup cost of the electrical node is given by

$$\langle c_{exc} \rangle = \frac{1}{N} \sum_{n=1}^{N} c_{exc,n}, \tag{5.64}$$

where $\langle P^k \rangle$ is the average number of ports per electrical node. Replacing (5.63) in (5.64), we obtain

$$\langle c_{exc} \rangle = \gamma_{e0} + \gamma_{e1} \tau \langle P^k \rangle, \tag{5.65}$$

where $\langle P^k \rangle$ is the average number of ports per electrical node.

In an electrical and optical bandwidth management system, the channels that go through the node are switched in the optical domain and the channels that are local to the node are processed in the electrical domain. The optical bandwidth management is performed at the optical channel (circuit) level (Fig. 5.8).

A plausible model for the cost of the optical switch is to consider a base cost plus a cost that increases proportionally with the number of ports, i.e.

$$c_{oxc,n} = \gamma_{o0} + \gamma_{o1} P_n^k, \tag{5.66}$$

where γ_{o0} is the base cost for the optical switch, γ_{o1} is the added cost per port, and P_n^k is the switch number of ports. In this case, the electrical switch is going to process only local traffic; therefore,

$$c_{exc,n} = \gamma_{e0} + \gamma_{e1}\tau\,(P_{ADD,n} + P_{DROP,n}) = \gamma_{e0} + 2\gamma_{e1}\tau\,P_{ADD,n}. \tag{5.67}$$

Considering the average values, we obtain for the optical switch

$$\langle c_{oxc}\rangle = \frac{1}{N}\sum_{n=1}^{N} c_{oxc,n}, \tag{5.68}$$

and using (5.66) in (5.68), we obtain

$$\langle c_{oxc}\rangle = \gamma_{o0} + \gamma_{o1}\langle P^k\rangle. \tag{5.69}$$

For the electrical switch, we have

$$\langle c_{exc}\rangle = \frac{1}{N}\sum_{n=1}^{N} c_{exc,n}, \tag{5.70}$$

and using (5.67) in (5.70), we have

$$\langle c_{exc}\rangle = \gamma_{e0} + 2\gamma_{e1}\tau\langle P_{ADD}\rangle. \tag{5.71}$$

5.1.2 Case Studies for Metro and Access Networks

Broadband Internet access is becoming a commodity product in the Western world. Most fixed broadband access networks currently rely on either twisted pair copper wires, using the DSL standards, or HFC cables, using the DOCSIS standards. In the last decade, in some countries, optical fibre access or FTTH networks have been extensively deployed, mainly in Asia (Japan, South Korea) and more recently in the United States. A lot of Western European countries, however, are lagging compared to the rest of the world. On average, only 1.5% of the broadband connections in Western Europe were served by fibre in 2008, corresponding to 3.37% of the 34 million worldwide FTTH users. It is expected, however, that DSL and HFC networks will run out of bandwidth in the (near) future, and an upgrade to a higher-capacity, most probably optical fibre–based, access network will be inevitable in the coming years.

Several implementations of an optical access network exist. Depending on the end point of the fibre path, they are referred to as fibre to the x (FTTx). In the case

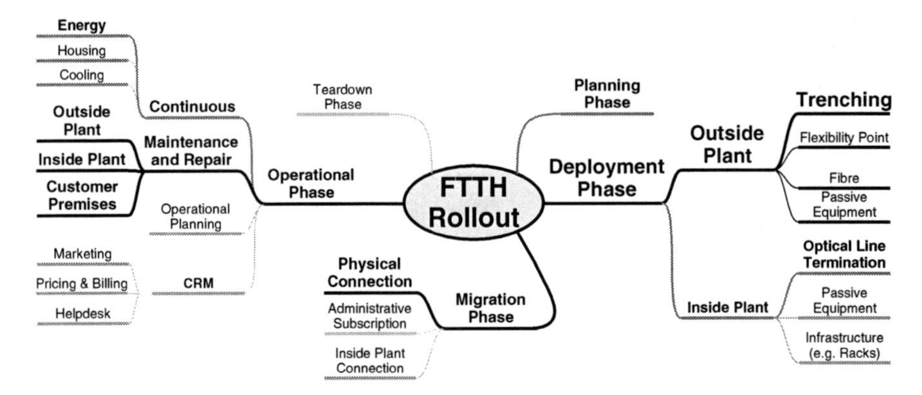

Fig. 5.9 Cost breakdown approach for the rollout cycle of an FTTH network (Casier 2008)

of FTTH or FTTB, the fibre reaches the user's house or building. FTTC, on the other hand, brings the fibre to a service node (e.g. a street cabinet) close to the end user. In the latter case, the fibre does not reach the user itself and the remaining distance has to be covered by another technology, either wired (e.g. VDSL or HFC) or wireless (e.g. WiMAX, WiFi, 3G).

There are two main categories of FTTx technologies, i.e. active and passive. AONs provide a (logical) P2P connection between the CO and each user. Most used AON topologies are home run (HR, with a dedicated fibre from the CO to each user, also known as P2P network) and active star (AS, with a switch or router installed between the CO and the user, e.g. Ethernet access switch in the street cabinet). More meshed topologies, however, could also be envisaged for an AON. PONs on the other hand, are P2MP networks, where the access fibre is shared by several users (e.g. 32, 64) through a branched tree topology. Nowadays, the most used PON configuration is a (power splitting) TDM PON, with GPON and EPON as the two most important standards. WDM PONs and hybrid WDM/TDM PONs can be considered as strong candidates for NG-PONs.

The main cost components for a fibre access network rollout are shown in Fig. 5.9 (Casier 2008).

This representation is based on a rollout cycle consisting of four phases (planning, deployment, migration and operational). The costs made in the planning and deployment phases mainly correspond to the CapEx, while most of the other costs (with the exception of some installation and equipment costs in the migration phase) correspond to the OpEx. The main capital expenditures are related to the outside plant, and especially the trenching cost for rolling out an optical fibre to every home. Another specific cost for a fibre network (and more generally a fixed network) is the physical termination at the user side. One of the most costly OpEx processes is failure reparation (Mas Machuca 2007). Basically, there are two failure-

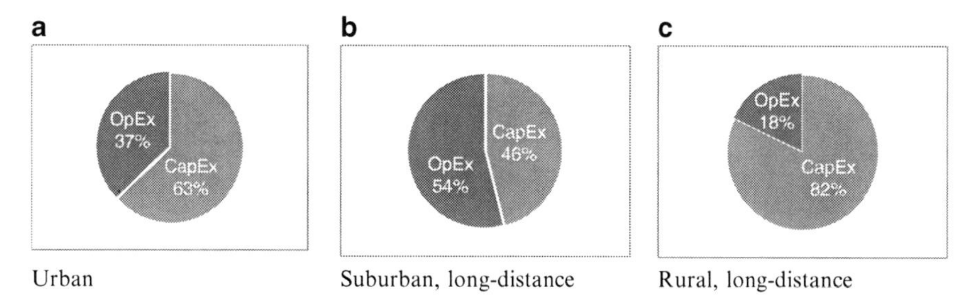

Fig. 5.10 Division between CapEx and OpEx for three considered scenarios, using an AON-HR (or P2P)

related operational costs, namely, failure reparation cost and penalty cost, which are proportional to the number of disconnected users for each failure of an unprotected component.

We have compared the costs of fibre access networks for several architectures and technologies based on some specific scenarios. To compare the cost of different fibre access deployments, three scenarios (urban, long-distance suburban, long-distance rural) have been analysed over a duration of 15 years. The first scenario is related to an FTTH rollout in the city centre of Ghent (Belgium) (Lannoo 2008). The considered network covers an area of 20 km², counting 90,000 inhabitants (i.e. ca. 43,000 households). The average fibre distance between the CO and the user is 1.5 km, and 80% of the fibres are buried underground. The other two scenarios consider a long-distance FTTH network (20 km between CO and user) in a suburban and a rural area, respectively (Wosinska 2008).

In the suburban scenario (also indicated as collective case), the length of the FF connecting the CO with the RN is assumed to be FF = 19.5 km, while the length of the DF between the CO and the RN is equal to DF = 0.5 km. In the rural scenario (dispersive case), FF = 15 km and DF = 5 km. The total number of users supported by one FF was assumed to be 256.

The CapEx–OpEx ratio is presented in Fig. 5.10 for an AON-HR (or P2P) architecture. Note that a discount rate of 10% is used to incorporate the time value of money. The urban and long-distance rural cases experience the highest CapEx, due to a high unit cost for trenching in an urban area vs. long distances, and a high amount of fibre and trenches in a rural area. Both factors are much smaller for the considered suburban case, leading to a lower CapEx. Using equal assumptions (e.g. underground trenching and comparable digging cost), on the other hand, should always favour the more densely populated areas, but this example shows that other parameters can play a significant role in this comparison. Note that in both long-distance cases, the absolute value of the OpEx is approximately the same and that the main difference is thus caused by the much-higher CapEx for a rural area.

For the two long-distance cases, Fig. 5.11 shows a relative comparison between three different FTTH architectures (AON-HR, AON-AS, PON), with AON-HR depicted as 100%. CapEx is the highest for AON-HR and the lowest for PON.

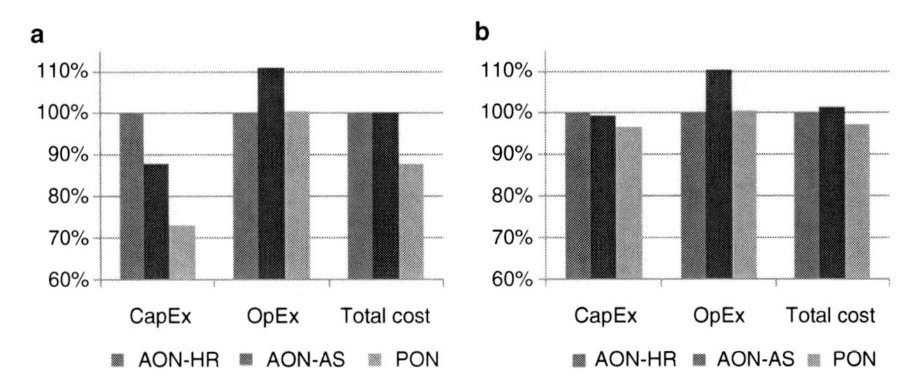

Fig. 5.11 Comparison between AON-HR, AON-AS, and PON for the suburban (**a**) and rural (**b**) long-distance scenarios

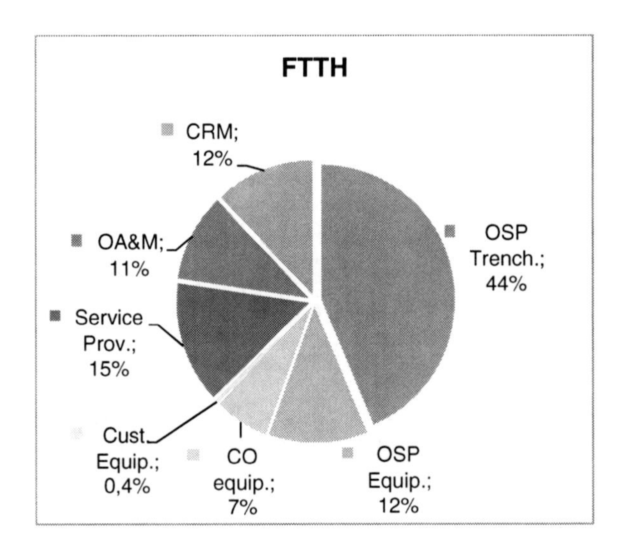

Fig. 5.12 Detailed CapEx and OpEx breakdown for urban FTTH scenario

OpEx is highest for AON-AS, containing a lot of active equipment in the field, and almost equal for PON and AON-HR. We see that the CapEx difference between the architectures decreases when going to a more expensive (rural) scenario, because the advantage of sharing fibres will decrease.

An overview of a more detailed division between CapEx and OpEx from the viewpoint of the operator is given in Fig. 5.12 for the urban case (cf. Fig. 5.10). The largest CapEx part is the cost for the OSP, consisting of digging and equipment (ducts, fibre) costs. Other CapEx are related to the CO and customer side. The OpEx division can be split between network-related (indicated as OA&M) and service-related (service provisioning and CRM, consisting of pricing and billing, help desk, marketing) costs.

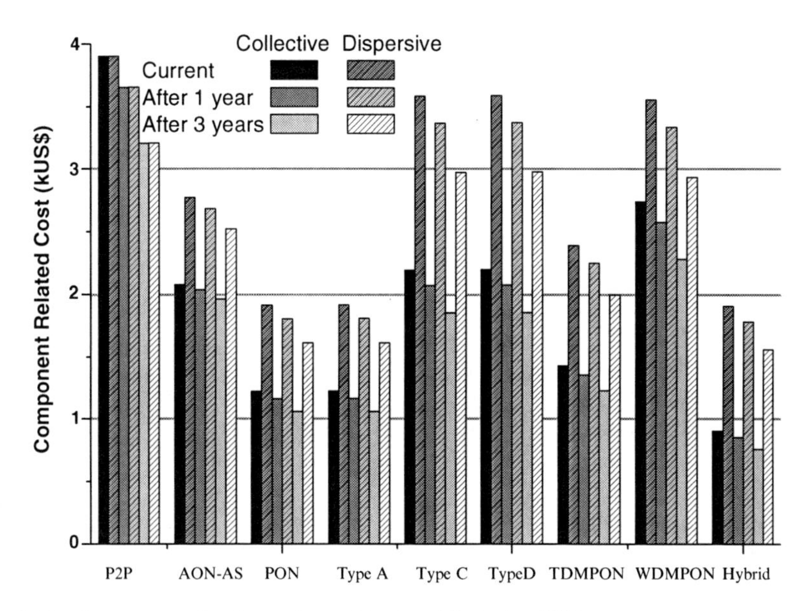

Fig. 5.13 Component-related cost changed in time (Wosinska 2008)

In the following sections, we present two case studies, namely, one where the component cost is changed in time and the second one with a variable take rate. Besides the comparison for the P2P or AON-HR, AON or AON-AS and PON schemes, we also present the comparison of CapEx and OpEx for several other representative architectures with protection schemes, namely, PONs with standard protection defined by ITU-T (G983.1 1998) (i.e. protection schemes Type A, C and D) and some cost-efficient protection schemes for TDM PON, WDM PON and Chen (2006), Chan (2003) and Chen (2007), respectively.

5.1.2.1 Case Study I: Changed Cost in Time

As mentioned before, the main capital expenditures are related to the outside plant, and especially the trenching cost for rolling out an optical fibre to every home. Besides, CapEx also includes component-related cost. The component-related cost consists of component and fibre cost, installation, housing and chassis. The prices of fibre and components are expected to decrease over time according to the Moore's law. In contrast, the cost related to the civil work, e.g. burying fibres, housing and chassis, is relatively stable in time. Furthermore, the installation cost increases since the salary for operator's employee is expected to increase in time. We assume 7% cost reduction per year for fibre and each component and 3% increase per year for the hourly rate of operator's employee, Chen (2010). Figure 5.13 shows component-related cost per ONU calculated for all the considered architectures in rural (dispersive) and suburban (collective) scenarios. Please notice that we consider cost per ONU and not a price that will be charged from the users, which may depend on pricing policy and other factors. It can be seen that the component-related cost is influenced strongly by cost reduction in time. In P2P and all the PON architectures, the decrease of the component-related cost is similar to the assumed change of component cost in time. However, in AON-AS (indicated as AON in Fig. 5.13) it is quite different, because the expensive housing for the Ethernet switch at RN, whose cost does not fluctuate in time, takes nearly 50% of the total component-related cost. However, the total CapEx per ONU for all the considered architectures does not change significantly with time, since the cost of burying fibre is the dominating component of the cost for the deployment of fibre-based access networks.

Typically the fibre infrastructure can have much longer lifetime than the other components/equipments in an access network, since it can be reused for future solutions. Therefore, it may be reasonable to charge a new user only for the component-related cost, and the remaining part as well as operational cost can be included in the subscription of services and paid, e.g. on a monthly basis. It can be observed in Fig. 5.13 that the component-related cost for basic PON, Type A, and both TDM PON and hybrid PON with neighbouring protection is relatively low (much lower than $2,500), which may be acceptable as an initial installation price for a new user. Furthermore, one can expect that after 3 years this cost per new user will be reduced by around 20%. Moreover, the component-related cost for WDM PON is relatively high, but on the other hand it can offer much higher capacity per user and scale better than other architectures.

Furthermore, the equipment cost variation will also have an impact on the time at which the spare component is bought for reparation. The later it is bought, the cheaper it is, but longer total time to repair and higher penalty costs can be expected.

5.1.2.2 Case Study II: Variable Take Rate

It is noticed that the CapEx per user is dependent on the take rate, referred to as the percentage of homes or buildings covered by the access network infrastructure that subscribe to the service. Infrastructure costs (e.g. enclosure construction and fibre deployment) may be incurred for all homes, even though they can only be recovered from fees paid by subscribers. In this case with respect to take rate, we consider two different strategies to bury distribution fibres. In the first one, fibres are buried to all homes or buildings from the beginning but the installation is made upon subscription. In the second strategy, distribution fibres are buried only to the homes or buildings that subscribe for the service. Obviously, when take rate is 100%, there is no difference between these two strategies. However, as shown in Fig. 5.14, when the take rate is lower than 100%, the cost per ONU varies more in the first strategy than in the second one since burying distribution fibres is expensive, in particular, in rural areas. Therefore, from a cost point of view, the second strategy to bury distribution fibres is recommended. However, in some cases the first strategy should

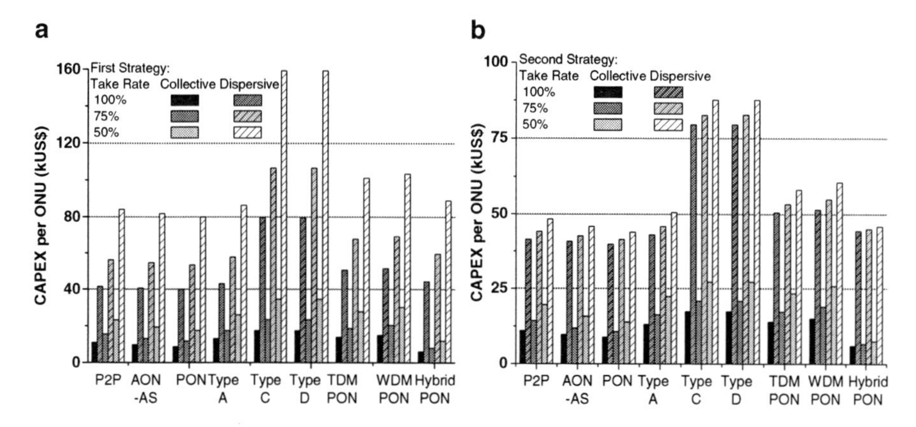

Fig. 5.14 Results for case study II (variable take rate): (**a**) CapEx for the first deployment strategy, where distribution fibres are buried to all the homes or buildings from the beginning but the installation is made upon subscription, and (**b**) CapEx for the second deployment strategy, where distribution fibres are buried only to the homes or buildings that subscribe for the service (J. Chen L. W. 2010)

be applied to reduce the installation time for new users after network deployment. Furthermore, if the network provider can expect that the take rate will increase soon, it might also be desirable that all distribution fibres are buried from the beginning.

The take rate also has an impact on the operational costs. Higher take rate implies more installed equipment and, therefore, higher number of failures. In general, the higher the take rate is, the higher the number of users and the lower the operational costs per user are (since there are more users with whom to share the costs).

5.2 The Benefit Side: The Economic Impact of Future Internet

The Internet is perhaps the most revolutionary technological development that has taken place during this generation. More recently, the advent and quick growth of broadband access has taken Internet connectivity to a new level and made a large portion of the world population aware of the potential of this technology and its impact on daily life. In some countries, this awareness has matured into a political persuasion (see the broadband-for-all slogan) that is directly or indirectly driving the development of broadband access. There are various more or less pronounced political ambitions behind increased broadband penetration. An important one is e-government, i.e. the ability to offer online interaction with the citizens, which is believed to simultaneously benefit democracy and significantly reduce the spending within the public administration through higher efficiency. Second, broadband penetration leads to ICT maturity, which is believed to benefit industry and trade, through the stimulation of front-edge service and product development, as well as more efficient production, logistics, resource management, etc. Finally, high broadband penetration is argued to lead to a more environmentally sustainable development (e.g. video conferences instead of travelling) and to lessening of migration from rural areas to large cities.

5.2.1 Socio-economic Gains (Acreo)

One of the perceived benefits of widespread broadband access is an increase in trade. On one hand, goods and services directly purchased over the Internet would be traded to a lower extent or not be traded at all if they needed to be physically shipped (e.g. multimedia content or IP telephony), whereas some goods and services that need to be physically shipped or provided (e.g. books, standard consumer electronics, as well as hotels and airline trips) see their cost of marketing, transportation and distribution reduced – thanks to improved communications, improved logistics and reduced need for intermediaries between producer and consumer.

One obvious way to examine the effect of broadband penetration on trade is to follow the evolution of trade and broadband penetration in time. This however has the disadvantage of not taking into account other factors that may have an impact on trade. Moreover, the statistical significance of the data would be compromised by the limited number of years during which broadband access has been around. Finally, a correlation between broadband penetration and trade is not per se sufficient to establish a causal link. A method (Korotky 2004) that solves the first two problems is cross-country regression analysis, in which bilateral trade between a large number of countries is measured, as well as indicators for number of factors which are suspected to have an impact on trade (among which broadband penetration). A general equation is then written to model the impact of these factors on trade, and when this equation is evaluated against the measured data for each country pair, the relative importance (if any) of each specific factor is estimated. This method has the advantage of separating the effect of broadband on trade from other factors, and moreover it allows collecting a very large number of samples. Our model is described by the so-called gravity equation (Korotky 2004), which predicts that the amount of trade between two countries T_{ij} be directly proportional to their respective GDP Y_i and Y_j, and inversely proportional to their geographical distance D_{ij}

$$D_{ij}T_{ij} = K_{ij}\frac{Y_i^{\beta_Y}Y_j^{\beta_Y}}{D_{ij}^{\beta_D}}; \tag{5.72}$$

the coefficients $\beta_Y > 0$ and $\beta_D < 0$ determine how important economic size and geographical distance are with respect to each other, and K_{ij} is a proportionality constant accounting for all other factors. Measurable factors likely to be correlated with trade include whether two countries speak the same language, whether they share a border, and broadband penetration: these factors are included in the study. Other factors such as common free-trade area and political relations can also have an impact but are neglected in the first version of the study. It is convenient to write the gravitation equation in logarithmic form:

$$\ln T_{ij} = \beta_0 + \beta_Y \cdot (\ln Y_i + \ln Y_j) + \beta_P \cdot (\ln P_i + \ln P_j) + \\ + \beta_D \cdot \ln D_{ij} + \beta_b \cdot b_{ij} + \beta_l \cdot l_{ij} + \varepsilon_{ij} \tag{5.73}$$

in which we have expanded the proportionality factor K_{ij} in terms of a common language binary variable, l_{ij}, a shared border binary variable, b_{ij}, and broadband penetration, P_i and P_j. The factor β_0 is the intercept for the linear regression (mathematically, the amount of trade predicted by the model when all other terms are zero, even if the model is not necessarily significant in that case), while ε_{ij} is the prediction error of the model for country pair *i,j*. Equation 5.73 also helps giving another interpretation to the β coefficients in terms of elasticity coefficients. Because a marginal change in the natural log of a variable is approximately the relative change in the underlying variable (mathematically, $d(\ln x) = dx/x$ or $\Delta(\ln x) \approx \Delta x/x$), one can rewrite Eq. 5.73 as $\Delta T_{ij}/T_{ij} \approx \beta_x \ \Delta x/x$, i.e. a 1% change in a certain variable x (keeping all other variables fixed) gives a β_x% change in the trade volume.

The model can be estimated against data if all the variables can be measured for a number of country pairs. Large databases with such measurements are available online. We used the World Bank's WDI database (Casier 2008) to obtain broadband penetration in terms of connections per thousand inhabitants, and Prof. Rose's online database at Berkeley University (Mas Machuca 2007) for the remaining data (trade and GDP are expressed in US dollars; broadband penetration in terms of number of connections per 100 inhabitants). The time chosen is the year 2000, which had the richest set of data (29 countries, giving more than unique 400 country pairs, or samples). An ordinary least square (ols) regression can then be run, producing a value for each β coefficient, which minimises the average of $|\varepsilon_{ij}|$. It is found that all parameters included in the model are significant, except for the shared border indicator, for which a T-statistic parameter of 0.2 is obtained (for a parameter to be considered significant, it is common to require T-stat to be above 2 (Lannoo 2008), which gives a 95% confidence interval). In particular, β_Y is found to be 0.92 (slightly less than one, which can be explained by the fact that smaller economies have to rely more on specialisation and therefore on trade than bigger ones), and it is highly significant (T-stat 30). Distance does have a large and negative impact ($\beta_D = -0.99$), and it is very highly significant (T-stat 21). Common language is found significant (T-stat 2.2) and positive ($\beta_D = 0.33$): Within the 95% confidence interval, speaking the same language statistically increases trade between two countries by 39% ($e^{0.33}$–1).

Finally, it is found that broadband penetration is significantly (T-stat 9) and positively correlated with trade ($\beta_P \approx 0.25$). This implies that, all things equal, a 1% increase in broadband penetration in both trading partners is correlated with a 0.5% increase in trade between the partners. This strong correlation can also be visualised graphically. Figure 5.15 shows bilateral trade (controlled for other

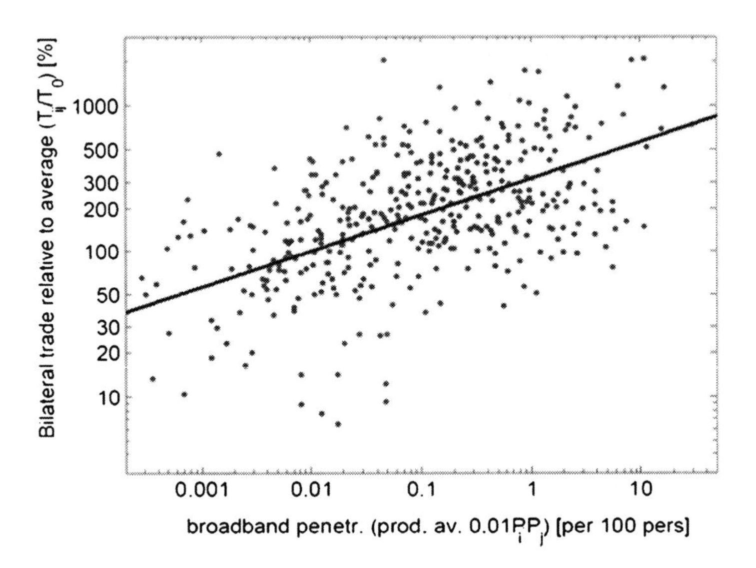

Fig. 5.15 Correlation between broadband penetration and trade for 400 country pairs

variables) between any two countries in the data set, versus broadband penetration in each country pair (400 dots). The model prediction (solid line) is also plotted for comparison. Naturally, the existence of a correlation is per se not enough to establish a link of causality. Three possibilities exist. The first is that broadband penetration causes an increase in trade; specifically, a 1% increase in broadband penetration in both trading partners leads to a 0.5% increase in trade between them (or, for large changes: a fourfold increase in broadband penetration in both trading partners leads to a doubling of trade). The second possibility is that larger volumes of trade lead to an increase in broadband penetration, specifically a doubling in trade causes a fourfold increase in broadband penetration in both trading partners. Finally, one can assume that there is an unknown factor causing both broadband penetration and trade to increase.

The last option cannot be safely rejected, and neither can the second one. One could argue that most Internet traffic is generally not directly linked to trade: peer-to-peer file-sharing applications such as Bit Torrent, Kazaa, Direct Connect, and E-Mule dominate Internet traffic volume (see for instance traffic measurements for a high-speed symmetric access network in Sweden (J. Chen L. W., 2010), while video services such as YouTube and social networks such as Facebook and MySpace are maintained to dominate number of Internet sessions (though no clear numbers are available in this respect); so the need to make trade easier does not seem to be a major driver for the demand of Internet connectivity. In summary, it is most reasonable (though by no means proven) to conclude that it is high broadband penetration that causes increased trade and not vice versa.

A next step would be a more direct study of the economic impact on economic wealth, which would be very interesting. However, this remains a difficult task. Although a correlation between, e.g. GPD per capita and broadband penetration is easy to find in the data, the strongest link of causality is arguably wealth-> broadband penetration. Proving the possible existence of a broadband-> wealth link of causality requires advanced regression techniques and may be hard to achieve.

5.3 Future Internet and New Business Models

5.3.1 The Open Network Model

The basic idea behind the open access model is to promote the highest degree of competition in order to maximise the freedom of choice for the end users, and avoid monopoly (Korotky 2004). An important underlying principle of open networks, however, is that of building an infrastructure for the society, not merely a revenue-driven system, and there is a strong political interest in open access networks – not only in Sweden but across Europe. Another important driver has been the failure of traditional large operators to provide broadband access at sustainable prices in remote areas. This is a reason why many rural communities across Sweden have deployed open access networks. Today it is estimated that 95% of the 173 municipality networks and 42% of the housing companies with FTTx currently operate according to an open access model (Casier 2008).

The traditional telecom model is based on "vertical integration," in which one entity delivers the service, operates the network, and owns the network infrastructure. Originally, the available services were mainly limited to telephony, radio, and television. These justified dedicated infrastructures each optimised to transmit information carried by a specific physical signal, and with inherently different traffic patterns. Technology has evolved dramatically since. Today the amount of available services is booming: from well-established ones such as telephony (mobile or fixed), web access, e-mailing and television (standard quality and HDTV), to rapidly growing ones such as video conferencing, video and music streaming and sharing, online gaming and e-health and to new and emerging ones such as 3D TV, grid computing, etc. For all these services, information is stored and transmitted digitally, and it is increasingly delivered using the IP protocol. Moreover, the end user is no longer just a consumer of contents but has also become a producer of, e.g. photos and video material using a variety of applications. A vertically integrated model with a dedicated network infrastructure for each service is therefore highly inefficient. Some degree of convergence has taken place during the past 15 years, but this has been a slow and incomplete process, hampered in great part by the resistance from the traditional vertical-integration business model. Ideally, there is no reason today why telecommunication services should be delivered by a network infrastructure

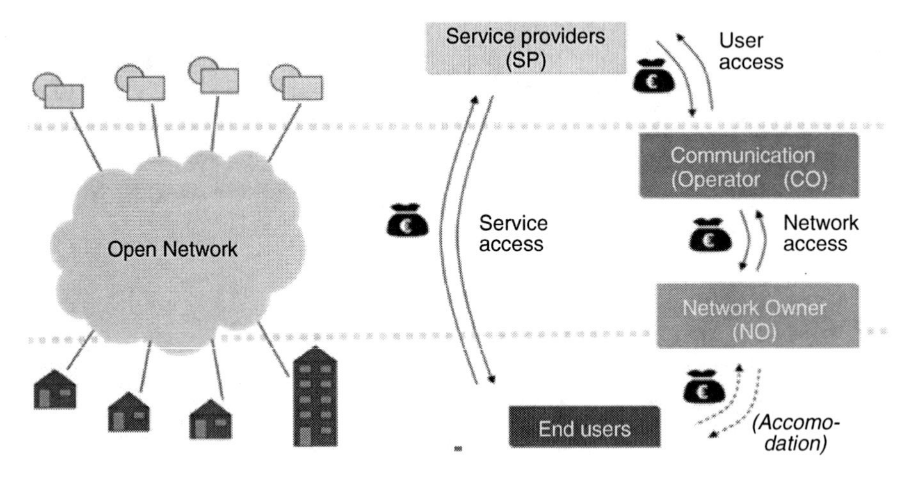

Fig. 5.16 The open network model and typical open access value chain

that is optimised to the type of end-user termination (urban vs. rural dwelling, heavy-vs.-light user, mobile vs. fixed, etc.) rather than the services being delivered.

The open network model, in which services are provided on a fair and non-discriminatory basis to the network users, is enabled by conceptually separating the roles of the service provider (SP) and the network and communication operator. Due to the different technical and economic nature of the different parts of the network, different roles and actors can be identified. A fibre access network broadly consists of a passive infrastructure (implying right-of-way acquisition, trenching, cable duct laying, local-office premises) and active equipment (transponders, routers and switches, control and management servers) (Fig. 5.16).

The passive infrastructure is typically characterised by high CapEx, low OpEx, and low economies of scale, and is highly local, hard to duplicate, and inherently subject to regulation. The active equipment is characterised by high OpEx and economies of scale, and is subject to limited regulation. These factors justify a further role separation between a NO, who owns and maintains the passive infrastructure (typically real estate companies, municipalities, utilities), and the CO, who operates (and typically owns) the active equipment (incumbent operators, new independent operators, broadband companies).

Depending on which roles different market actors take up, the network will be open at different levels and different business models will arise, as illustrated in Fig. 5.17. A single actor may act as NO and CO (a), in which case the network is open at the service level. If the roles of CO and NO are separate (c, d), then openness at infrastructure level is achieved. Generally, one NO operates the infrastructure, while one or several CO can be allowed to operate the active infrastructure generally over a fixed period of time, at the end of which the contract may or may not be renewed (in which case a new CO is designated and active equipment may need to be replaced). Most often, economies of scale make it impractical to have a truly

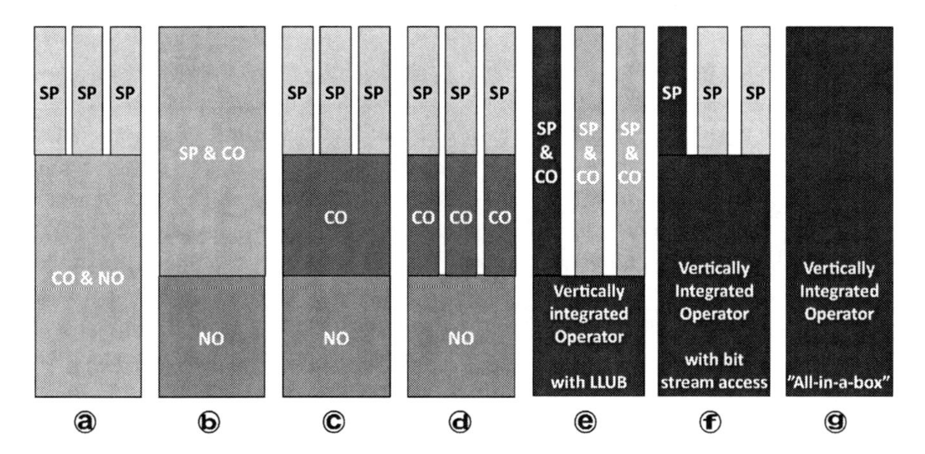

Fig. 5.17 Access network business models

multi-CO network (although larger networks may assign the operation of different geographical parts of the network to different CO). Independent of the specific model, however, the CO should offer different service providers access to the network (and therefore the users) on non-discriminatory conditions. The end users typically purchase services directly from the service providers. The CO receives revenue from the SP and pays a connection fee to the NO for network access.

If the CO also acts as SP (b), the network cannot be described as being really open according to the definitions here, but it is still "more" open than the conventional vertically integrated model in (g) which most incumbents worldwide follow today. In case of LLUB, a vertically integrated operator is still present, but there can be multiple actors working as combined CO and SP. In case of bit stream access, the vertically integrated operator assumes the role of CO, but there can be multiple SPs offering their services in the networks.

Observe that some of the roles in Fig. 5.17 can be subdivided into more roles and that the limit between the different roles is not always as clear as in the figure. However, the figure should give a pretty good idea of which kinds of business models (in particular with respect to open access) are used in Sweden today. The definition of "open" may vary a lot depending on the type of actor and which level in the network is regarded. It may for instance be claimed that Fig. 5.17b is also a truly open network. On the other hand, one could argue that if there is only one entity owning the fibre infrastructure, the network is not open – despite the fact that all COs get access to the fibre on equal terms.

Many of the Swedish open networks have started as a locally driven effort to build an infrastructure for the society, and often further motivated by the lack of interest of traditional large operators to provide broadband access at sustainable prices in remote areas. However, even national operators are increasingly finding open access networks profitable and stepping up their business with such networks. In the following, we give an overview of Swedish access networks operating different business models.

Stokab. In Stockholm the passive fibre network is owned and administered by Stokab, a company owned by the City of Stockholm. Stokab was founded in 1994 and was one of the first companies worldwide who embraced the idea of an operator-neutral ICT infrastructure in a larger scale. The historical reason for this has been twofold: to avoid uncoordinated trenching in the city and to create a foundation for the future IT society. It is today a huge dark fibre network comprising more than one million fibre kilometres. Stokab also owns fibre in parts of the Stockholm metropolitan area which include neighbour municipalities and the Stockholm archipelago. The network supports both public and local administrations, enterprises, and operators. Stokab's fibre network is operator neutral in the sense that all operators get access to the dark fibre, points of presence, and antenna sites on equal terms. The operators include both communication operators serving an open access network as well as vertically integrated operators. Apart from owning the dark fibre network, Stokab also acts as a communication operator for the city administration and internal communication. Stokab would correspond to the role of the NO in Fig. 5.17. All the models a, b, c, and d coexist in Stokab's network.

Svenska Bostäder, Stockholmshem, Familjebostäder. These three companies are 100% owned by the city of Stockholm and are (if added together), with more than 90,000 homes, Sweden's largest housing companies (Lannoo 2008). By the end of 2010, all apartments will have access to a fibre-optic broadband connection. Svenska Bostäder (SB) has chosen an open access model with different service providers and different services to choose from for the end users. SB is connected to Stokab's fibre network but owns the passive infrastructure within the MDU. This is typically single mode fibre between the communications room (serving one or several MDUs) and the apartments. The active equipment in the access network, the communication rooms, and the gateway in the apartments are owned and operated by a communication operator, and the money flow is according to Fig. 5.16, where the dashed line indicates that the tenants are paying rent to the housing company which is also the network owner. Competition at CO level is achieved by dividing Stockholm into three geographical zones, and each zone has its own CO. That is, only one CO operates in one zone. The CO has a contract for a limited number of years, which can be renegotiated by the end of the period. Many other municipal housing companies across Sweden are using a similar model. However, only SB deploys more than one COs in its network. The business model of Svenska Bostäder corresponds to Fig. 5.17d where Svenska Bostäder has the role of NO, and with multiple COs and SPs being contracted. Generally, a housing company would be the NO in Fig. 5.17c.

Mälarenergi Stadsnät. Mälarenergi Stadsnät is a company owned by Mälarenergi (Wosinska 2008), a local power utility owned by the city of Västerås. It is now a large municipality network and among the most successful in Sweden, with 40,000 homes connected. Mälarenergi Stadsnät owns the fibre infrastructure and acts as a communication operator in the network, a role which covers being a service broker. Interestingly, ME Stadsnät has recently entered the municipality network of neighbouring Eskilstuna as CO (G983.1 1998). Referring to the business models in Fig. 5.17, Mälarenergi is the combined NO and CO (Fig. 5.17a) in its own network in Västerås, but it acts as the CO (Fig. 5.17c) in Eskilstuna

Säkom. Another example of a municipality network following model (c) is Säffle municipality (in Värmland County) with a population of 15,800, a significant part of which is scattered over a large rural area. The municipality only owns the fibre infrastructure (through the wholly owned company Säkom (Chen and Chen 2006) acting as NO), while a separate company acts as CO with a 5-year contract. What makes Säkom interesting is that currently the incumbent TeliaSonera acts as CO. It is noteworthy that TeliaSonera, a former monopolistic telecom operator, exclusively operates as CO without owning the network infrastructure and without delivering its own services. TeliaSonera acts as a pure CO in a few other municipalities in Sweden.

TeliaSonera. TeliaSonera is Sweden's incumbent and originally operated as a typical vertically integrated operator only. However, TeliaSonera also embraces the open access model in different ways. The company functions as communication operator in some networks, as in the Säkom case and also in one of the three zones of Svenska Bostäder as described above. TeliaSonera also acts as a service provider in several municipality and housing company networks (e.g. Mälarenergi and in Svenska Bostäder's downtown part of its network), in competition with other service providers. That is, TeliaSonera is active in all the business models (b) to (g) in Fig. 5.17.

5.3.2 *Blue Ocean Strategies*

The strong emergence of the Internet into the public domain has completely changed the competitive landscape in the telecom industry. The telecom industry started witnessing a massive transformation in the structure and the rules of the game for competition. Services are converging, market barriers are collapsing, products are getting standardised, and companies are merging and building partnerships and alliances. When a set of alternative products become similar in functionality and usage, they turn into commodities. The competition will purely be based on price, and companies will fight to get a bigger slice of the same pie. This pushes telecom companies into head-to-head competition that erodes profitability for companies and diminishes the attractiveness of the entire industry.

The huge growth in Internet traffic and video-based applications has rendered legacy systems obsolete. Operators have to spend significant capital to keep up with the traffic demand and they face many options in the process. Incremental upgrades reduce the financial burden but could place the company at a competitive disadvantage. Massive transformation of the network might leverage the competitiveness of the company but could also bring a big financial risk if the investment does not

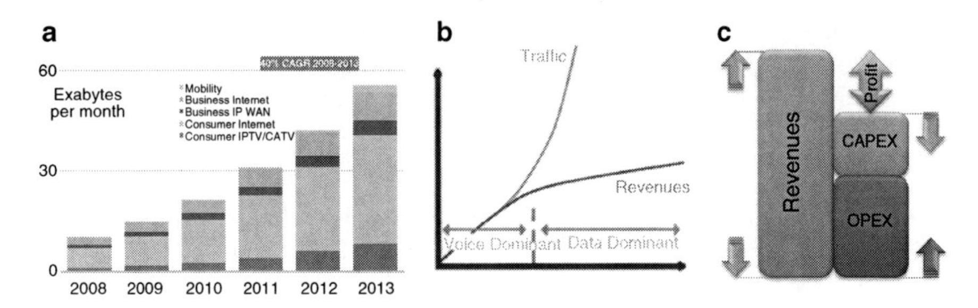

Fig. 5.18 (**a**) IP traffic growth (PTS-ER-2009:29, 2009), (**b**) gap between traffic and revenues, and (**c**) the profitability challenge

pay off. The challenge that operators face is to try to monetize the investment they make in their network. This is especially challenging with the trends in the services offered where operators have to make up for declining revenues from voice traffic and generate proportional revenues from their investments to keep up with the huge growth in the data traffic (Fig. 5.18) (Index, 2008).

The trends seen in Fig. 5.18a clearly show an exponential growth in the IP traffic with a breakdown of the different services. The services that have been enjoying the highest growth are the mobile data, IPTV, and consumer Internet. However, the growth in the IP traffic is not accompanied by a proportional growth in revenues for service providers. As shown in Fig. 5.18b, since the traffic moved from being TDM-based mostly for voice services to become IP-based, the traffic started growing at an accelerated rate while the growth rate for revenues was diminishing. This causes a big dilemma for service providers since they have to upgrade their network to keep up with the growing demand, but at the same time there is clearly an un-captured value illustrated by the gap between the growth of the traffic and the associated revenues. However, what matters the most is the profit, which is a function of revenues and cost in the form of CapEx and OpEx.

$$\mathit{Profitability} = f\left(\frac{\textbf{Revenues}}{\textbf{CAPEX} + \textbf{OPEX}}\right)$$

So if the name of the game is profitability, the good news is that if the revenue per bit is dropping, the cost per bit has been dropping as well. So the mission of the service provider is to translate this explosion in the IP traffic into an opportunity and capitalise on it by providing differentiated services at reduced costs.

In researching the technical, economical and strategic factors faced by the management in the telecom industry, we have found that adapting some of the teachings in the BOS to the telecom industry provides operators the best competitive position. A technology like FTTH, for example, provides a great opportunity to create a blue ocean by focusing on value innovation to make the competition irrelevant.

Fig. 5.19 Comparison between Red Ocean Strategy and Blue Ocean Strategy

BOS is a fundamental shift from the traditional strategy of SCP (Correia 2008). SCP focuses on analysing the industry structure and positioning the company accordingly, which determines the profitability of the company. BOS teaches to reconstruct market boundaries, revisit all activities in the value chain to reach beyond existing demand, and simultaneously pursue cost and differentiation strategy. Figure 5.19 shows a comparison between ROS and BOS. It is clear from the comparison that the companies in the telecom industry are competing in a red ocean. The only way for service providers to avoid this mutually destructive competition is by reinventing their business model to enable them to translate technology innovation into value innovation. The following sections will show the key benefits of FTTH and how to implement it to create a blue ocean of opportunities. The reason we chose FTTH in our study is because current access networks in most operators' networks are in desperate need for upgrade, and FTTH is a platform that enables the operators to completely reinvent their business model. Without driving the fibre deeper in the loop, preferably all the way to the home, the operators will be restricted in their options to provide differentiated services on a scalable and converged platform that can support future growth.

5.3.2.1 Technical, Economical, and Strategic Benefits of FTTH

The focus in this study will be on FTTH implementation in a PON, which has been the technology of choice for most tier-1 operators. However, the strategic benefits are applicable to other FTTH implementations such as point-to-point and active Ethernet. As shown in Fig. 5.20, optical fibre is used to directly connect the customer network element to the CO. The distribution network is completely passive, and CO

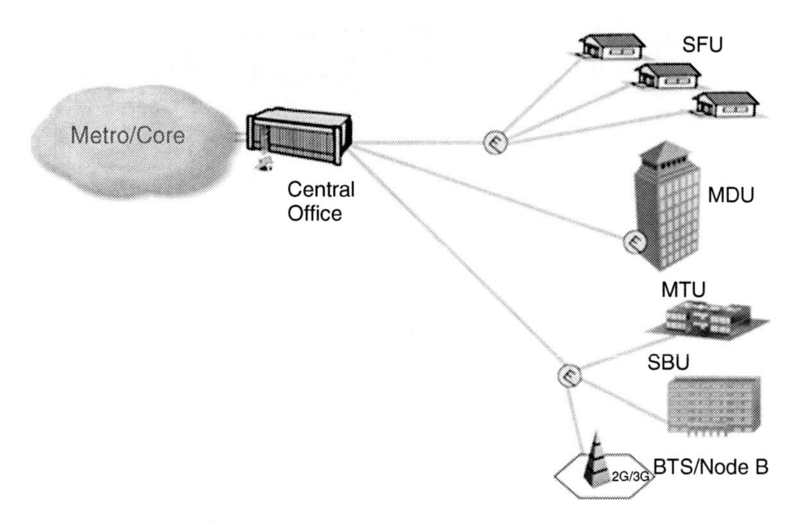

Fig. 5.20 FTTH PON implementation

equipment is shared through passive optical splitters. Fibre has virtually unlimited bandwidth, and driving fibre all the way to the home enables the operators to support new services and have a future-proof solution.

The passive nature of the distribution network makes it more reliable and eliminates the need for power. Also, supporting multiple customers with a single fibre reduces the capital cost per home and significantly reduces the power and space required in each CO. Finally, compared to copper-based solutions, FTTH enables operators to have more customers per CO, reduces the number of COs needed, and makes the network intelligent to monitor, report, and assist in troubleshooting and fault isolation. An exchange serving 15,000 customers, for example, would require 900 racks of copper equipment and 800 KW of power, while FTTH would only take one rack and 100 W of power (Mas Machuca 2007).

Customers and services can be supported from a single FTTH access platform. Single family units, multi-dwelling units, shopping plazas, enterprises, and even mobile base stations can all be connected to the same distribution network. They would only differ with the network termination element and the interface required from that element. Current commercial PON technologies are based on a shared TDM/TDMA pipe (time division multiplexing/time division multiple access). Once the network is deployed, higher data rates can be introduced by inserting additional wavelengths to support new customers or existing customers who need the service and are willing to pay for it. Eventually, the network can be transformed into a pure WDM-PON network where every customer enjoys a virtual point-to-point connectivity to the central office with a dedicated wavelength (Fig. 5.21). Changes will be minimal in the distribution network where optical power splitter may be replaced with a coloured device, such as AWG, to separate the wavelengths without having to change the outside network. TDM and WDM networks can coexist since

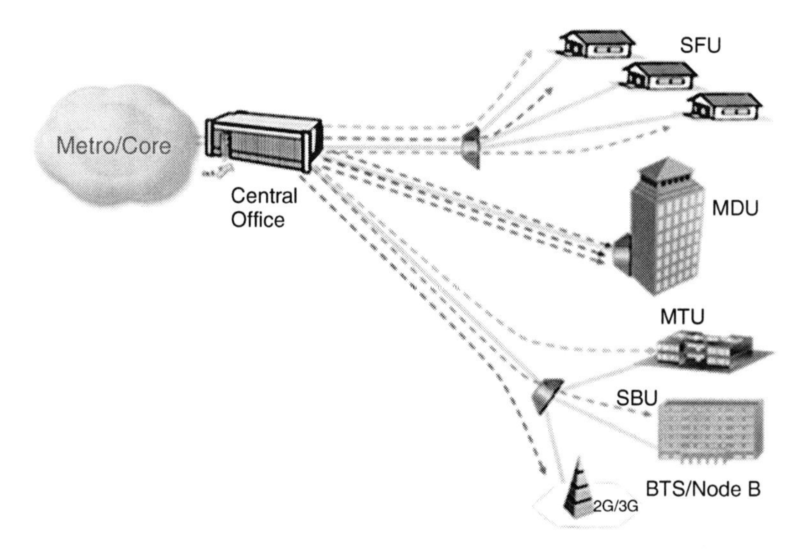

Fig. 5.21 FTTH upgrade with WDM

WDM-PON may use the rest of the optical spectrum (i.e. C-band), which is not used from the previously installed technologies (as in GPON networks that use IPTV instead of video overlay). This provides a nice upgrade path for current deployment to ensure that it is future-proof and that the return on the invested capital is maximised.

5.3.2.2 Utilising FTTH to Reinvent the Service Providers' Business Model

FTTH is not a new technology. It had several false starts dating back to 1977 and was considered too expensive for mass deployment. It looks like it is finally happening with the several deployments worldwide (B. Lannoo, Economic Benefits of a Community Driven Fibre to the Home Rollout, 2008). However, when we examine these deployments, we see more focus is on technology innovation instead of value innovation with hardly any differentiation in the product offering and the value perceived by the customer.

In order to translate all the technical and economical advantages of FTTH into a strategic competitive advantage, operators need to focus on value innovation. This enables them to profit from the great potential that FTTH brings. When operators create value for their customers, they end up with real opportunities (Chen 2007). In order for them to do so, the following strategic activities need to be considered and implemented.

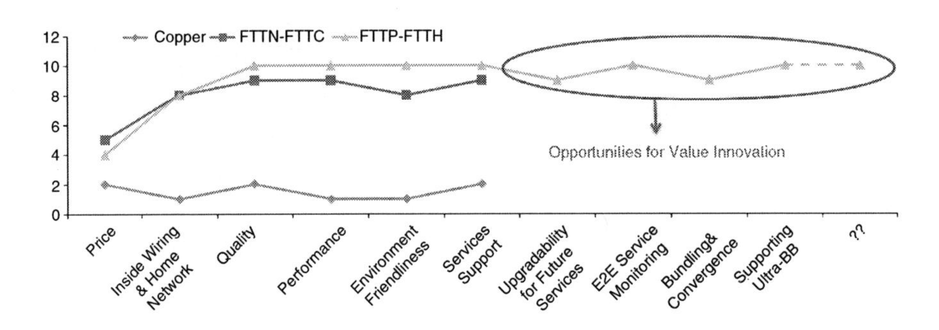

Fig. 5.22 FTTH strategy canvas

5.3.2.3 Reconstructing Market Boundaries

Most operators are no longer monopolistic entities who provide a single utility service in a static market and industry. The business model will have to be completely reshaped to evaluate all the old assumptions and consider the current key competitive drivers (such as customer preference, product quality, price, industry standard, etc.). A "strategy canvas" needs to be created to graphically display these factors. A possible FTTH strategy canvas is shown in Fig. 5.22.

When we look at cost, FTTH has a lower CapEx and OpEx than copper, but the investment required in brown fields, where a legacy network already exists, is pretty substantial. However, even in brown field applications, the significant OpEx savings will enable operators to still provide comparable prices to what is offered on existing networks (G983.1 1998) (Chen and Chen 2006). In fact, if we factor the OpEx savings only for the first 4 years of operation, the TCO is estimated to be 37% lower with FTTH (Chan 2003). The real value for customers will come from the quality of service and the ability to get new services that cannot be realised with copper-based solutions.

Operators need to transform their business model to provide services and solutions instead of getting caught in providing a fatter pipe. This requires a complete change in the mindset of the company. The variety of options and the complexity involved can easily confuse and turn off customers. The best way to retain customers and create opportunities for them and for the operator is to engage with them as a partner and as a solution provider. This is mainly applicable to enterprises and small–medium businesses. Outside providing turnkey solutions, opportunities exist in hosting services, managed services, cloud computing, and other network-based services. Furthermore, pushing network elements inside customers' home network provides tremendous opportunities to introduce many new services that never existed in the traditional business models.

5.3.2.4 Focus on the Big Picture, Not the Numbers

It is very easy to get lost in numbers that may or may not show that FTTH has a valid business model. Business case numbers were analysed extensively in the past, and it was never easy to justify the amount of investment required for FTTH (Chen 2007). However, when we look at the big picture, we see a completely different picture. Operators who took the strategic decision to pursue FTTH were pleasantly surprised by the improving economics. If we take the case of Verizon, they chose to deploy FTTH, branded as FiOS, for strategic reasons due to the fierce competition from their cable rivals (J. Chen L. W. 2010). That was the largest project in the company's history, and Wall Street punished Verizon for taking that risk. However, not only Verizon was able to benefit strategically by getting into a blue ocean away from their cable rivals, but the economics of the business case were improving quickly as the company travelled through the experience curve and the technology matured. The company started enjoying a drop of 80% in the network report rate and the dispatching calls. The cost per home passed dropped from $1,400 in 2004 to $700 in 2009, and the cost per home connected dropped from $1,200 to $640 during the same time period. Revenues from FiOS had a 56% growth rate in 2009, and the company received top rankings in surveys for customer satisfaction from several rating agencies (Larsen 2009).

Although Verizon felt that strategically they had to go to FTTH regardless of what the business case numbers said, they ended up with huge savings in OpEx, a big boost in revenues, and top customer satisfaction rate. The portfolio of services they offer still does not fully use the potential of the FTTH network, but it is considered superior to what their competitors offer. Verizon focused on the big picture and was able to identify an opportunity that enabled them to create value for their customers and gain a huge competitive advantage.

5.3.2.5 Reach Beyond Existing Demand

Instead of focusing on current customers and services, operators need to use trends they see in the industry to identify possible opportunities. Real growth lies beyond existing demand. It is very important to make sure that today's network can support tomorrow's applications. We have never been good in predicting the future. Nobody predicted the emergence of YouTube and social networking, and no one knows what other applications and products will emerge in the future. An introduction of a new popular device, like the iPhone, could drain the network resources and cause a PR nightmare for the operator. Operators need to ensure that their network can be transparently upgraded with minimum interruptions to existing customers. A platform like FTTH has this capability. Once the fibre is in place, additional wavelengths can be added and the traffic can be upgraded without having to change the optical distribution network (Fig. 5.23).

Fig. 5.23 An ecosystem at home with myriad of differentiated services

The telecom industry is extremely dynamic, and the potential is huge to introduce new products and services, or at a minimum, facilitate and stimulate such innovations by having the right platform deployed. A platform like FTTH not only enables operators to provide access to higher data rates, but also enables them to build a complete ecosystem to provide differentiated services. The possibilities are unlimited and include building management services, smart home solutions, user-generated contents, security solutions, infotainment, telecommuting, e-health, e-education, security, and many personalised and interactive video-based services.

5.3.2.6 Executing the Strategy

Technological innovation does not necessarily lead to "value innovation." The product usefulness and value need to be assessed to see how customers will buy it and use it. We should determine why customers will buy it and if it brings exceptional utility. In general, customers do not care how their bits are delivered to them. The satisfaction rate for fibre users is extremely high, but few of the non-fibre users even know about FTTH (Obstfeld 2000). It will be the task of marketing to perform mass education of the customers on the benefits of FTTH and continue

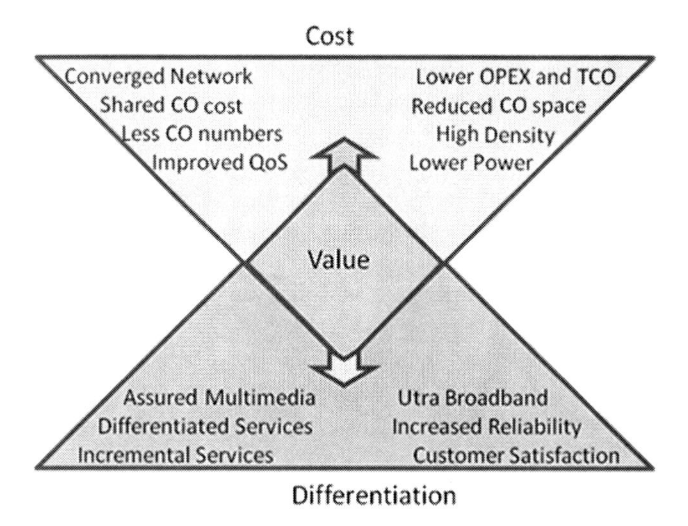

Fig. 5.24 Innovating value by increasing differentiation and reducing cost

to introduce innovative services at affordable prices to continue to create value for the customers. Another value to the customer is created by the inherent reliability of the fibre network and the ability to make the network intelligent. Monitoring and diagnostics can be performed at every layer and for every service in the network. Deterioration in the network and degradation in the performance can be detected, and problems can be resolved before they become service affecting. This enables operators to provide unprecedented service quality to their customers to gain their loyalty and retain them for long term (Fig. 5.24).

The bottom line of venturing in FTTH or any other technology is to improve profitability, for both short and long terms. Cost reduction directly improves the bottom line. As discussed earlier, FTTH enables the operators to reduce the total cost of ownership. At the same time, it helps consolidate the different networks that operators currently run into one converged network. One thing to keep in mind is that operators need to get the most out of their investment and not let the likes of Google and Apple "free ride" on their investment while they carry all the risks. The price per bit is dropping every year, and the IP traffic is increasing exponentially, but the revenues per the IP traffic are not increasing proportionally. If that gap cannot be narrowed, at least it needs to be kept under control to maximise the benefit from the traffic that runs through the operator's network. Operators should not treat all bits equally but should rather focus on the profitable ones. Different services require different bit rates and have different quality of service requirement. The focus should be on introducing new profitable services and ensure the quality of experience for these services.

Finally, operators must resolve any internal departmental differences. People will question and resist any new strategy, especially if it affects their position and area of expertise. Implementing a new strategy successfully starts and ends with the employees. With fibre replacing copper and TDM networks getting phased out with services converging to IP, it is important that the skills of the employees keep up with the changes in the technology and the industry. Management should make sure that these internal issues do not affect the overall performance of the organisation or impact the quality of the services delivered to customers.

The FTTH strategy should be built into the company's ongoing processes. Since it involves risks, it is important to build trust among the key stakeholders to generate an extra effort from a unified team. Innovation should be embedded in the operator's processes to continue to identify new and innovative products and services that create new opportunities to stay ahead and avoid clashes that would adversely impact the operator's business model. The increased competition in the telecom industry and the declining fixed telephony revenues are forcing operators to look for new opportunities to make up for lost revenues and to capitalise on the huge growth in the IP traffic. FTTH provides operators a blue ocean of opportunities. The possibilities are unlimited once FTTH is in place as long as the operators focus on introducing innovative services and solutions that create value for their customers. FTTH provides a better technical performance, a more economical solution in terms of the CapEx and OpEx, and a strategically superior platform that delivers differentiated services for a lower cost.

References

Casier, K., Verbrugge, S.: Techno-economic evaluations of FTTH roll-out scenarios. NOC (2008)

Chan, V.T., Chan, V.C., et al.: A self-protected architecture for wavelength division multiplexed passive optical networks. IEEE Photonic Technol. Lett. **15**(11), 1660–1662 (2003)

Chen, J., Chen, B.: Self-protection scheme against failures of distributed fibre links in an Ethernet passive optical network. OSAJ. Opt. Networks **5**(9), 662–666 (2006)

Chen, J., Wosinska, Lena: Analysis of Protection schemes in PON compatible with smooth migration from TDM-PON to Hybrid WDM/TDM PON. OSA J. Opt. Network **6**, 514–526 (2007)

Chen, J., Machuca, C.M., Wosinska, L., Jaeger, M.: Cost vs. Reliability performance study of fiber access network architectures. IEEE Comm. Mag. **48**(2), 56–65 (2010)

Correia, A.: Planeamento e dimensionamento de redes opticas com encaminhamento ao nível da camada optica. Master Degree Thesis, University of Aveiro (2008)

ITU-T Recommendation G983.1: Broadband optical access systems based on passive optical network (PON) (1998)

Korotky, S.: Network global expectation model: a statistical formalism for quickly quantifying network needs and costs. J. Lightwave Technol. **22**(3), 703–722 (2004)

Lannoo, B., Casier, K.: Economic benefits of a community driven Fibre to the home rollout. IEEE Broadnets (2008)

Larsen, C.P., Forzati, M.: A study of broadband access: Policy & status in Sweden, and global economic impact. OFC/NFOEC (2009)

Machuca, C. Mas., Moe., O.: Modeling of OpEx in network and service life-cycles In ECOC 2007 OpEx workshop, Berlin, September 2007 (2007)

Obstfeld, M., Krugman, P.R.: International Economics: Theory and Policy. Addison Wesley, Boston (2000)

Wosinska, L., Chen. J.: Reliability performance analysis vs. deployment cost of fibre access networks. COIN (2008)

Index

A. Teixeira and G.M.T. Beleffi (eds.), *Optical Transmission: The FP7 BONE Project Experience*, Signals and Communication Technology, DOI 10.1007/978-94-007-1767-1,

CPSIA information can be obtained at www.ICGtesting.com
Printed in the USA
LVOW070232140312

272992LV00005B/41/P